Fundamentals of Rocket Propulsion

Fundamentals of Rocket Propulsion

D.P. Mishra

CRC Press
Taylor & Francis Group
Boca Raton London New York

CRC Press is an imprint of the
Taylor & Francis Group, an **informa** business

CRC Press
Taylor & Francis Group
6000 Broken Sound Parkway NW, Suite 300
Boca Raton, FL 33487-2742

Printed on acid-free paper

International Standard Book Number-13: 978-1-4987-8535-8 (Hardback)

Library of Congress Cataloging-in-Publication Data

Names: Mishra, D. P. (Aerospace engineer), author.
Title: Fundamentals of rocket propulsion/D.P. Mishra.
Description: Boca Raton: CRC Press, 2017. | Includes bibliographical
references and index.
Identifiers: LCCN 2016056000 | ISBN 9781498785358 (hardback: acid-free paper)
Subjects: LCSH: Airplanes--Rocket engines--Textbooks. | Rocket
engines--Combustion--Textbooks.
Classification: LCC TL709.M57 2017 | DDC 621.43/56--dc23
LC record available at https://lccn.loc.gov/2016056000

Three stalwarts who steered the Indian Space Programme: Professor Vikram Ambalal Sarabhai, Professor Satish Dhawan, and Dr. A. P. J. Abdul Kalam

Contents

Preface

This book is designed and developed as an introductory text on the fundamental aspects of rocket propulsion for both undergraduate and graduate students. It is believed that the practicing engineers in the field of space engineering can benefit from the topics covered in this book. A basic knowledge of thermodynamics, combustion, and gas dynamics is assumed.

I have been teaching courses in aerospace propulsion to undergraduate and graduate students for the last 20 years. I felt a need to codify my accumulated lecture notes, which have undergone considerable modifications, in a textbook form for the benefit of students who have completed the course. The main motivation for writing this book was to emphasize the basic principles of rocket propulsion, which may encourage students to take up this subject, while inculcating in students confidence in their innate capabilities.

Chapter 1 starts with a brief introduction to rocket propulsion, covering its application both in aerospace and nonaerospace branches. A bird's-eye view of nonchemical rocket engines is provided in this chapter so that students can gauge the entire gamut of rocket propulsion. Subsequently, Chapter 2 covers aerothermodynamics, which is essential for an analysis of rocket engines. The basic principles of rocket propulsion, fundamentals of thermodynamics, chemistry and gas dynamics are discussed briefly in this chapter. The elements of rocket propulsion are discussed in Chapter 3. Performance parameters that are useful in characterizing rocket engines, namely, specific impulse, impulse-to-weight ratio, specific propellant flow rate, mass flow coefficient, thrust coefficient, characteristic velocity, and propulsive efficiencies, are defined and discussed. The main purpose of a nozzle in a rocket engine is to expand the high-pressure hot gases generated by the burning of a propellant to a higher jet velocity for producing the requisite thrust. This important component of the rocket propulsion system is covered in Chapter 4.

The performance of space flight is covered in Chapter 5. Several flight regimes, namely, atmospheric flight, near-space flight, and deep-space flight, are considered. Besides this, the atmospheric flight regime is also considered for air–surface missile and sounding rocket applications, while near-space flight is considered for satellites, space labs, and so on.

In Chapter 6, various kinds of chemical propellants consisting of fuel and oxidizer along with certain additives are discussed elaborately. Chapter 7 deals with solid-propellant rocket engines (SPREs), covering all aspects of their design and development. Design aspects covering a wide range of applications in the form of various propulsive devices, namely, spacecraft, missiles, aircraft, retro-rockets, and so on, and gas-generating systems are also included in this book. The liquid-propellant rocket engine (LPRE) is covered in Chapter 8. Several kinds of LPRE with their relative merits and demerits are included in this chapter. Analyses of LPREs for their various components are developed, which can be used for the design and development of rocket engines. Chapter 9 is devoted to the hybrid propellant rocket engine (HPRE) with relative merits and demerits compared to SPREs and LPREs. Chapter 10 covers the injection systems and atomization processes involved in spray formation of liquid propellants, which find applications in LPREs and HPREs. Chapter 11 discusses nonchemical rocket engines, which are used in recent times for space applications.

Worked-out examples are provided at the end of each chapter to demonstrate the fundamental principles and their applications to engineering problems. More emphasis is given to delineate the fundamental processes involved in chemical rocket engines. Adequate problems are given such that students can grasp the intricate aspects of rocket propulsion, which will help them design and develop rocket engines for peaceful purposes.

Several individuals have contributed directly or indirectly during the preparation of the manuscript. I appreciate the help rendered by many students at IIT Kanpur and other institutes. I am indebted to my graduate students, Drs. Swarup Jejurkar, P.K. Ezhil Kumar, S.J. Mahesh, and Manisha B. Padwal for suggesting corrections in the manuscript. I am also thankful to my other graduate students, Malena, Rohan, Deepthy, Pranav, Shruti, and Abhishek among others for helping me in several ways. The official support provided by Pankaj and Mohit is highly appreciated. Finally, the persistent support of my family in this time-consuming project is highly appreciated.

D.P. Mishra

Author

Dr. D.P. Mishra is a professor in the Department of Aerospace Engineering at the Indian Institute of Technology (IIT), Kanpur, India, where he was instrumental in establishing a combustion laboratory. He currently holds the Indian Oil Golden Jubilee Professional Chair in IIT Kanpur. He was a visiting professor in 2002 at the Tokyo-Denki University, Japan. His areas of research interest include combustion, computational fluid dynamics, atomization, and so on. He was the recipient of the Young Scientist Award in 1991 from the Ministry of New and Renewable Energy, Government of India. He was conferred the INSA-JSPS Fellowship in 2002. In recognition of his research, Dr. Mishra received the Sir Rajendranath Mookerjee Memorial Award from the Institution of Engineers (India). He is a recipient of the Samanta Chandrasekhar Award for his contributions to science and technology. For technological contribution to the common people, he has been conferred with the Vikash Prerak Sanman in 2010. He had served as an assistant editor of *International Journal of Hydrogen Energy*, Elsevier, USA. Besides this, he also serves as an editorial board member of the *Journal of the Chinese Institute of Engineers*, Taylor & Francis, and the *International Journal of Turbo and Jet Engines*, De Gruyter. He holds few Indian patents and has published around 200 research papers in refereed journals and in conference proceedings. He authored a textbook titled *Fundamentals of Combustion*, published by Prentice Hall of India, New Delhi. He has written two other textbooks, *Experimental Combustion* and *Engineering Thermodynamics*, published by Taylor & Francis, USA, and Cengage India Pvt Ltd., New Delhi, respectively. He also published a textbook titled *Gas Turbine Propulsion* (MV Learning, New Delhi/London). He also serves as a managing trustee of the International Foundation of Humanistic Education.

CHAPTER 1

Introduction

> The rockets can be built so powerful that they would be capable of carrying a man aloft.
>
> HERMAN OBERTH, 1923

1.1 INTRODUCTION

I enjoy looking at the star-studded sky particularly at night. Sometimes, I hum the childhood rhyme "twinkle twinkle little star, how I wonder what you are." I am sure most of you must feel happy gazing at a dark clear sky. Besides, when you relax in the moonlight away from the din and bustle of a hectic life in a remote village during your vacation, you may look at the sky and feel like going to the moon in your imagination. You may wonder how to travel to outer space. It is not that you are attracted by the beauty of the moon or stars. Rather the charisma of celestial objects like the sun, stars, planets, and moon has impressed several poets and writers to create volumes of literary work since ages. Humans from time immemorial have been baffled by the exotic beauty of the celestial bodies in space. The desire of modern humans to know more about these bodies in space has led to the invention of the telescope, which has been instrumental in unraveling a plethora of knowledge about space. Today, humans are able to send spacecraft even to distant celestial bodies like the moon and mars. If one wants to travel above 25 km into space, one cannot use an air-breathing turbojet because of the nonavailability of sufficient oxygen. One must consider using a non-air-breathing propulsive device, which can carry both fuel and an oxidizer. Such a device is known as a rocket engine. Thus, a rocket engine is a non-air-breathing jet propulsive device that produces the required thrust by expanding high-temperature and pressured gas in a convergent–divergent (CD) nozzle. This book is devoted to the rocket engine, particularly one that uses chemical fuel and an oxidizer. However, a brief account of a nonchemical rocket engine is provided in Chapter 11. Interested readers can refer to other advanced books on rocket engines [1–3,6,7]. Let us now look at the basic principle of propulsion.

1.2 BASIC PRINCIPLE OF PROPULSION

Recall that propulsion is a method by which an object is propelled in a particular direction. The word "propulsion" stems from the Latin word *propellere*, where *pro* means forward or backward and *pellere* means drive or push. In addition, we know that the verb "propel" means to drive or cause to move an object in a specified direction. Hence, for the study of propulsion we will have to concern ourselves with this propelling force, the motion thereby caused, and the bodies involved. The study of propulsion is not only concerned with rocket engines but also with vehicles such as aircraft, automobiles, trains, and ships. We may recall that the principle of Newton's laws of motion is the basis for the theory of jet propulsion. Jet propulsion can be expounded mainly by the second and third laws of motion. For example, a spacecraft is flying vertically at uniform speed. The resultant force in the vertical direction must be zero to satisfy Newton's second law of motion, according to which an unbalanced force acting on the body tends to produce an acceleration in the direction of the force which is proportional to the product of mass and acceleration. In other words, the spacecraft must produce thrust which must be equal to the drag force caused due to the fluid motion over the body of this spacecraft and the gravitational force. For accelerating the spacecraft, one needs to supply higher thrust than that of drag forces and gravitational force acting on it. According to Newton's third law of motion, we know that for every acting force, there is an equal and opposite reacting force. The acting force is the force exerted by one body on another, while the reacting force is exerted by the second body on the first. Although these forces have equal magnitude and occur in opposite directions, they never cancel each other because these forces always act on two different objects.

1.3 BRIEF HISTORY OF ROCKET ENGINES

Interestingly, the Chinese Han Dynasty that prevailed around 200 BC had developed rockets which were used, of course, for fireworks at that time.

But the early invention of the basic principle for a jet engine goes back to the Hero of Alexandria (around AD 67), an Egyptian mathematician and inventor who had invented several machines utilizing water, air, and steam. Figure 1.1 shows a schematic of the aeropile of Hero, which is considered to be the first device in the world to illustrate the reactive thrust principle much before Newton, who established the third law of motion. The name of this device derives from the two Greek words *aeolos* and *pila*, which mean the ball (pila) of *Aeolus*, the Greek god of the wind. This device consists of a metal boiler, a connecting pipe, and rotating joints that carry two opposing jets. It may be noted in Figure 1.1 that the heat from the burning fuel is utilized to convert water into steam. Two tubular pipes attached to the head of the boiler carry the steam to two nozzles. The steam that issues from the two nozzles forms two opposing jets, which can make the system rotate. It is really an interesting device for demonstrating the principle of reactive thrust, which is the basis of rocket propulsion. It is also believed

FIGURE 1.1 Schematic of the aeropile by Hero. (From Treager, I.E., *Aircraft Gas Turbine Engine Technology*, 3rd edn., McGraw-Hill Inc., New York, 1995.)

that around the same period the Chinese had developed windmills based on the principle of reactive thrust.

However, the real rocket was invented by the Chinese around the tenth century AD while experimenting with gunpowder and bamboo. The gunpowder was discovered in the ninth century AD by a Taoist alchemist. Subsequently, Feng Jishen managed to fire a rocket using gunpowder and bamboo, and this is considered to be the first rocket engine to leave the ground. In the beginning, a bamboo tube closed on one side is filled with gunpowder with a small opening in the tail end as shown in Figure 1.2a. On igniting the gunpowder, high-pressure hot gas, which had moved around in an erratic manner, is ejected from the small opening. Later, a bamboo stick was attached to the gunpowder-loaded bamboo rocket (see Figure 1.2b) to give stability to the rocket, which is similar to our present-day fireworks rocket. All these developments were used to make beautiful displays of color and light in the dark sky. It is also believed that around this time, a Chinese scholar, Wan Hu, had developed a rocket sled (see Figure 1.3), which comprised of a series of rockets attached to the seat. Unfortunately, he died while operating this device due to an explosion of the rockets. But later on, the concept behind these toy rockets was developed further for use as a deadly weapon in war. During the Japanese invasion around 1275, Kublai Khan used gunpowder artillery to win the war. Around the

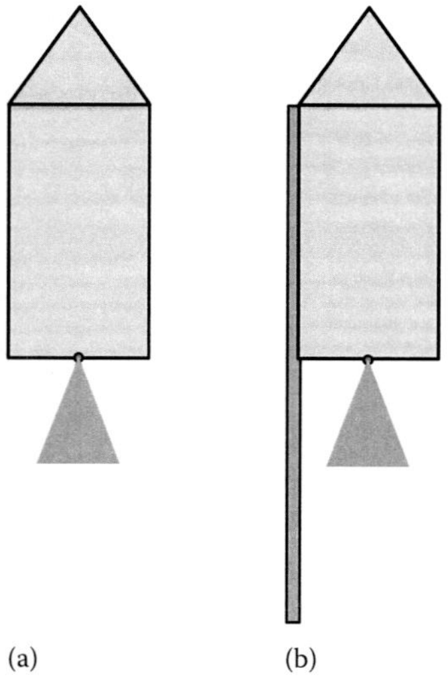

FIGURE 1.2 Schematic of (a) a bamboo rocket and (b) a bamboo rocket with a stick.

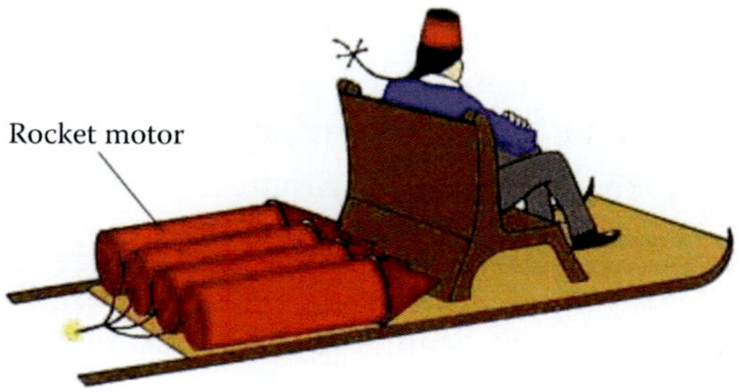

FIGURE 1.3 Rocket sled by Wan Hu. (From Treager, I.E., *Aircraft Gas Turbine Engine Technology*, 3rd edn., McGraw-Hill Inc., New York, 1995.)

thirteenth century, rockets were used as bombardment weapons even in Western countries like Spain and was generally brought by the Arabs and Mongol hordes. Later on, similar devices made of metal were used in India during wars mainly by Mughals and Marathas. Subsequently, Hyder Ali and his son Tipu Sultan created havoc in the British army by using military rockets during several battles in the seventeenth century. Tipu Sultan had 5000 rocketeers in his army, which was about one-seventh of his total army's strength. Learning from the battle experience with Tipu Sultan, the British army led by Sir William Congrieve began developing a series of barrage rockets ranging in weight from 8 to 136 kg. Congrieve-designed rockets were used at sea against Napoleon's army in 1812.

Rocket technology based on a solid propellant was developed by various researchers across the globe up to the twentieth century. During the First World War, the work on rocket engines was expedited by various countries to win the war. For the first time, rockets were fired from aircraft to shoot down hydrogen gas–filled balloons that were used for moving armies. During this period, it was found that solid propellants are difficult to handle and hence scientists across the globe started serious experiments for developing liquid fuel–fired rocket engines. Konstantin Tsiolkovsky was the first person to propose the idea of a liquid-propellant rocket engine (LPRE). Two young scientists named Robert H. Goddard in America and Wernher von Braun in Germany were successful in designing and developing LPREs. Goddard had initiated his experiments in rocket engines while studying for his doctorate in Clark University in Worcester, United Sates. In 1919, he published a paper titled "A Method of Reaching Extreme Altitudes," in which he presented a mathematical analysis of his ideas on rocketry and suggested how to fly to the moon. Around 1927, Goddard launched a liquid-powered rocket, designed and developed by him at his

Aunt Effie Goddard's farm in Auburn, Mass, which flew only 46 m. In 1923, Hermann Oberth wrote a book titled as *The Rocket into Interplanetary Space*, which attracted the attention of many youngsters who had dreams of space flight. In 1925, Wernher von Braun came across this interesting book and got into rocketry. Five years later, von Braun had joined Oberth to assist him in his rocket experiments. Subsequently, around 1932, the German Army, accepting von Braun's idea of rocketry, took many initiatives for developing rockets to win the war. By December 1934, von Braun scored his first success with an A2 rocket using ethanol and liquid oxygen. Between 1934 and 1941, Von Braun and his colleagues conducted several experiments on new designs with incremental success. With plenty of initial hiccups, he and his team managed to successfully launch the famous A-4 rocket engine, known in history as V-2 (vengeance weapon number two), on October 3, 1942, from Peenemuende. The rocket followed its programmed trajectory perfectly, and landed on target 193 km away. This launch is considered to mark the beginning of space age. Wernher von Braun was the first person along with his research group who had successfully launched a long-range ballistic missile. Of course, after World War II, the Americans and Russians scooped out the technology of the V2 rocket engine, based on which their space program flourished at the time. After 10 years, army colonel Sergei Korolev, who worked in Germany during World War II, became the chief designer of spacecraft and was responsible for developing the Vostok, Voshkod, and Soyuz spacecraft, which had carried all Soviet cosmonauts into orbit. Even the entire research team of von Braun moved to the United Sates and provided leadership in the development of an ambitious American space program. In 1956, the Army Ballistic Missile Agency was established at Redstone Arsenal under von Braun's leadership to develop the Jupiter intermediate-range ballistic missile. Subsequently, von Braun, along with the entire Army Ballistic Missile Agency, joined NASA to augment the ambitious space program of America. Both Americans and Russians abandoned the V2 rocket design of Germany and developed their own design for the next generation of rocket engines. The Russians launched their first satellite known as Sputnik and sent the first man, Yuri Gagarin, into space on April 12, 1961. After several years of struggle, American engineers and scientists landed on the moon on July 20, 1969, as part of the NASA Apollo II mission and created history in space exploration. Subsequently, several countries and agencies, namely, the European Space Agency, France, Japan, China, Israel, and India, developed their own launch vehicles.

1.4 CLASSIFICATION OF PROPULSIVE DEVICES

In the last century, several propulsive devices based on various principles were designed and developed, which are used in modern aircraft and spacecraft [4,5]. The propulsive devices can be broadly divided into two categories: air-breathing and non-air-breathing engines (see Figure 1.4). In the case of air-breathing engines, ambient air is used as an oxidizer for burning fuel. Air-breathing engines can be broadly divided into two categories: constant-pressure combustion and constant-volume combustion. The first category is based on the Brayton cycle, which can be further divided into two categories: gas turbine and ramjet. There is another exotic jet engine known as the *scramjet* engine. The gas turbine engine can be classified broadly into three categories: turbojet, turbofan, and turboprop/turboshaft.

In non-air-breathing engines, the oxidizers are not taken from the atmosphere but are rather carried onboard the vehicle. Most non-air-breathing engines are generally termed as rocket engines/motors. Note that the term rocket used in the literature is more broad-based, and its definition can be conceived as "the device in which mass contained in the vehicle is ejected

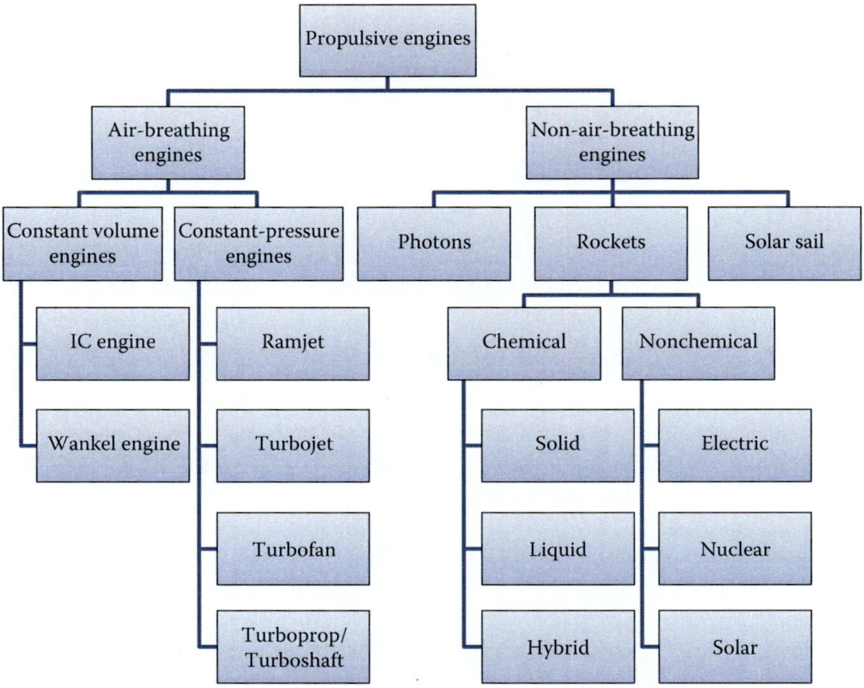

FIGURE 1.4 Classification of propulsive devices.

rearward by which thrust is produced for its propulsion." Hence, we will use rocket engine in place of rocket motor in this book. Rocket engines are broadly divided into two categories: chemical and nonchemical. Based on the type of energy used for propulsion, nonchemical rockets are divided into solar, electrical, and nuclear engines. The nonchemical engines can be further classified into various engines, which will be discussed in Chapter 11. Chemical rockets, based on the type of propellant used, are divided into three categories—solid, liquid, and hybrid—as shown in Figure 1.4.

In this book, we are mainly concerned with chemical propulsive devices in which chemical energy resulting from the breaking or forming of chemical bonds is used to impart the requisite force for propulsion to take place. Other forms of energy—electrical, nuclear, plasma, and so on—have been used to propel rocket engines, which will be discussed briefly as a separate chapter.

1.4.1 Comparison of Air-Breathing and Rocket Engines

We know that for propelling an aircraft, piston and gas turbine engines use air from the atmosphere as these vehicles are confined within the earth's atmosphere. Besides, ramjet engines, employed mostly for missile application, come under the category of air-breathing engines. We know that an air-breathing engine's propulsive efficiency drops down to low values when the flying altitude exceeds 25 km. Therefore, one must think of a non-air-breathing propulsive device that can carry both fuel and oxidizer along with it. Such a device is known as a rocket engine, as mentioned earlier. Thus, a rocket engine is a non-air-breathing jet propulsive device that produces the required thrust by ejecting high-pressure and high-temperature gas from a propellant burning through a CD nozzle. Let us compare both air-breathing and non-air-breathing (rocket) engines as enumerated in Table 1.1.

TABLE 1.1 Comparison of Air-Breathing and Rocket Engine

Air-Breathing Engine	Rocket Engine
1. It uses atmospheric air for combustion.	It carries its own oxidizer and fuel.
2. It cannot operate in space (vacuum).	It can operate in atmosphere and space.
3. Its performance is dependent on altitude and flight speed.	Its performance does not depend on altitude and flight speed.
4. The thrust developed by this engine is dependent on flight speed.	The thrust of this engine is independent of flight speed.
5. It cannot operate beyond supersonic speed ($M = 5.0$).	It can operate at any flight velocity (Mach number: $M = 0$–25).
6. Rate of climb decreases with altitude.	Rate of climb increases with altitude.
7. Flight speed is always less than jet velocity.	Flight speed is not limited.

1.5 TYPES OF ROCKET ENGINES

In the previous section, we have learnt that the rocket engine can be broadly classified into two categories: (1) chemical rockets and (2) nonchemical rockets. Based on the physical form of the chemical propellant, chemical rocket engines can be divided into three categories: (1) solid propellant, (2) liquid propellant, and (3) hybrid propellant. But the nonchemical engines based on type of energy are further classified into three categories: (1) nuclear rockets, (2) electrical rockets, and (3) solar rockets. In this book, we will be mostly discussing chemical rocket engines. However, a brief account of other nonchemical rocket engines will be provided in Chapter 11. Both chemical and nonchemical rocket engines are described in the following section.

1.5.1 Chemical Rocket Engines

In case of chemical rocket engines, chemical energy released during the burning of fuel and oxidizer is used to raise the temperature and pressure of the gas which is expanded in a CD nozzle to produce thrust. Generally, the hot gases at high pressure are accelerated to high supersonic velocities in the range of 1500–4000 m/s for producing thrust. It may be noted that both fuel and oxidizer are being carried along with the engine unlike in air-breathing engines. Based on the physical state of the propellant (fuel and oxidizer), as discussed earlier, chemical rocket engines can be broadly divided into three categories: (1) solid propellant, (2) liquid propellant, and (3) hybrid propellant. These engines are described here briefly.

1.5.1.1 Solid-Propellant Rocket Engines

We know that solid-propellant rocket engine (SPRE) is one of the oldest non-air-breathing engines as it is believed to have been used in China as early as the thirteenth century onward for war purposes. The solid propellant composition, which was initially black powder, underwent a series of changes with time. Currently, solid propellants have found a wider application in various propulsion and gas-generating systems. It has a wide range of thrust levels ranging from a few N (Newton) to several hundred N. Besides having a solid form, this propellant can be stored in the combustion chamber ready for use for a longer period of time, on the order of 10–20 years, provided they are hermitically sealed. Compared to other types of chemical rocket engines, these are economical, reliable, and simple. Hence, these engines find a wider range of both civilian and military applications.

Let us consider a simple SPRE as shown in Figure 1.5a which basically consists of the major components that are a solid propellant, a combustion chamber, an igniter, and a nozzle. Note that the propellant, which mainly consists of fuel, oxidizers, and various additives, is entirely stored within the combustion chamber in the form of blocks of definite shape called grain and is supported by the walls or by special grids, traps, or retainers (not shown in Figure 1.5a). Note that this grain contributes to around 80%–95% of the total mass of an SPRE. Hence, the performance of this kind of engine and its payload capability are dependent on the optimal design of the grain, which will be discussed in Chapter 7. The igniter initiates the combustion process on the surface of the propellant when actuated with the help of an electrical switch, whose details will be discussed in Chapter 7. As a result, the propellant grains will start burning and filling the empty combustion chamber, hence building up the chamber pressure. Subsequently, the high-temperature and high-pressure gases are expanded in the supersonic nozzle to produce the requisite thrust. Generally, these nozzles are made of high-temperature materials, namely, metals with graphite coating, and are ablative materials that can take a high thermal load with minimal

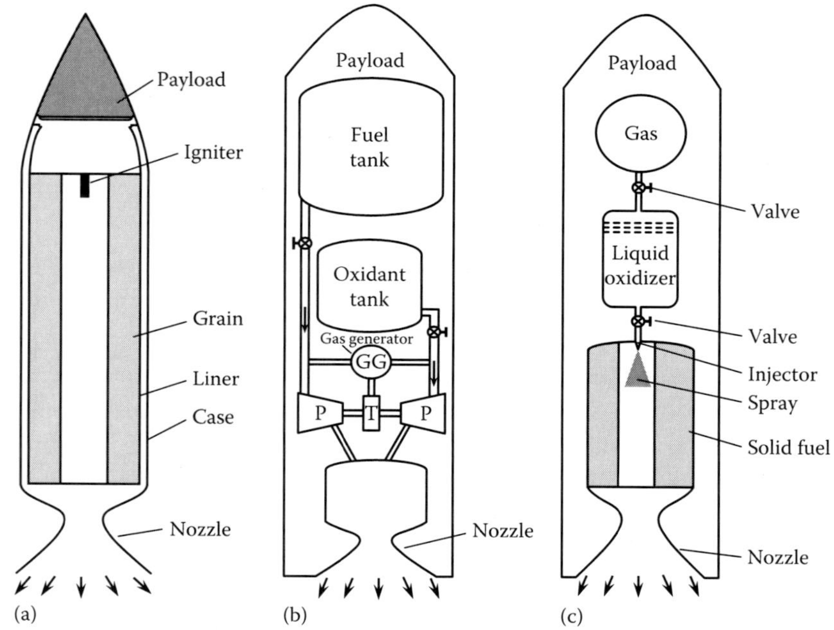

FIGURE 1.5 Three types of chemical rocket engines: (a) solid propellant, (b) liquid propellant, and (c) hybrid propellant.

corrosion. Generally, a fixed nozzle is preferred in SPREs as shown in Figure 1.5a. Hence, a solid rocket engine is considered to be a non-air-breathing vehicle without any moving parts. But in recent times, the gimbaled nozzle is being used for controlling the direction of thrust.

As discussed earlier, the main characteristic of a solid propellant rocket engine (SPRE) is its simplicity. In view of its simplicity, the SPRE is particularly well suited for developing very high thrust within a short interval of time, particularly in the booster phase. With recent advancements in propellant chemistry, it can be used for fairly long burning time (sustainers).

Advantages

- It is simple to design and develop.
- It is easier to handle and store unlike liquid propellant.
- Detonation hazards of many modern SPREs are negligible.
- Better reliability than Liquid Propellant Rocket Engine (LPRE) (>99%).
- Much easier to achieve multistaging of several motors.
- The combination pressure in SPREs is generally higher than in LPREs since it is not subject to the limitation of a feed system.
- Development and production cost of SPREs is much smaller than that of LPREs, especially in the high-thrust bracket.

Disadvantages

- It has lower specific impulse compared to LPREs and hybrid propellant rocket engines (HPREs).
- It is difficult to turn off its operation unlike in an LPRE.
- Transport and handling of solid propellants are quite cumbersome.
- It is difficult to use the thrust vector control and thrust modulation.
- The cracks on the propellant can cause an explosion.
- Careful design of the nozzle is required as active cooling cannot be used.
- The erosion of the throat area of the nozzle due to high-temperature solid particles can affect its performance adversely.

1.5.1.2 Liquid Propellant Rocket Engines

Recall that around 1927, an American professor, Robert Goddard, had designed and developed an LPRE, which had flown only 46 m. Subsequently, the Germans took this technology to a mature level that culminated in the famous V2 rocket engine. Currently, the LPRE has found a wider application in various propulsion and gas-generating systems. It has a wide range of thrust levels ranging from a few N (Newton) to several hundred N. In addition to having a liquid form, this propellant can be stored in a separate tank and can be controlled easily, and hence thrust can be varied easily unlike in an SPRE. As LPREs are stored in separate tanks unlike SPRE, one can achieve a higher level of thrust and is thus considered to be more powerful than an SPRE. Therefore, it is preferred for large spacecraft and ballistic missiles. However, the design of an LPRE is quite complex and requires specialized nozzles. Compared to other types of chemical rocket engines, LPREs are compact, light, economical, and highly reliable. Hence, they have a wider range of both civilian and military applications.

Let us consider a simple LPRE as shown in Figure 1.5b, which basically consists of major components, namely, a propellant feed system, a combustion chamber, an igniter system, and a nozzle. Note that both fuel and oxidizer propellants are stored separately in special tanks at high pressure. Of course the propellant feed system along with the propellant mass contributes significantly to the mass of the engine but it is significantly less compared to the total mass of an SPRE. In fact, sometimes, the mass of the nozzle for deep-space applications is comparable to the propellant mass and its feed system in the case of an LPRE.

The pressurized liquid propellants are converted into spray consisting of arrays of droplets with the help of atomizers as shown in Figure 1.5b. Of course, an igniter is used to initiate the combustion process on the surface of the propellant. As a result, the propellant will start burning and fill up the empty thrust chamber, thereby building up pressure in the chamber similar to that of other chemical rocket engines. Subsequently, these high-temperature and high-pressure gases are expanded in a CD nozzle to produce the requisite thrust. As mentioned earlier, these nozzles are made of high-temperature materials, namely, metals with graphite coating and ablative materials that can take a high thermal load with minimal corrosion. It may be noted that propellant feed lines have several precision valves with whose help the operations of such kinds of rocket engine can be started and shut off at will, and hence repetitive operation is possible for this engine unlike the SPRE. The advantages and disadvantages of an LPRE are enumerated as follows.

Advantages

- An LPRE can be reused.
- It provides greater control over thrust.
- It can have higher values of specific impulse.
- In case of emergency its operation can be terminated very easily.
- It can be used on pulse mode.
- It can be used for long-duration applications.
- It is easy to control this engine as one can vary the propellant flow rate easily.
- The heat loss from the combustion gas can be utilized for heating the incoming propellant.

Disadvantages

- This engine is quite complex compared to the SPRE.
- It is less reliable as there is a possibility of malfunctioning of the turbopump injectors and valves.
- Certain liquid propellants require additional safety precaution.
- It takes much longer to design and develop.
- It becomes heavy, particularly for short-range application.

1.5.1.3 Hybrid Propellant Rocket Engines

In order to achieve better performance, elements from SPREs and LPREs are combined to devise a new engine known as the HPRE. Note that this engine can use both solid and liquid types of propellants. All permutation and combination of propellants can be used for this kind of engine. But the most widely used propellant combination is a liquid oxidizer along with a solid propellant. Let us consider a simple HPRE as shown in Figure 1.5c, which basically consists of major components, namely, a propellant feed system, a combustion chamber, a solid fuel grain, an igniter system, and a nozzle. Note that only the oxidizer propellant in the present example is stored in a special tank under high pressure. The pressurized propellants are converted into spray consisting of arrays of droplets with the help of atomizers as shown in Figure 1.5c. Some of the propellant evaporates due to the recirculation of hot gases and comes into contact with the gaseous fuel that emanates from the solid fuel grains due to pyrolysis. The combustion products start burning and fill the empty thrust chamber, thereby building up pressure inside the chamber similar to that of other chemical rocket engines. In a similar manner, thrust is produced due to the expansion of these high-temperature and high-pressure gases in a supersonic nozzle. It may be noted that the liquid propellant feed line has a few valves with which the operation of such rocket engines can be controlled at will. Hence, it can find applications in missions that need throttling, restart, and long range. It has similar features to an LPRE, namely, compact, light, economical, and highly reliable. Besides, these engines may have better performance compared to both solid and liquid engines. Hence, these engines may find a wider range of both civilian and military applications, although these are still in research stages. Let us learn the advantages and disadvantages of an HPRE.

Advantages

- An HPRE can be reused.
- It provides greater control over thrust.
- It has relatively lower system cost compared to the LPRE.
- It can have higher values of average specific impulse compared to the SPRE.
- It has higher density of specific impulse than that of the LPRE.
- It has higher volume utilization compared to the LPRE.
- It has start–stop–restart capability.

Disadvantages

- This engine is quite complex compared to the LPRE.
- Its mixture ratio varies to some extent and hence it is quite difficult to achieve steady-state operation.

- It has lower density of specific impulse compared to SPRE.
- There is underutilization of solid fuel due to a larger sliver of residual grain at the end of the operation.
- Certain liquid propellants require additional safety precaution.
- It takes much longer to design and develop.
- It becomes heavy, particularly for short-range application.
- It has an unproven propulsion system for large-scale applications.

1.5.2 Nonchemical Rocket Engines

We use chemical rocket engines liberally for most applications, but nonchemical rocket engines are being designed and developed for quite some time now as for certain missions it is undesirable to use chemical rocket engines due to their higher propellant mass per unit impulse. In some deep-space applications, nonchemical rocket engines are much sought after. Of course, this book is mostly concerned with chemical rocket engines, but for the sake of completeness, we will briefly discuss nonchemical engines in this section. Based on the source of energy, these engines can be broadly divided into three categories: (1) electrical rocket engines, (2) nuclear rocket engines, and (3) solar rocket engines. Of course, there are several other alternative rocket engines which are not being discussed in this section due to their limited uses. We will discuss nonchemical rocket engines in detail in Chapter 11.

1.6 APPLICATIONS OF ROCKET ENGINES

In comparison to other air-breathing engines, rocket engines have very specific applications, which can be broadly divided into four types: (1) space launch vehicle, (2) spacecraft, (3) missile, and (4) miscellaneous.

1.6.1 Space Launch Vehicle

Space launch vehicles are meant for launching satellites or certain payloads into space or into spacecraft. They are generally designed and developed for certain specific applications. Starting from the first launch vehicle Sputnik in 1957 by the Russians, several space launch vehicles have been built by various agencies across the globe. Note that each space launch vehicle is designed for a specific mission, such as placing a satellite into a certain orbit, a mission to the moon, or space exploration. Space launch vehicles have several kinds of payloads, namely, military, civilian satellite, space exploration, commercial, tourist, and so on. The military applications span over command and control satellites, reconnaissance satellites, and so on. Similarly, civilian launch vehicles are used for weather forecasting, mapping of seismic zones, geo-positioning satellites (GPS), communicating satellites, and so on. Space exploration is another area for which space launch vehicles are being developed. In addition, satellites for commercial and touristic purposes are being launched in recent times. Of course, the configuration of each space launch vehicle for these applications can differ from the other. Even for the same mission requirement, the configuration of the space launch vehicle will depend on the prior experiences of designers, available resources, the involved agency in the development of such vehicles, and so on. Based on the number of stages, these can be classified as single-stage, double-stage, triple-stage, and so on. The number of stages depends on the specific space trajectory, types of maneuvers, types of propellant, and so on. Generally, the first stage is called the booster as it provides the initial thrust to overcome the initial inertia of the launch vehicle. Subsequently, this stage is separated from the moving vehicle. By staging launch vehicles, its payload capability increases significantly. Such kind of staging is essential for placing a satellite into a higher orbit. It is a must for manned space mission. Hence, the vehicles may be labeled manned or unmanned vehicles. As mentioned earlier, based on the types of propellant, they are classified as solid, liquid, or hybrid launch vehicles. Of course, in real space launch vehicles, both solid and liquid engines can be used either together or individually in a single stage. Moreover, space launch vehicles can use both non-cryogenic and cryogenic engines as in the case of the geo-synchronous launch vehicle (GSLV) of India. In recent times, a reusable single-stage device to orbit the launch vehicle is being designed and developed across the globe which may be adopted in the future.

1.6.2 Spacecraft

A spacecraft is a vehicle that is designed and developed to travel in space. It can be piloted like a space shuttle or be unpiloted. Spacecraft can be employed for various applications, namely, earth observation, meteorology, navigation, planetary exploration, and space colonization. Depending on its application, it can be classified as interplanetary, manned/unmanned spacecraft, or trans-solar vehicle. Some of the primary functions of a spacecraft are orbit insertion and orbit change maneuvers and space flight. Besides this, the spacecraft has several secondary functions such as attitude control, spin control, momentum wheel and gyro unloading, and stage separation. In order to execute these functions, a spacecraft can carry a number of rocket engines. Pulsed small rockets with short bursts of thrust are also being used for attitude control. It may be noted that most spacecraft use an LPRE along with a solid rocket engine as a booster at the time of their launch. Generally, electric rocket engines are used for both primary and secondary functions during the flight of spacecraft in the space region as they can enable long-duration flights.

1.6.3 Missile

A missile is a self-propelled guided weapon system that is propelled by a rocket engine. The first missile known as the V1 flying bomb was built and used by the Germans during World War II. A modern missile has four major components: a rocket engine, targeting, guidance, and warhead systems. Several kinds of missile have been developed in various countries for winning wars and colossal amounts of taxpayers' money are being misused in fear of war. In other words, science and technological knowledge are being abused while misusing natural resources that could have been used for the development of humanity on this beautiful earth. A missile can be broadly divided into two categories: (1) ballistic and (2) cruise. A ballistic missile follows a ballistic flight path to deliver the warheads to a particular target. Its flight path is mainly guided by gravity except in the initial period briefly after its launch. Ballistic missiles are preferred for long-range and land attack missions where accuracy is not very important. These missiles can be launched from mobile launchers, ships, submarines, and underground silos. In recent times, long-range intercontinental ballistic missiles (ICBMs) are being designed and developed by several countries for carrying nuclear warheads. If some of these missiles are used by militants just imagine what could happen to life on this beautiful earth. I wish and hope that designers and scientists keep in mind the onset of a possible holocaust during their research work.

In order to win a war, several kinds of cruise missiles are being designed and developed that can deliver warheads accurately to the predetermined target as their flight paths are guided and controlled unlike ballistic missiles. Based on their launch and application, several kinds of cruise missiles, namely, surface-to-surface, air-to-surface, surface-to-air, and air-to-air, are being evolved which are propelled by chemical rocket engines. Most of the missiles use SPREs as they are quite simple to design and develop. Of course, LPREs are also used in missile, as in Bramos (India).

1.6.4 Other Civilian Applications

Rocket engines have been used for several other civilian applications depending on the imagination of people. Rocket engines that are used routinely for collecting weather predictions are commonly known as "sounding rockets." Rocket engines are also used for research in airplanes, in rocket-assist takeoff, and for providing lifelines to ships under distress. Besides this, they are considered for use in developing propulsion belts. Rockets can be used in the future for providing relief materials to inaccessible places during natural calamities.

REVIEW QUESTIONS

1. Explain the concept of jet propulsion.
2. Recount the history of rocket propulsion.
3. What was the progress made in the development of rocket engines during AD 1900–2000?
4. What are the differences between a rocket and a gas turbine engine?
5. What is meant by a photon engine? How is it different from a jet engine?
6. What is the difference between rocket engine and photon engine?
7. What do you mean by solar sail? What is the basic principle of this engine? How is it different from photon engine?
8. Enumerate the relative advantages and disadvantages of solid-propellant and liquid-propellant engines?
9. What are the relative advantages and disadvantages of chemical and non-chemical rocket engines?
10. What are the different applications of rocket engines?

REFERENCES AND SUGGESTED READINGS

1. Sutton, G.P. and Ross, D.M., *Rocket Propulsion Elements*, 5th edn., John Wiley & Sons, New York, 1975.
2. Barrere, M., Jaumotte, A., de Veubeke, B.F., and Vandenkerckhove, J., *Rocket Propulsion*, Elsevier Publishing Company, New York, 1960.
3. Treager, I.E., *Aircraft Gas Turbine Engine Technology*, 3rd edn., McGraw-Hill Inc., New York, 1995.

4. Mishra, D.P., *Gas Turbine Propulsion*, MV Learning, London, U.K., 2015.
5. Timnat, Y.M., *Advanced Chemical Rocket Propulsion*, Academic Press, London, U.K., 1987.
6. Turner, M.J.L., *Rocket and Spacecraft Propulsion*, Springer Verlag, Heidelberg, Germany, 2001.
7. Hill, P.G. and Peterson, C.R., *Mechanics and Thermodynamics of Propulsion*, Addison Wesley Publishing Company, Reading, MA, 1965.

CHAPTER 2

Aerothermochemistry of Rocket Engines

> Great spirits have always encountered violent opposition from mediocre minds.
>
> ALBERT EINSTEIN

2.1 INTRODUCTION

The knowledge of thermochemistry and gas dynamics is essential for the understanding of rocket engine. Some of these topics might have been covered in your previous courses. However, it is important to recapitulate these topics for better understanding of rocket propulsion. Hence, the purpose of this chapter is to review briefly the rudiments of thermochemistry and gas dynamics from the perspective of rocket engine. Firstly, we will review the basics of classical thermodynamics, starting with its definition and its various terminologies.

2.2 BASIC PRINCIPLES OF CHEMICAL THERMODYNAMICS

The term thermodynamics is a combination of two Greek words, namely, *therme*, meaning "heat," and *dynamics*, meaning "work." Thus, thermodynamics deals with work and heat and their interaction between system and surrounding. Recall that thermodynamics is based mainly on three concepts: energy, equilibrium, and entropy. The detailed treatments of thermodynamics are given in textbook [1] and certain fundamental aspects are discussed briefly here.

2.2.1 Basic Definitions

We know that a system is regarded as a collection of materials with fixed identity in a region of space with certain boundaries. The system boundary can be flexible or rigid, across which energy in the form of heat or work can be transferred. Note that the system boundary can change its size, shape, or position, as in a piston cylinder arrangement, when the system interacts with its surrounding. The boundary need not necessarily be real; it may even be an imaginary one. Everything external to the system is known as the surrounding. Usually, the term surrounding means those things outside the system boundary that interact in certain ways with the system. A system and its surrounding are known as a universe. The system can be conveniently classified into three categories: (1) closed system, (2) open system, and (3) isolated system. If a system consists of a particular quantity of matter such that no matter will cross its boundary, we can call it a closed system. But in an open system, both matter and heat energy can flow across the system boundary. Note that no matter but heat energy is transferred across the boundary of a closed system. In case of isolated system, neither mass nor energy is transferred through the system boundary. The rocket engine can be considered as an open system in which propellants after getting burnt leave as exhaust gas. Hence the control volume approach rather than control mass approach is used for analyzing the various components of rocket engine, namely, combustion chamber and nozzle in the present book.

During the aerothermochemistry analysis of any component of a rocket engine, we have to deal with the macroscopic properties of gas such as pressure, temperature and volume, density, and mass fraction. For gaseous phase system, we need to invoke the ideal gas law. Besides this, the concept of equilibrium needs to be invoked during the thermodynamic analysis of rocket engine. Note that a system is said to be in the state of thermodynamic equilibrium only when its mechanical, thermal, and chemical equilibrium are satisfied.

2.3 THERMODYNAMIC LAWS

The various components of rocket engines can be analyzed using the laws of thermodynamics. The first, second, and third laws of thermodynamics are briefly discussed in the following. For the details of these laws and derivations, one can refer to some standard thermodynamic books [1–3].

2.3.1 First Law of Thermodynamics

The first law of thermodynamics states that energy can be neither created nor destroyed but only transforms from one form to another. In other words, whenever energy transfer takes place between a system and its surrounding, there will be a change in the system. To illustrate this, let us consider gas confined in a piston cylinder as a closed mass system, as shown in Figure 2.1. Let δQ be the heat added to the system and δW be the work done by the system on the surrounding by the displacement of boundary. The heat added to the system and work done by the system cause a change in energy in the system. According to the first law of thermodynamics, this is given by

$$dE_t = \delta Q - \delta W; \quad dE_t = dU + dKE + dPE \tag{2.1}$$

where

dE_t is the total energy content of the system

E_t is the internal energy (U) + kinetic energy (KE) + potential energy (PE)

Note that dE_t is the exact differential that exists for state function and its value depends on the initial and final states of the system. In contrast, δQ and δW depend on the path followed by the process. Keep in mind that the first law of thermodynamics is an empirical relation between heat, work, and internal energy, first put forward by Joule in 1851 through conducting a series of experiments. Of course, it has been verified thoroughly by subsequent experimental results. To date, nobody has managed to disprove this law.

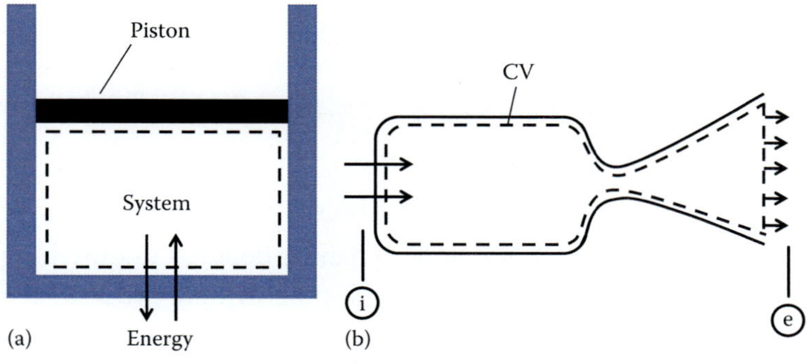

FIGURE 2.1 Schematic of thermodynamic systems: (a) piston cylinder arrangement and (b) rocket engine.

2.3.2 First Law for Control Volume

In order to analyze various components of rocket engine, we need to consider a control volume (CV), as shown in Figure 2.1b. Hence the control volume approach rather than control mass approach is used for analyzing the various components of rocket engine, namely, combustion chamber and nozzle in the present book. Before applying the first law of thermodynamics, let us consider the assumptions being made for deriving this law:

- The control volume is fixed with respect to the coordinate system.
- No need to consider the work interactions associated with the moving body.
- Uniform flow conditions over the inlet and the outlet flow areas.

By invoking these assumptions, the first law of thermodynamics for the CV is given by

$$\frac{dE}{dt}=\left[\dot{m}_i\left(h_i+V_i^2/2+gZ_i\right)-\dot{m}_o\left(h_o+V_o^2/2+gZ_o\right)\right]+\dot{Q}-\dot{W}_{sh} \tag{2.2}$$

where

$\dot{m}$ is the mass flow rate
h is the enthalpy
V is the velocity
Z is the height
E is the total energy
$\dot{Q}$ is the energy transfer rate as heat
$\dot{W}_{sh}$ is the shaft power

This form of the first law of thermodynamics for CV is very useful for the analysis of components in the rocket engine. The reader can refer to certain standard thermodynamics books [3–5] to get acquainted with this important thermodynamic law.

2.3.3 Second Law of Thermodynamics

We know that the first law of thermodynamics does not say anything about the feasibility of direction in which the process may proceed. However, the second law of thermodynamics stipulates the direction of the process. According to the famous scientist Kelvin Plank, the second law of thermodynamics states that it is impossible to construct a cyclically operating device such that it produces no other effects than absorbing energy as heat from a single thermal reservoir and performs an equivalent amount of work. In other words, it is impossible to have a heat engine with thermal efficiency of 100%. Further, the second law of thermodynamics defines an important property of a system, known as entropy, which is expressed as

$$dS = \frac{\delta Q}{T} \quad \text{(reversible process)} \tag{2.3}$$

where dS is the change of entropy of the system during an incremental reversible heat exchange δQ, when the system is at temperature T. The entropy being a state variable can be used either for reversible or irreversible process. An alternative, more general, relation for the entropy is

$$dS = \frac{\delta Q}{T} + dS_{irrev} \quad \text{(irreversible process)} \tag{2.4}$$

where

δQ is the actual amount of heat added to the system during which entropy change

dS_{irrev} is generated due to the irreversible, dissipative phenomena

Note that irreversible processes are caused due to friction, heat transfer with finite temperature gradient, and mass transfer with finite concentration gradient. These irreversible processes due to their dissipative nature always result in increase of the entropy as given in the following:

$$dS_{irrev} \geq 0 \tag{2.5}$$

Combining Equations 2.4 and 2.5, we have

$$dS \geq \frac{\delta Q}{T} \tag{2.6}$$

For adiabatic process ($\delta Q = 0$), Equation 2.6 becomes

$$dS \geq 0 \tag{2.7}$$

Note that this expression derived from the second law of thermodynamics (Equation 2.7) indicates the direction in which the process can proceed. A process can proceed either in the direction of increasing entropy or constant entropy of the system and its surrounding. The process cannot proceed in the direction of decreasing entropy. Engineers attempt to reduce the irreversibility to a large extent to enhance the performance of any practical device. The change in specific entropy of a system, ds, can be determined by assuming the heat interaction to be taking place in reversible manner by using the following relation:

$$\delta q = Tds \tag{2.8}$$

where

q is the heat per unit mass
s is the entropy per unit mass

Substituting Equation 2.8 in the energy equation (Equation 2.1), we get

$$Tds = du + Pdv \tag{2.9}$$

From the definition of enthalpy, we have

$$dh = du + Pdv + vdP \tag{2.10}$$

For thermally perfect gas, we can assume $dh = C_p\, dT$. Substituting this relation in Equation 2.10 and using Equation 2.9, we get

$$ds = C_p \frac{dT}{T} - R\frac{dP}{P} \tag{2.11}$$

For an isentropic (reversible and adiabatic) process, $ds = 0$, we can integrate Equation 2.11 between state (1) and (2):

$$\frac{P_2}{P_1} = \left(\frac{T_2}{T_1}\right)^{C_p/R} \tag{2.12}$$

However, for a calorically perfect gas, specific heat can be expressed in terms of specific heat ratio γ, as follows:

$$\frac{C_p}{R} = \frac{\gamma}{\gamma - 1} \tag{2.13}$$

Using Equation 2.13, we may express Equation 2.12 in the form

$$\frac{P_2}{P_1} = \left(\frac{T_2}{T_1}\right)^{\frac{\gamma}{\gamma-1}} \tag{2.14}$$

By employing the perfect gas law, we can write the isentropic relation as

$$\frac{P_2}{P_1} = \left(\frac{\rho_2}{\rho_1}\right)^{\gamma} = \left(\frac{T_2}{T_1}\right)^{\frac{\gamma}{\gamma-1}} \tag{2.15}$$

We will use this equation very often while dealing with rocket engine. This expression for isentropic process can be easily applicable for the flow outside the boundary layer. As in the case of nozzle flow, since the thickness of boundary layer is very thin in comparison to entire flow domain, one can easily analyze it by assuming the flow to be isentropic except in its boundary layer. Thus, this isentropic relation (Equation 2.15) can be used for the analysis of a wide range of practical problems in rocket engine.

2.4 REACTING SYSTEM

In a rocket engine, propellant is burnt to produce high-temperature and high-pressure gas while undergoing chemical reactions. In the chemical reaction, mass balance describing exactly how much oxidizer has to be supplied for complete combustion of certain amount of fuel is generally termed as stoichiometry. The details of the reacting system discussed in specialized books on combustion [2,3] are reviewed in this section.

2.4.1 Stoichiometry

The word "stoichiometry" basically originates from the Greek *stoikhein*, which means element, and *metron*, which means measure. In other words, the term "stoichiometry" literally means the measure of elements. Hence, an elemental balance is to be carried out to estimate stoichiometric ratio for a chemical reaction. In other words, stoichiometry is the relationship between the mass of the reactants and the products of a chemical reaction. Let us consider an example in which certain amount of methane is burnt in the presence of certain amount of oxygen undergoing chemical reaction and liberating certain amount of heat energy. The chemical reaction that represents such a reaction is shown in the following:

$$H_2 + 1/2O_2 \rightarrow H_2O + \Delta H_R \tag{2.16}$$

$$(2\ g) + (16\ g) \rightarrow (18\ g)$$

It can be observed from this reaction that 1 mol of hydrogen is reacted with 1/2 mol of oxygen to produce 1 mol of water. Note that this is a balanced equation in which numbers of chemical elements are balanced. It is interesting to note that 2 g of hydrogen can react with 16 g of oxygen to produce 18 g of water. That means the mass in the left side of reaction is the same as that in the right side, thus stating that mass is conserved. This is the underlying principle of chemical reactions that no mass is created or destroyed in a chemical reaction. In contrast, number of moles need not be conserved. Such a reaction is known as stoichiometric reaction. Hence the quantity of oxidizer just sufficient to burn a certain quantity of fuel completely in a chemical reaction is known as **stoichiometry**. It is known as the ratio of oxidizer to fuel that is sufficient to burn fuel, leading to the formation of complete products of combustion. In the example cited, the **stoichiometry**, (m_{ox}/m_{fuel}) would be 8. In combustion, oxidizer/fuel ratio always may not to be in stoichiometric proportion. On several occasions, excess oxidizers are supplied to ensure complete combustion of fuel in practical devices. When more than a stoichiometric quantity of oxidizer is used, the mixture is known as fuel lean or **lean** mixture. In contrast, if less quantity of oxidizer than a stoichiometric quantity is present, then the mixture is known as fuel-rich or **rich** mixture. On several occasions, hydrocarbon fuels are burnt in the presence of air. For a hydrocarbon fuel represented by C_xH_y, the stoichiometric relation is given as

$$C_xH_y + a(O_2 + 3.76N_2) \rightarrow xCO_2 + (y/2)\, H_2O + 3.76a\, N_2 \qquad (2.17)$$

In this reaction, the air is assumed to be consisting of 21% O_2 and 79% N_2 by volume for simplicity and this will be considered throughout this book. When the reaction is balanced, "a" turns out to be $x + y/4(a = x + y/4)$. Then, the stoichiometric air–fuel ratio would be

$$\left(A/F\right)_{stoic} = \left(m_{ox}/m_{fuel}\right) = 4.76\,a\left(MW_{air}/MW_{fuel}\right) \qquad (2.18)$$

where MW_{air} and MW_{fuel} are the molecular weights of air and fuel, respectively. The stoichiometric air-fuel ratio can be obtained using Equation 2.18 for any hydrocarbon fuels such as for methane, propane, butane, and so on. It can be noted that the stoichiometric air–fuel ratio for methane is 17.16, while for higher hydrocarbons such as propane and butane, it is around 15. Also, it must be appreciated that many more times of oxidizer than fuel has to be supplied for complete combustion of fuel even theoretically. But in practice, one has to supply the oxidizer in a larger proportion than the stoichiometric ratio to ensure complete combustion. When the ratio other than stoichiometric air–fuel ratio is used, one of the useful quantities known as equivalence ratio is employed to describe the air–fuel mixture. The **equivalence ratio φ** is defined as

$$\varphi = \frac{(F/A)}{(F/A)_{stoic}}; \quad \Rightarrow (F/A) = \varphi (F/A)_{stoic} \tag{2.19}$$

Note that the equivalence ratio is a nondimensional number in which fuel–air ratio is expressed in terms of mass. This ratio describes quantitatively whether a fuel–oxidizer mixture is rich, lean, or stoichiometric. From Equation 2.19, it is very clear that the stoichiometric mixture has an equivalence ratio of unity. For fuel-rich mixture, the equivalence ratio is greater than unity ($\boldsymbol{\varphi} > 1$). Lean mixture has less than unity ($\boldsymbol{\varphi} < 1$). By lean and rich mixture, we mean basically the extent of fuel in the mixture. In several combustion applications, the equivalence ratio is one of the most important parameters that dictate the performance of the system. Another parameter used very often to describe relative stoichiometry is the **percent stoichiometric air**, which can be related to the equivalence ratio as

$$\%\text{stoichiometric air} = \frac{100\%}{\varphi} \tag{2.20}$$

The other useful parameter to define relative stoichiometry of mixture is **percent excess air**:

$$\%\text{excess air} = \frac{(1-\varphi)}{\varphi} \times 100 \tag{2.21}$$

Example 2.1

Ethyl alcohol is burnt with oxygen. The volumetric analysis of products on dry basis is CO_2 = 92.51%, O_2 = 6.76%, CO = 0.73%. Determine (i) *F*/*A* ratio, (ii) equivalence ratio, and (iii) % stoichiometric air.

Solution

$$XC_2H_5OH + aO_2 \rightarrow 92.51CO_2 + 0.73CO + 6.76O_2 + bH_2O$$

By mass balance, we can have

$\mathbf{C}: 2X = 92.51 + 0.73; \rightarrow X = 46.62$

$\mathbf{H}: 6X = 2b; \rightarrow b = 139.86$

$\mathbf{O}: 2a + X = 92.51 \times 2 + 0.73 + 2 \times 6.76 + 139.86$

$a = 146.25$

Then, the balance equation becomes

$46.62C_2H_5OH + 146.25O_2 \rightarrow 92.51CO_2 + 0.73CO + 6.76O_2 + 139.86H_2O$

Let us recast the preceding chemical reaction in terms of 1 mol of fuel, as follows:

$C_2H_5OH + 3.14(O_2 + 3.76N_2) \rightarrow 1.98CO_2 + 0.016CO + 0.15O_2 + 3H_2O$

Then the fuel/air ratio by mass becomes

$$\left(\frac{F}{A}\right)_{actual} = \frac{m_F}{m_A} = \frac{(12 \times 2 + 6 + 16)}{(3.14 \times 32)} = 0.457$$

In order to calculate the equivalence ratio, we will have to obtain the stoichiometric fuel–air ratio by considering a balance equation, as follows:

$$C_2H_5OH + 3O_2 \rightarrow 2CO_2 + 3H_2O$$

Then, the stoichiometric fuel–air ratio becomes

$$\left(\frac{F}{A}\right)_{Stoic} = \frac{m_F}{m_A} = \frac{(12 \times 2 + 6 + 16)}{3 \times 32} = \frac{46}{96} = 0.479$$

Note that the stoichiometric air–fuel ratio is 15.05. Then, the equivalence ratio becomes

$$\varphi = \frac{(A/F)_{stoic}}{(A/F)} = \frac{(F/A)}{(F/A)_{stoic}} = \frac{0.457}{0.479} = 0.95$$

Hence, this mixture is a slightly rich one. Then, let us calculate percent of stoichiometric air, as follows:

$$\%\text{stoichiometric air} = \frac{100\%}{\varphi} = \frac{100}{0.95} = 105.26\%$$

2.4.2 Ideal Gas Mixture

We have learnt that the fuel and oxidizer can occur in three forms, namely, solids, liquids, and gases. The actual measured properties of solid and liquid substances can be used, which can be expressed in the form of the thermodynamic equation of state. But one might assume that such kind of equation of state may be quite restrictive in nature. But the thermodynamic equation of state for ideal gaseous fuel and oxidizer is quite broad and can be applied for combustion systems even though it deals with a mixture of gases. Note that the properties of a mixture can be found out by assuming it to be an ideal gas. We know that an ideal gas obeys equation of state, which is given in the following:

$$PV = nR_uT \tag{2.22}$$

where

P is the pressure
V is the volume of the gas
T is the temperature of the gas
n is the number of moles of gas
R_u is the universal gas constant (R_u = 8.314 kJ/kmol K)

But when we are dealing with a mixture of gases, we can find out the properties of the mixture from individual gases by applying the Gibbs–Dalton Law.

Let us consider a container C, which contains multicomponent mixture of gases composed of m_A grams of species A, m_B grams of species B, and m_i grams of ith species. Then, the total number of moles, m_{tot}, in the container would be given by

$$m_{tot} = m_A + m_B + \cdots + m_i \tag{2.23}$$

By dividing Equation 2.23 by total mass m_{tot}, we will get

$$1 = Y_A + Y_B + Y_C + \cdots + Y_i = \sum Y_i \tag{2.24}$$

where $Y_A(= m_A/m_{tot})$ is the mass fraction of species A. It can be noted that the sum of all mass fraction of individual species in a mixture is equal to unity. Similarly, we can have a relationship for total number of moles:

$$n_{tot} = n_A + n_B + \cdots + n_i \tag{2.25}$$

By dividing Equation 2.25 by total number of moles, n_{tot}, we will get

$$1 = X_A + X_B + X_C + \cdots + X_i = \sum X_i \tag{2.26}$$

where $X_A(= n_A/n_{tot})$ is the mole fraction of species A. It can be noted that the sum of all mole fraction of individual species in a mixture is equal to unity.

The mole fraction X_i and mass fraction Y_i can be related easily, as follows:

$$Y_i = \frac{m_i}{m_{mix}} = \frac{n_i MW_i}{n\, MW_{mix}} = X_i \frac{MW_i}{MW_{mix}} \tag{2.27}$$

where MW_i is the molecular weight of ith species. The mixture molecular weight can be easily estimated by knowing either species mole or mass fractions, as follows:

$$MW_{mix} = \sum_i X_i MW_i \tag{2.28}$$

$$MW_{mix} = \frac{1}{\sum_i Y_i / MW_i} \tag{2.29}$$

We can express the pressure of a mixture of gases such as species $A, B, C, \ldots,$ as follows:

$$P = \frac{n_A R_u T}{V} + \frac{n_B R_u T}{V} + \frac{n_C R_u T}{V} + \cdots + \frac{n_i R_u T}{V} = p_A + p_B + p_C + \cdots + p_i = \sum_i p_i \tag{2.30}$$

where p is the partial pressure of individual species by definition. Keep in mind that in this case, we have assumed that individual species are in the same temperature and volume of that of the mixture. This is known as the Dalton's law of partial pressure, which is as follows:

> The pressure of gaseous mixture is the sum of the pressure that each component would exert if it alone occupies the same volume of mixture at the same temperature of the mixture.

It must be kept in mind that the partial pressure of a species can be determined by knowing mole fraction of the corresponding species and total pressure of the mixture by the following relation:

$$p_i = X_i P \tag{2.31}$$

The specific internal energy u of the mixture can be determined from a knowledge of their respective values of the constituent species by invoking the Gibb's theorem, which states the following.

> The internal energy of a mixture of ideal gases is equal to the sum of the internal energy of individual component of the mixture at same pressure and temperature of the mixture.

Then, the specific internal energy u of mixture can be obtained by either mole fraction or mass fraction weighted sum of individual component's specific internal energy, given as follows:

$$\bar{u}_{mix} = \sum_i X_i \bar{u}_i \tag{2.32}$$

$$u_{mix} = \sum_i Y_i u_i \tag{2.33}$$

where

$\bar{u}_i$ is the internal energy of ith species per unit mole

u_i is the internal energy of ith species per unit mass

Similarly, the specific enthalpy h of the mixture can be obtained by similar relations, given as follows:

$$\bar{h}_{mix} = \sum_i X_i \bar{h}_i; \quad h_{mix} = \sum_i Y_i h_i \tag{2.34}$$

The enthalpy of a species can be expressed as

$$h_{i,T}^0(T) = h_{f,298.15}^0 + \int_{298.15}^{T} C_{P,i} \, dT \tag{2.35}$$

Note that the enthalpy of any species at particular temperature is composed of two parts: (1) the heat of formation, which represents the sum of enthalpy due to chemical energy associated with chemical bonds; and (2) sensible enthalpy, as it is associated with temperature. The other specific properties of the mixture such as entropy s, Gibbs free energy g, specific heat C_P, can be obtained by similar relation from the individual species.

2.4.3 Heats of Formation and Reaction

In the combustion process, several chemical reactions take place simultaneously. In some reactions, heat will be evolved and in others, heat will be absorbed. Hence, it is important to evaluate the heat liberated or absorbed in a chemical reaction. Let us take an example of a burner in which 1 mol of propane is reacting with 5 mol of oxygen to produce 3 mol of carbon dioxide and 4 mol of water, as per the following chemical reaction:

$$C_3H_8 + 5O_2 \rightarrow 3CO_2 + 4H_2O \tag{2.36}$$

We need to determine the amount of heat liberated during this reaction, for which we must know heat of formation of each participating species. In other words, heat of formation of each species must be known to evaluate the heat of reaction. The **heat formation** of a particular species can be defined as the heat of reaction per mole of product formed isothermally from elements in their standard states. It must be kept in mind that the heat of formation of elements in their standard states is assigned a value of zero as per international norms. For example, nitrogen gas at standard temperature and pressure is the most stable whose heat of formation at standard state is zero. In contrast, the **heat of reaction** can be defined as the difference between the enthalpy of the products and enthalpy of reactants at the specified states. Note that the heat of reaction is stated in the standard state. If one of the reactants happens to be fuel and other one is the oxidizer, then combustion takes place, liberating certain amount of heat, then the heat of reaction is known as the **heat of combustion**. The heat of formation of some important pure substances is given in Table 2.1, at standard conditions. The main advantage of standard heat of formation is that one only needs to keep track of heat of formation of only few numbers of species to calculate heat of reaction of several reactions.

In combustion calculation, we need to deal with thermochemical systems that can be handled easily invoking the Hess Law, which is also known as constant heat summation. The Hess Law states that the resultant

TABLE 2.1 Heat of Formation of Some Important Species

Chemical Formula	Species Name	State	Standard Heat of Formation (kJ/mol)
O_2	Oxygen	Gas	0.0
O	Element oxygen	Gas	247.4
H_2	Hydrogen	Gas	0.0
H	Element hydrogen	Gas	218.1
OH	Hydroxyl	Gas	42.3
H_2O	Water	Gas	−242.0
H_2O	Water	Liquid	−286.0
C	Graphite	Solid	0.0
CO	Carbon monoxide	Gas	−110.5
CO_2	Carbon dioxide	Gas	−394.0
CH_4	Methane	Gas	−74.5
C_3H_8	Propane	Gas	−103.8
C_4H_{10}	Butane(n)	Gas	−124.7
C_4H_{10}	Butane(iso)	Gas	−131.8
C_2H_2	Acetylene	Gas	226.9
N_2	Nitrogen	Gas	0
H_2O	Water	Gas	−242.0
H_2O	Water	Liquid	−272.0

heat evolved or absorbed at constant pressure or constant volume for a given chemical reaction is the same whether it takes place in one or many steps. In other words, it does not depend on the intermediate paths that may occur between the reactants and products. Hence we can manage to add or subtract thermochemical equations algebraically for obtaining final heat of reaction by using the Hess Law.

Example 2.2

Estimate the higher and lower heating values at 298 K of LPG (Propane = 70%, Butane = 30%) per mole and per kilogram of fuel.

Solution

$$0.7C_3H_8 + 0.3C_4H_{10} + a(O_2 + 3.76N_2) = xH_2O + yCO_2 + zN_2$$

$$\mathbf{C}: 0.7 \times 3 + 0.3 \times 4 = y; \rightarrow y = 3.3$$

$$\mathbf{H}: 0.7 \times 8 + 0.3 \times 10 = 2x; \rightarrow x = 4.3$$

$$\mathbf{O}: 2a = 24 + x$$

$$\rightarrow a = 5.45$$

$$0.7C_3H_8 + 0.3C_4H_{10} + 5.45(O_2 + 3.76N_2) = 4.3H_2O + 3.3CO_2 + 3.76 \times 5.45N_2$$

Energy balance:

$$Q = \sum_{i=p} n_i \bar{h}^0_{f,f} - \sum_{i=R} n_i \bar{h}^0_{f,i}$$

$$Q = 3.3\bar{h}^0_{f,CO_2} + 4.3\bar{h}^0_{f,H_2O} + 3.76 \times 5.46\,\bar{h}^0_{f,N_2} - 0.7\bar{h}^0_{f,C_3H_8} - 0.3\bar{h}^0_{f,C_4H_{10}} - 5.45\bar{h}^0_{f,O_2} - 3.76 \times 5.46\bar{h}^0_{f,N_2}$$

$$Q = 3.3\bar{h}^0_{f,CO_2} + CO_2 + 4.3\bar{h}^0_{f,H_2O} - 0.7\bar{h}^0_{f,C_3H_8} - 0.3\bar{h}_{f,C_4H_{10}}$$

$$\bar{h}^0_{f,CO_2}(g) = -393.522 \text{ kg/mol}$$

$$\bar{h}_{f,H_2O}(l) = -286.000 \text{ kJ/mol for } (\text{HHV})$$

$$\bar{h}^0_{f,H_2O}(g) = -241.826 \text{ kJ/mol for LHV}$$

$$\bar{h}^0_{f,C_3H_8} = -103.92 \text{ kg/mol}$$

$$\bar{h}_{f,C_4H_{10}} = -124.733 \text{ kJ/mol}$$

By mole basis, we can get

$$\text{HHV} = -3.3 \times 393.522 - 4.3 \times 286 + 72.744 + 37.4199$$

$$= -1298.6226 - 1229.8 + 72.744 + 37.4199$$

$$\text{HHV} = -2418.25 \text{ kJ/mol}$$

$$\text{LHV} = -3.3 \times 393.522 - 4.3 \times 241.826 + 0.7 \times 103.92 + 0.3 \times 124.733$$

$$-1298.6226 - 1039.85 + 72.744 + 37.4199 - 2228.30 \text{ kJ/mol}$$

$$= -2228.30 \times 10^3 \text{ kJ/kmol}$$

By mass basis, we can have

$$MW_{mix} = \sum_i MW_i$$

$$= n_{C_3H_8} \cdot MW_{C_3H_8} + n_{C_4H_{10}} \cdot MW_{C_4H_{10}} = 0.7 \times 44 + 0.3 \times 58$$

$$= 48.2 \text{ kg/kmol}$$

$$\text{LHV} = \frac{-2228.30 \times 10^3}{48.2} \text{kJ/kg} = -46.23 \text{ MJ/kg}$$

$$\text{HIV} = \frac{-2418.25 \times 10^3}{48.2} \text{kJ/kg} = -50.17 \text{ MJ/kg}$$

2.4.4 Adiabatic Flame Temperature

We need to know the theoretical/ideal flame temperature during combustion process for a particular fuel–air ratio, provided no heat liberated can be transferred from its system boundary. Let us consider the combustor in which certain amount of fuel and air in certain ratio is reacted, which leads to completion (equilibrium state) by constant pressure and adiabatic process. Then, the final temperature attained by the system is known as **adiabatic flame temperature** T_{ad}. Note that T_{ad} depends on initial pressure P, initial unburnt temperature T_u, and composition of the reactants. Only when the final compositions of the products are known, then the first law of thermodynamics is sufficient to determine the adiabatic flame temperature T_{ad}. Let us consider the physical boundary of the combustor as the control volume. The process is considered to be adiabatic as it is insulated perfectly. Under this condition, the first law of thermodynamics turns out to be

$$H_P\left(T_{ad}P\right) = H_R\left(T_uP\right) \tag{2.37}$$

where

H_p is the total enthalpy of products at adiabatic temperature T_{ad} and pressure P

H_R is the total enthalpy of reactants at initial temperature T_u and ambient pressure

H_p and H_R are expressed in terms of heat of formation and sensible enthalpy of participating species, given as follows:

$$H_R = \sum_{i=R} n_i\bar{h}_i = \sum_{i=R} n_i\left[\bar{h}^0_{f,i} + \int_{T_u}^{T} C_{pi}\left(T\right)dT\right] \tag{2.38}$$

$$H_p = \sum_{i=P} n_i\bar{h}_i = \sum_{i=P} n_i\left[\bar{h}^0_{f,i} + \int_{T_u}^{T_{ad}} C_{pi}\left(T\right)dT\right] \tag{2.39}$$

where

$\bar{h}^0_{f,i}$ is the heat of formation of ith species, which is available in Table 2.1

n_i is the number of moles of ith species

C_{pi} is the specific heat of ith species that are dependent on temperature

TABLE 2.2 Adiabatic Flame Temperature of Typical Fuels at Stoichiometric Mixture

System at T_u (K)	P (MPa)	T_{ad} (K)
CH_4–air, 300 K	0.1	2200
CH_4–air, 300 K	2.0	2278
CH_4–air, 600 K	2.0	2500
CH_4–O_2, 300 K	0.1	3030
C_4H_{10}–air, 300 K	0.1	2246
H_2–air, 300 K	0.1	2400

Note that the appropriate value of the specific heats of the products must be chosen judiciously to get the correct adiabatic temperature. In Equation 2.39, the final composition of the product must be known. Otherwise, we cannot evaluate the adiabatic flame temperature T_{ad}. Generally, the final equilibrium composition is used for the evaluation of adiabatic temperature T_{ad}. Unfortunately, the final equilibrium composition is dependent on the final temperature T_{ad}. Hence, one has to resort to an iterative technique for determining both T_{ad} and equilibrium composition. Let us consider an example for the determination of adiabatic temperature T_{ad}, in which the final compositions are known.

The adiabatic flame temperature data for stoichiometric mixture of certain fuel–oxidizer system at their respective initial temperature and pressure are shown in Table 2.2. It can be noted that adiabatic flame temperature does not depend much on the nature of hydrocarbon fuel as long as air is used as oxidizer. It can also be observed from Table 2.2 that the flame temperature of hydrocarbon–air is around 2200 + 100 K. However, if the air is replaced by oxygen, the flame temperature increases by 500–600 K. Besides this, an increase in initial temperature enhances the flame temperature by almost the same amount.

Example 2.3

Determine the adiabatic flame temperature of the stoichiometric C_4H_{10}–air mixture at 298 K, 0.1 MPa, assuming no dissociation of the products for the following two cases: (1) evaluating C_P of each species at 298 K, and (2) evaluating C_P of each species at 2000 K.

Solution

$$C_4H_{10} + \frac{13}{2}\left(O_2 + 3.76N_2\right) \rightarrow 4CO_2 + 5H_2O + 24.44N_2$$

Applying the first law of thermodynamics for CV, we have from energy balance

$$\int_{298}^{Tad} \Sigma n_i C_{pi} dT = \left[\sum_{i=R} n_i \bar{h}_{fi,298} - \sum_{i=P} n_i \bar{h}_{fi,298} \right]$$

$$= \bar{h}_{f,C_4H_{10}} + 6.5\bar{h}_{f,O_2} + 24.44\bar{h}_{f,N_2} - 4\bar{h}_f, w_2$$
$$-5\bar{h}_{f,H_2O} - 24.44\bar{h}_{f,N_2}$$
$$= -131.8 - \left(4 \times (-394.0) + 5 \times (-247)\right) = 2654.2 \text{ kJ} \tag{E1.2}$$

For standard heat of formation see Reference 1

1. We can consider C_p at 298 K

$$C_{P,CO_2} = 37.129 \text{ J/kmol K}$$
$$C_{P,H_2O} = 35.59 \text{ J/kmol K}$$
$$C_{P,N_2} = 29.124 \text{ J/kmol K}$$

Substituting these values in LHS of Equation E1.2

$$(4 \times 37.129 + 5 \times 35.590 + 24.44 \times 29.12)(T_{ad} - 288) = 2654.2 \times 10^3$$

$$T_{ad} = 2844.64 \text{ K}$$

2. By considering the C_p value at 2000 K, we can have

$$C_{P,CO_2} = 60.35 \text{ J/kmol K}$$
$$C_{P,H_2O} = 51.18 \text{ J/kmol K}$$
$$C_{P,N_2} = 35.97 \text{ J/kmol K}$$

Substituting these values in Equation E1.2, we can determine the adiabatic flame temperature T_{ad} as

$$(4 \times 60.350 + 5 \times 51.180 + 24.44 \times 35.971)(T_{ad} - 298) = 2654.2 \times 10^3$$
$$\rightarrow T_{ad} = 2226.41 \text{ K}$$

Note that this estimated adiabatic temperature of 2226.4 K is closer to the adiabatic temperature obtained from the equilibrium calculation (see Table 2.2), as compared to 2844.6 K obtained for case (a). Hence, it is important to choose specific heat values at appropriate temperature.

2.4.5 Chemical Equilibrium

In Section 2.4.4, we have analyzed the chemical process by assuming the chemical reactions to be capable of going to completion. However, in rocket engine, gases being at high temperature and pressure, the saturated molecules, namely, O_2, H_2, H_2O, CO_2, undergo dissociation reactions, as follows:

$$H_2 \Leftrightarrow H + H \tag{2.40}$$

$$O_2 \Leftrightarrow O + O \tag{2.41}$$

$$2H_2 + O_2 \Leftrightarrow 2H_2O \tag{2.42}$$

Note that dissociation is highly endothermic in nature, which will be affecting the adiabatic flame temperature when evaluated using this procedure. It can be observed from these reactions that the opposite of dissociation, known as recombination, does take place. The extent of dissociation and recombination can be determined by invoking the criteria for chemical equilibrium.

We can explore the criteria of chemical equilibrium by considering the example of reaction system containing H_2, O_2, and H_2O as chemical compositions at a particular pressure and temperature, as given earlier. Note that the chemical composition of this mixture would change only when there is a change in pressure and temperature. Hence, let us consider a reacting system at a particular fixed temperature and pressure for the development of a general criterion for chemical equilibrium. One might be tempted to invoke the increase in entropy principle relationship stipulated by the second law of thermodynamics as the criterion for chemical equilibrium. Unfortunately, this entropy principle cannot be applied for rocket engine as it is inherently a nonadiabatic system. Therefore, we will have to develop another criterion of equilibrium for nonadiabatic reacting system.

Let us consider a simple compressible system of fixed mass at particular temperature T and pressure P. By invoking the first and second laws of thermodynamics, we can have

$$dU + PdV - TdS \leq 0 \tag{2.43}$$

We know that the Gibbs function is given by

$$G = H - TS \tag{2.44}$$

By differentiating the Gibbs function at constant pressure and temperature, we can have

$$(dG)_{T,P} = dH - TdS - SdT = (dU + PdV + VdP) - TdS$$
$$= dU + PdV - TdS \tag{2.45}$$

Since $dT = 0$ and $dP = 0$, then, from Equations 2.43 and 2.45, we can get $(dG)_{T,P} \leq 0$. This indicates that a process cannot proceed when $(dG)_{T,P} > 0$. In other words, the process can only proceed in the direction of decreasing Gibbs function until it reaches its minimum value, as shown in Figure 2.2. As per this criterion of minimum Gibbs free energy, a chemical reaction cannot proceed in the direction of increasing Gibbs function as it violates the second law of thermodynamics. In other words, no further change in chemical composition can take place at the minimum value of Gibbs function.

Let us now consider a homogeneous system involving chemical reactions for four chemical compounds, A, B, C, and D, at a given pressure and temperature. Let n_A, n_B, n_C, and n_D be the number of moles of respective

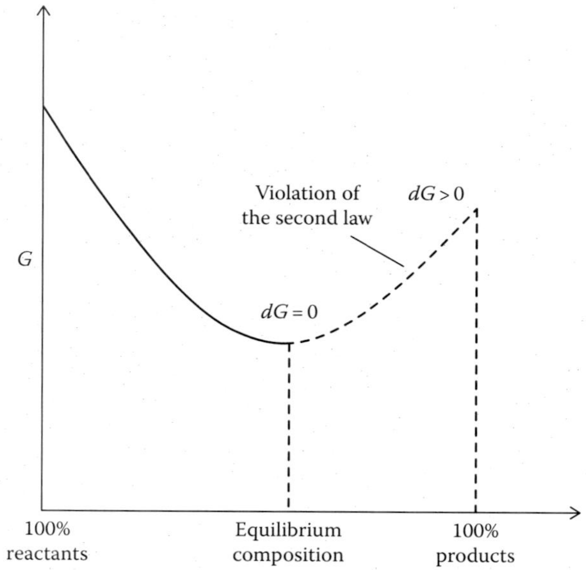

FIGURE 2.2 Illustration of the criterion of chemical equilibrium at a specified P and T.

compounds. The chemical reaction among these four compounds can be represented by the following global reaction:

$$\nu_A A + \nu_B B \rightleftharpoons \nu_C C + \nu_D D \tag{2.46}$$

where ν are stoichiometric coefficients. We can designate degree of reaction by $d\varepsilon$ and change in the number of moles of any species during a chemical reaction can be expressed as

$$dn_A = -\nu_A d\varepsilon; \quad dn_B = -\nu_B d\varepsilon; \quad dn_C = \nu_C d\varepsilon; \quad dn_D = \nu_D d\varepsilon; \tag{2.47}$$

Note that the minus sign indicates a decrease with the progress of the reaction. Similarly, the positive sign indicates an increase with the progress of the reaction. Now we can evaluate the change in the Gibbs function to evaluate the chemical equilibrium for the preceding reaction as

$$\begin{aligned}(dG)_{T,P} &= \Sigma(dG_i)_{T,P} = \Sigma(\bar{g}_i dn_i)_{T,P} \\ &= \bar{g}_C dn_C + \bar{g}_D dn_D - \bar{g}_A dn_A - \bar{g}_B dn_B = 0\end{aligned} \tag{2.48}$$

where $\bar{g}$ is the molar Gibbs function of respective species at specified temperature and pressure. Substituting the values of change in the number of moles of each component from Equation 2.47 in Equation 2.48, we will get

$$\bar{g}_C \nu_C + \bar{g}_D \nu_D - \bar{g}_A \nu_A - \bar{g}_B \nu_B = 0 \tag{2.49}$$

Considering the mixture of gases stated earlier to be an ideal gas, we can derive a criterion of chemical equilibrium by using the above Gibbs function relation. We know that the Gibbs function values are dependent on both temperature and pressure. But the Gibbs values at a fixed reference pressure P_0 (0.1 MPa) are generally as a function of temperature. The Gibbs function of ith component of ideal gas mixture at its partial pressure p_i and temperature T will become

$$\bar{g}_i(T,p_i) = \bar{g}_i^0(T) + R_u T \ln p_i \tag{2.50}$$

where

$\bar{g}_i^0(T)$ is the Gibbs function of ith component at 0.1 MPa and temperature T

p_i is the partial pressure of ith component in the mixture

Substituting the relation for the Gibbs function (Equation 2.50) for each species in Equation 2.49, we can get

$$\begin{aligned}\nu_C \bar{g}_C^0 + \nu_D \bar{g}_D^0 - \nu_A \bar{g}_A^0 - \nu_B \bar{g}_B^0 &= \Delta G_T^0 \\ &= -R_u T\left[\nu_C \ln\left(\frac{p_C}{P_0}\right) + \nu_D \ln\left(\frac{p_D}{P_0}\right) - \nu_A \ln\left(\frac{p_A}{P_0}\right) - \nu_B \ln\left(\frac{p_B}{P_0}\right)\right]\end{aligned} \quad (2.51)$$

By defining ΔG_T^0 as the standard state Gibbs function change, we can rewrite Equation 2.51 as

$$\Delta G^0 = -R_u T \ln \frac{\left(\frac{p_C}{P_0}\right)^{\nu_C}\left(\frac{p_D}{P_0}\right)^{\nu_D}}{\left(\frac{p_A}{P_0}\right)^{\nu_A}\left(\frac{p_B}{P_0}\right)^{\nu_B}} = -R_u T \ln K_P \quad (2.52)$$

where K_P is the chemical equilibrium constant based on pressure, which is dependent on the temperature for a particular reaction. For an ideal gas mixture, we can express it as

$$K_P = e^{\left(-\frac{\Delta G_T^0}{R_u T}\right)} = \frac{\left(\frac{p_C}{P_0}\right)^{\nu_C}\left(\frac{p_D}{P_0}\right)^{\nu_D}}{\left(\frac{p_A}{P_0}\right)^{\nu_A}\left(\frac{p_B}{P_0}\right)^{\nu_B}} \quad (2.53)$$

Hence, the equilibrium constant K_P of an ideal gas mixture at a fixed temperature can be easily computed by knowing the change in the standard Gibbs function. The values of K_P for certain important reactions are listed in Table B.1. By knowing the equilibrium constant K_P, we can evaluate the composition of reacting ideal gas mixture, which is illustrated in Example 2.4.

We know that the partial pressure of the species in a mixture can be expressed in terms of their respective mole friction as given by

$$p_i = X_i P = \left(\frac{n_i}{n_{tot}}\right) P \quad (2.54)$$

where

P is the pressure of the mixture

n_{tot} is the total number of moles present in the mixture

Substituting the partial pressure of each species in Equation 2.53, we will get

$$K_p = e^{\left(-\frac{\Delta G_T^0}{R_u T}\right)} = \frac{X_C^{\nu_C} X_D^{\nu_D}}{X_A^{\nu_A} X_B^{\nu_B}} \left(\frac{P}{P_0}\right)^{\Delta\nu} \quad \text{where } \Delta\nu = \nu_C + \nu_D - \nu_A - \nu_B \qquad (2.55)$$

The preceding expression given for K_P has been derived for a simple reaction involving two reactants and two products. However, similar expression can be obtained for any reaction involving more number of reactants and products. Note that the value of K_P depends on the temperature for a particular chemical reaction. For a negative value of Gibbs free energy change ($\Delta G_T^0 = -\text{ve}$), K_p will be more than one, indicating that products will be favored. On the other hand, when the change in Gibbs free energy is positive, the reactants will be favored at equilibrium.

Example 2.4

At 0.1 MPa, hydrogen gas is dissociated into H by $H_2 \rightleftarrows 2H$. Determine the equilibrium constant, K_P at (1) 3000 K and (2) 298 K.

Solution

1. By using Equation 2.55, we can determine the equilibrium constant K_P as

$$K_P = e^{\left(-\frac{\Delta G_T^0}{R_u T}\right)}$$

However, by using data from Appendix B of combustion book [2], the change in Gibbs function at 3000 K can be evaluated as

$$\Delta G_T^0 = 2\bar{g}_{\text{H}}^0 - \bar{g}_{\text{H}_2}^0 = 2 \times 46.007 - 0 = 92.014 \text{ kJ/mol}$$

Then the equilibrium constant K_P can be evaluated as

$$K_P = e^{\left(-\frac{\Delta G_T^0}{R_u T}\right)} = e^{\left(-\frac{92014}{8.314 \times 3000}\right)} = 0.0249$$

2. By following similar procedure, the equilibrium constant K_P can be evaluated at 298 K as

$$K_P = e^{\left(-\frac{\Delta G_T^0}{R_u T}\right)} = e^{\left(-\frac{406,580}{8.314\times 298}\right)} = 5.37\times 10^{-72}$$

It indicates that the equilibrium constant K_P for hydrogen dissociation reaction at 298 K is quite small as compared to the equilibrium constant K_P at 3000 K. Hence, equilibrium compositions will be negligibly small at 298 K as compared to that of the 3000 K case.

2.4.5.1 *Evaluation of Equilibrium Composition of Simultaneous Reactions*

In practical systems, reactive mixture involving more than one reaction is routinely encountered. For example, 5 mol of hydrogen is reacted with 2 mol of oxygen at 2500 K and 1.8 MPa. We know that when chemical equilibrium is reached, certain species will be formed. We need to guess possible species at equilibrium. Let us identify possible species, namely, H_2, H, O, O_2, OH, H_2O at equilibrium. Then, the following reaction can be expressed as

$$5H_2 + 2O_2 \rightarrow aH_2 + bH + cO_2 + dO + eOH + fH_2O \tag{2.56}$$

In order to determine the final equilibrium composition, we can invoke (1) equilibrium constant and (2) conservation of elements of the reactants. For this purpose, we can identify the possible elementary reaction steps, as follows:

$$H_2 \Leftrightarrow 2H \quad K_p = \frac{X_H P^0}{X_{H_2}} \tag{2.57}$$

$$O_2 \Leftrightarrow 2O \quad K_{P_2} = \frac{X_O^2 P^0}{X_{O_2}} \tag{2.58}$$

$$H + O \Leftrightarrow OH \quad K_{P_3} = \frac{X_{OH}}{X_O X_H P^0} \tag{2.59}$$

$$H_2 + O \Leftrightarrow H_2O \quad K_4 = \frac{X_{H_2O}}{X_{H_2} X_O P^0} \tag{2.60}$$

It can be noted that the same components, for example, O, H, and H_2 are involved in four reactions. Each species would not take part to the same extent in each reaction. In other words, it is very much needed to apply the equilibrium criterion to all equations in the present example. In general, the equilibrium criterion must be applied to all possible reactions along with conservation of mass for each chemical species. As a result, a system of simultaneous equation are formed that must be determined to evaluate the chemical composition. Of course, the chemical equilibrium criterion $(dG)_{T,P} = 0$ must be satisfied for each and every reaction involved in the reacting system. We can now strike an element conservation balance for each element, as follows:

$$5H_2 + 2O_2 \rightarrow aH_2 + bH + cO_2 + dO + eOH + fH_2O \tag{2.61}$$

$$\mathrm{H}:\ 2n_{H_2} + n_H + n_{OH} + 2n_{H_2O} = N_H \tag{2.62}$$

$$\mathrm{O}:\ 2n_{O_2} + n_O + n_{OH} + n_{H_2O} = N_O \tag{2.63}$$

Note that the total numbers of elements for H and O in Equation 2.61 are equal to 10 and 4, respectively ($N_H = 10$; $N_O = 4$). By dividing Equation 2.62 by the total number of moles, we can have

$$2X_{H_2} + X_H + X_{OH} + 2X_{H_2O} = \frac{N_H}{N_{tot}} \tag{2.64}$$

Similarly, from Equation 2.63 we can get

$$2X_{O_2} + X_O + X_{OH} + X_{H_2O} = \frac{N_O}{N_{tot}} \tag{2.65}$$

Divide Equation 2.64 by Equation 2.65 and we can have

$$\frac{2X_{H_2} + X_H + X_{OH} + 2X_{H_2O}}{2X_{O_2} + X_O + X_{OH} + X_{H_2O}} = \frac{N_H}{N_O} \tag{2.66}$$

We need to get another equation as six equations are required to find out solutions for six unknown variables. For this let us invoke one more conservation equation, as follows:

$$\sum X_i = 1, \quad \text{then } X_{H_2} + X_H + X_{OH} + X_{H_2O} + X_O + X_{O_2} = 1 \tag{2.67}$$

Thus, we have a well-constituted set of six simultaneous equations to get equilibrium composition. But the solution is not straightforward, as K_p involves nonlinearly. Thus, an iterative technique like the Newton–Raphson method can be used to solve the equation to arrive at equilibrium composition. Besides this, one needs to know the adiabatic flame temperature. The following procedure is to be adopted for estimating the adiabatic flame temperature.

Given data: P, T_{in}, initial mixture ratio

Step 1: Assume a value of T_{ad}^*

Step 2: Compute equilibrium composition at P, T_{ad}^*

Step 3: Estimate $\Delta H_R = \sum_{i=p} n_i \bar{h}_{f,i}^0 - \sum_{i=R} n_i \bar{h}_f^0$

Step 4: Estimate new temperature T_{ad}

Step 5: If $\begin{array}{l} |T_{ad} - T_{ad}^*| \le \varepsilon(\text{error}) \quad T_{ad} = T_{ad}^{\lambda} \quad (\text{stop/end}) \\ |T_{ad} - T_{ad}^*| > \varepsilon(\text{error}) \quad T_{ad}^* = T_{ad} + \alpha(T_{ad} - T_a) \quad (\text{Go to step 2}) \end{array}$

where α is the under relaxation parameter.

For a reacting mixture involving more number of species and relations, the number of simultaneous equations to be solved is quite large. Then, one has to go for iterative numerical method using a computer. Nowadays, several powerful numerical techniques are being developed for estimating chemical equilibrium composition. Among these programs, the computer code developed by NASA-SP-273 is noteworthy, which is based on the minimization principles of Gibbs function. As mentioned earlier, iterative techniques like Newton–Raphson method can be used to solve the equations to arrive at equilibrium composition. Besides this, one needs to know the adiabatic flame temperature. Note that both adiabatic flame temperature and equilibrium composition of any fuel–oxidizer system can be determined simultaneously by the following iterative calculation procedure. For more details, readers can refer books on combustion [6].

2.5 BASIC PRINCIPLES OF GAS DYNAMICS

In the analysis of rocket engine we will be dealing with the gas dynamics aspect of the flow field. Gas dynamics is a highly developed subject and cannot be covered in this chapter. The main objective of this section is to review certain aspects of gas dynamics that can be used for analyzing various components of rocket engine. Interested readers can refer to the standard textbooks [4–6] for an exhaustive treatment.

2.5.1 Conservation Equations

Let us recall that the gas dynamics is governed by the equations of continuity, momentum, and energy, along with the equation of state. In case of rocket engine, the gas flow can be safely assumed to be perfect in nature. In this section, we will derive the conservation equations of flow for a calorically perfect gas.

2.5.2 Steady Quasi-One-Dimensional Flow

The flow across various components of rocket engine is inherently three-dimensional and unsteady in nature. However, it is quite difficult to handle three-dimensional characteristics of fluid flow for the flow analysis of rocket engine. Hence, we need to simplify the actual problems with certain assumptions so that it can be handled easily. One can assume the flow in various components of rocket engine to be a quasi-one-dimensional steady flow. For example, flow in nozzle can be assumed to be quasi-one-dimensional steady flow as it is basically a varying area duct that changes in a gradual manner, as shown in Figure 2.3, along its length. If this rate of increase of area for subsonic flow is quite small, then we can assume flow properties to be function of x-direction only which remain uniform across any given stream tube. Such kind of flow is known as quasi-one-dimensional flow. We can apply the equations of mass, momentum, and energy, along with the equation of state for perfect gas to a quasi-one-dimensional differential CV, as shown in Figure 2.3, with the following assumptions:

- Steady quasi-one-dimensional flow
- Inviscid flow
- No body forces
- Ideal gas with constant thermophysical properties

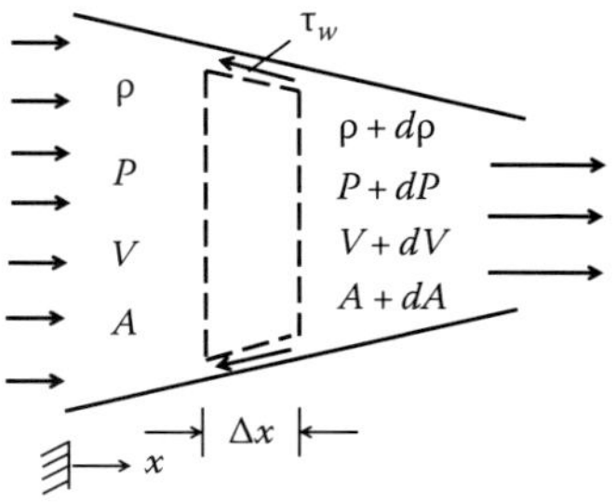

FIGURE 2.3 Control volume for one-dimensional steady flow.

The integral form of mass conservation equation for steady flow in CV is given by

$$\iint \rho(V \cdot n)\,dA = 0 \tag{2.68}$$

For steady one-dimensional flow in the CV, as shown in Figure 2.3, preceding Equation 2.68 becomes to

$$\frac{d\rho}{\rho} + \frac{dA}{A} + \frac{dV}{V} = 0 \tag{2.69}$$

Similarly, the integral form of momentum equation for steady one-dimensional flow in CV can be expressed as

$$\iint \rho V(V \cdot n)\,dA = \sum F_x \tag{2.70}$$

where $\sum F_x$ is the sum of forces due to pressure, body force, wall shear, and so on. From this equation, the differential form of momentum equation for quasi-one-dimensional flow can be derived as follows:

$$\rho V \frac{dV}{dx} = -\left(\frac{dP}{dx} + \frac{\tau_\omega C}{dA}\right) \tag{2.71}$$

where

τ_ω is the wall shear stress
C is the duct circumference

In the same way, the integral form of the energy equation can be deduced for the present case:

$$dh + VdV = dq - dw \tag{2.72}$$

where q and w are the heat and work transfer per unit mass. Assuming the gas to be perfect, the equation of state (Equation 2.73) will be

$$P = \rho RT \tag{2.73}$$

Generally, these equations cannot be solved easily. However, the closed form solution to these equations can be obtained for some special cases. We discuss isentropic flow through varying area duct in the following.

2.5.3 Isentropic Flow through Variable Area Duct

The isentropic flow through a variable area duct is analyzed in this subsection. By using momentum equation (Equation 2.71), we can derive expression for quasi-one-dimensional steady flow as follows:

$$\rho V dV + dP = 0 \tag{2.74}$$

Combining continuity (Equation 2.69) and momentum equations (Equation 2.74), we can get

$$dP + \rho V^2\left(-\frac{dP}{\rho a^2} - \frac{dA}{A}\right) = 0 \tag{2.75}$$

As per the definition of speed of sound, we know that

$$d\rho = \frac{dP}{a^2} \tag{2.76}$$

By rearranging Equation 2.75, we can have

$$dP\left(1 - M^2\right) = \rho V^2 \frac{dA}{A} \tag{2.77}$$

where M is the Mach number. Note that Equation 2.77 describes the effect of the Mach number on compressible flow through a duct of varying cross sections. For subsonic flow ($M < 1$), the term in the left side of Equation 2.77, $(1 - M^2)$, is positive. Hence, the pressure increases with cross-sectional area (dA = +ve). We can also derive an expression relating area with velocity using the momentum equation (Equation 2.74), as follows:

$$\frac{dA}{A} = \left(M^2 - 1\right)\frac{dV}{V} \tag{2.78}$$

This is known as the area–velocity relation. This important relation is quite helpful to understand various regimes of flow. For subsonic flow ($0 < M < 1$), the velocity decreases in divergent duct and increases in convergent duct. For supersonic flow ($M > 1$), the velocity increases in divergent duct and decreases in convergent duct, which happens to be opposite to that of the subsonic flow. For sonic flow ($M = 1$), Equation 2.78 indicates that change

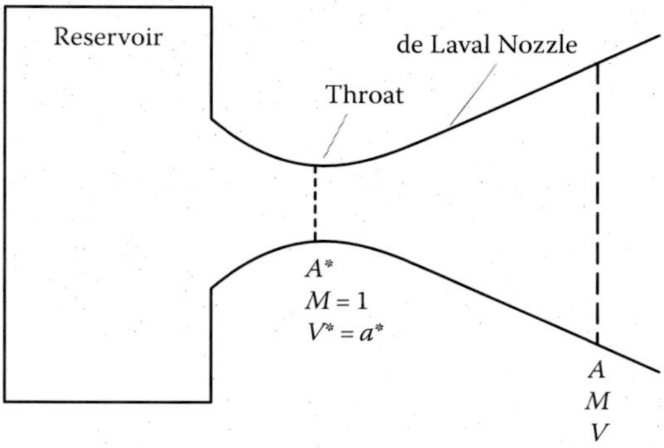

FIGURE 2.4 Schematic of de Laval nozzle.

in area becomes zero ($dA = 0$). In order to accelerate the flow to high subsonic speed from sonic condition, one has to use a convergent duct, as shown in Figure 2.4. Note that convergent–divergent duct is also known as de Laval nozzle, which is used extensively in rocket engine. Note that this de Laval nozzle was designed, and developed for the first time, by the Swedish engineer Carl G.P. de Laval for a single-stage steam engine used in marine applications. The minimum area of this nozzle is known as throat.

2.5.4 Mass Flow Parameter

In the last section, we discussed briefly the expansion process in a CD nozzle. We will now discuss the mass flow parameter, which will be useful for designing various components of rocket engine. Besides this, we will derive a relation between ratio of duct area to the sonic throat area and Mach number, which will be useful for designing a CD nozzle. For this purpose, let us consider a varying area duct as shown in Figure 2.4. Considering isentropic quasi-steady one-dimensional flow, we can invoke the continuity equation as follows:

$$\dot{m} = \rho VA = \rho^* V^* A^* = \text{Constant} \tag{2.79}$$

where ρ, V, and A are density, velocity, and area, respectively. Note that asterisk mark, (*) corresponds to sonic flow condition. Then, by using Equation 2.79, we can derive an expression for mass flux, as follows:

$$\frac{\dot{m}}{A} = \rho V = \rho M\sqrt{\gamma RT} = \frac{\rho_t}{\left(1+\frac{\gamma-1}{2}M^2\right)^{\frac{1}{\gamma-1}}} M \sqrt{\frac{\gamma RT_t}{1+\frac{\gamma-1}{2}M^2}} \tag{2.80}$$

By simplifying Equation 2.80 and expressing density in terms of pressure and temperature, we can have

$$MPR = \frac{\dot{m}\sqrt{T_t}}{AP_t} = \frac{\sqrt{\gamma}}{R} M \left(\frac{1}{1 + \frac{\gamma - 1}{2} M^2} \right)^{\frac{\gamma+1}{2(\gamma-1)}} = f(M) \tag{2.81}$$

Note that $\dot{m}\sqrt{T_t}/AP_t$ is known as mass flow parameter (MFP), which is unique function of Mach number and specific heat ratio γ for calorically perfect gas. The variation of mass flow parameter with Mach number for air is plotted in Figure 2.5. It can be observed that maximum MFP occurs at sonic condition for given stagnation condition, which is expressed as

$$MFP_{maa} = \frac{\dot{m}\sqrt{T_t}}{A^* P_t} = \frac{\sqrt{\gamma}}{\sqrt{R}} \left(\frac{2}{\gamma + 1} \right)^{\frac{\gamma+1}{2(\gamma-1)}} \tag{2.82}$$

Dividing Equation 2.82 by Equation 2.81, we can get

$$\frac{A}{A^*} = \frac{1}{M} \left[\frac{2}{\gamma + 1} \left(1 + \frac{\gamma - 1}{2} M^2 \right) \right]^{\frac{\gamma+1}{2(\gamma-1)}} \tag{2.83}$$

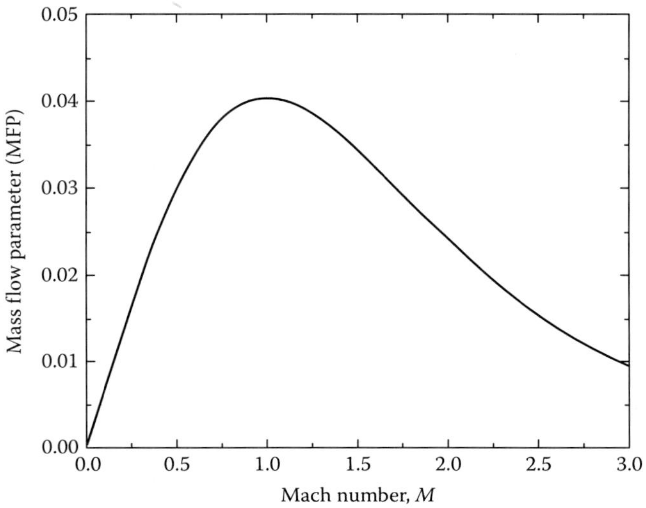

FIGURE 2.5 Variation of mass flow parameter (MFP) with Mach number.

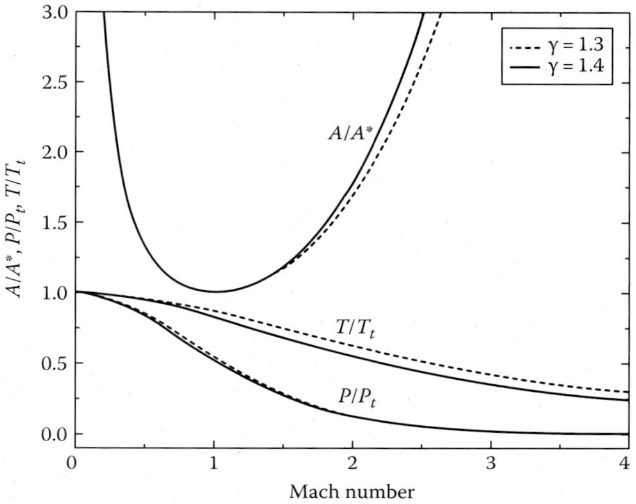

FIGURE 2.6 Variation area ratio A/A^*, temperature ratio T/T_t, and pressure ratio P/P_t with Mach number for air ($\gamma = 1.4$ and 1.3).

This is known as the Area–Mach number relation. It can be noted that the Mach number at any location of the duct is a function of ratio of local cross-sectional area to sonic throat area and specific heat ratio γ. Note that for a particular area ratio ($A/A^* > 1$), two values of Mach number, namely, (1) subsonic and (2) supersonic values can be obtained from Equation 2.83, depending on the pressure ratio between the inlet and exit of the duct. The variation of A/A^* is plotted in Figure 2.6 with Mach number for air ($\gamma = 1.4$). It can be observed that for subsonic flow, A/A^* decreases with increase in Mach number, indicating that the duct is convergent. Note that at $M = 1$, A/A^* attains unity value, indicating it to be a throat. On the other hand, for supersonic flow, A/A^* increases with Mach number, indicating the duct to have a divergent section. By knowing the local Mach number, one can determine the variation of pressure ratio P/P_t and temperature ratio T/T_t, as shown in Figure 2.6.

Example 2.5

A CD nozzle is to be designed for exit Mach number of 3.5. The combustion products with $\gamma = 1.25$ are entering from combustion chamber at pressure 12 MPa and temperature 2500 K. If the exit diameter of nozzle is 1.45 mm, determine (1) throat and exit area ratio and (2) maximum mass flow rate.

We can estimate area throat and exit area ratio as follows:

$$\frac{A}{A^*} = \frac{1}{M}\left[\frac{2}{\gamma+1}\left(1+\frac{\gamma-1}{2}M^2\right)\right]^{\frac{\gamma+1}{2(\gamma-1)}}$$

$$= \frac{1}{2}\left[\frac{2}{1.25+1}\left(1+0.125\times 3.5^2\right)\right]^{\frac{1.25+1}{2(1.25-1)}} = 19.22$$

$$\Rightarrow A^* = \frac{A}{19.22}$$

The exit area of nozzle is estimated as

$$A = \frac{\pi}{4}d^2 = \frac{3.14}{4}(1.45)^2 = 1.65 \text{ m}^2$$

$$\Rightarrow A^* = \frac{1.65}{19.22} = 0.086 \text{ m}^2$$

We can estimate maximum mass flow rate through the nozzle as follows:

$$\dot{m}_{\text{max}} = \frac{\sqrt{\gamma}}{\sqrt{RT_t}}\left(\frac{2}{\gamma+1}\right)^{\frac{\gamma+1}{2(\gamma-1)}}$$

$$P_t A^* = \frac{0.658}{\sqrt{287\times 2500}} 12\times 10^6 \times 0.086 = 801.7 \text{ kg/s}$$

2.5.5 Normal Shocks

A rocket engine does experience shock and expansion waves during its flight as it flies at supersonic speed. Recall that the shock wave is a compression wave caused by supersonic motion of body in the medium. It can occur either in internal or external supersonic flow field. The proper understanding of shock behavior is very important in designing various components of rocket engine. Hence, the next section is devoted to analyzing the flow containing normal shock. Before analyzing the normal shocks, it is essential to get a clear physical picture of the shock itself. Note that the thickness of the shock is very thin, around 0.3 μm. We know that the finite changes in fluid properties such as temperature, pressure, and velocity lead to large gradients across the small distance of the shock, making the process to internally irreversible. Hence, we are interested in this analysis to

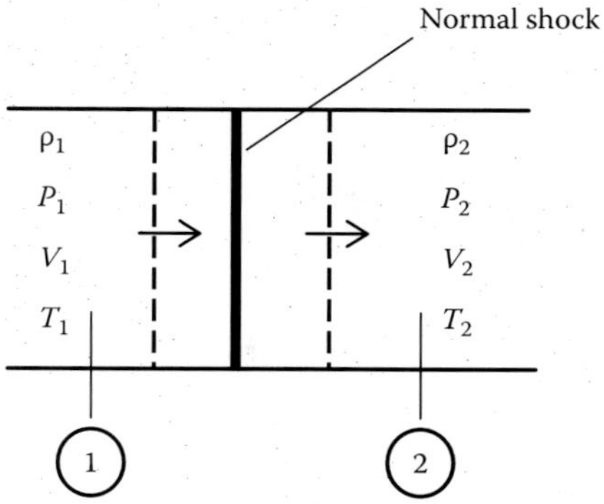

FIGURE 2.7 Schematic of CV for analysis of normal shock.

predict the changes that occur across the shock rather than its detailed structure. Let us apply the basic governing equations to the thin control volume containing a normal shock, as shown in Figure 2.7. The following assumptions are made for this analysis:

1. Steady flow
2. Uniform flow at each section
3. $A_1 = A_2 = A$ (as the shock is quite thin)
4. Negligible friction at channel wall as the shock is quite thin
5. Negligible gravity force
6. Adiabatic reversible flow
7. Ideal gas with constant thermodynamic properties

By mass conservation between station (1) and (2), we can have

$$\rho_1 V_1 = \rho_2 V_2 \tag{2.84}$$

Similarly by invoking momentum conservation between station (1) and (2), we can have

$$P_1 + \rho_1 V_1^2 = P_2 + \rho_2 V_2^2 \tag{2.85}$$

By energy conservation between stations (1) and (2), we can have

$$h_1 + \frac{V_1^2}{2} = h_2 + \frac{V_2^2}{2} \Rightarrow h_{t2} = h_{t2} \tag{2.86}$$

Note that the total energy of the flow remains constant as there is no energy addition. For an ideal gas, the equation of state is given by

$$P = \rho RT \tag{2.87}$$

Let us first obtain property ratios in terms of M_1 and M_2. The temperature ratio can be expressed as

$$\frac{T_2}{T_1} = \frac{T_2}{T_{t2}} \cdot \frac{T_{t2}}{T_{t1}} \cdot \frac{T_{t1}}{T_1} \tag{2.88}$$

For calorifically perfect gas, the total enthalpy equation becomes

$$h_{t1} = h_{t2} \Rightarrow T_{t1} = T_{t2} \quad \left(\text{for an ideal gas}\right) \tag{2.89}$$

Note that the stagnation temperature remains constant across the shock. Then static temperature ratio can be expressed in terms of Mach number using isentropic relationship between total and static temperature, as follows:

$$\frac{T_2}{T_1} = \frac{1+\frac{\gamma-1}{2}M_1^2}{1+\frac{\gamma-1}{2}M_2^2} \tag{2.90}$$

We can find a relationship for velocity ratio, as

$$\frac{V_2}{V_1} = \frac{M_2 a_2}{M_1 a_1} = \frac{M_2\sqrt{\gamma RT_2}}{M_1\sqrt{\gamma RT_1}} = \frac{M_2}{M_1}\sqrt{\frac{T_2}{T_1}} \tag{2.91}$$

From Equations 2.90 and 2.91, we have

$$\frac{V_2}{V_1} = \frac{M_2}{M_1}\left[\frac{1+\frac{\gamma-1}{2}M_1^2}{1+\frac{\gamma-1^2}{2}M_2^2}\right]^{\frac{1}{2}} \tag{2.92}$$

From the continuity equation, a relationship between the density ratios can be obtained, as follows:

$$\frac{\rho_2}{\rho_1} = \frac{V_1}{V_2} = \frac{M_1}{M_2}\left[\frac{1+\frac{\gamma-1}{2}M_2^2}{1+\frac{\gamma-1}{2}M_1^2}\right]^{\frac{1}{2}} \tag{2.93}$$

Similarly, from the momentum equation, we can obtain a relationship between pressure ratio, as follows:

$$P_1\left(1+\frac{V_1^2}{RT_1}\right) = P_2\left(1+\frac{V_2^2}{RT_2}\right) \tag{2.94}$$

As $V^2 = \gamma RTM^2$, we can have

$$\frac{P_2}{P_1} = \frac{1+\gamma M_1^2}{1+\gamma M_2^2} \tag{2.95}$$

Now we will have to find a relationship between M_1 and M_2. For this, let us use equation of state and express the temperature ratio as

$$\frac{T_2}{T_1} = \frac{P_2}{P_1}\frac{\rho_1}{\rho_2} = \left[\frac{1+\gamma M_1^2}{1+\gamma M_2^2}\right]\frac{M_2}{M_1}\left[\frac{1+\frac{\gamma-1}{2}M_1^2}{1+\frac{\gamma-1}{2}M_2^2}\right]^{\frac{1}{2}} \tag{2.96}$$

Using Equation 2.96 and the expression for static temperature ratio, we get

$$\left[\frac{1+\frac{\gamma-1}{2}M_1^2}{1+\frac{\gamma-1}{2}M_2^2}\right]^{\frac{1}{2}} = \left[\frac{1+\gamma M_1^2}{1+\gamma M_2^2}\right]\frac{M_2}{M_1} \tag{2.97}$$

Squaring this equation, we can get

$$\frac{1+\frac{\gamma-1}{2}M_1^2}{1+\frac{\gamma-1}{2}M_2^2} = \frac{M_2^2}{M_1^2}\left[\frac{1+\gamma M_1^2}{1+\gamma M_2^2}\right]^2 \tag{2.98}$$

By solving this equation, two solutions can be obtained as follows:

$$M_2 = M_1 \quad \text{(trivial solution)} \tag{2.99}$$

$$M_2^2 = \frac{M_1^2 + \dfrac{2}{\gamma - 1}}{\dfrac{2\gamma}{\gamma - 1} M_1^2 - 1} \tag{2.100}$$

This result can be plotted in Figure 2.8 for $\gamma = 1.25$. It indicates that for supersonic flow at inlet ($M_1 > 1$), the Mach number M_2 at downstream of shock will be subsonic ($M < 1$) and a compression shock will be produced. But, when inlet flow is subsonic ($M_1 < 1$), then an expansion wave will be formed ($M_1 > 1$).

We can find the relationship between stagnation pressure across the shock wave, as given in the following:

$$\frac{P_{t2}}{P_{t1}} = \frac{P_{t2}}{P_2} \cdot \frac{P_2}{P_1} \cdot \frac{P_1}{P_{t1}} = \frac{P_2}{P_1} \left[\frac{1 + \dfrac{\gamma - 1}{2} M_2^2}{1 + \dfrac{\gamma - 1}{2} M_1^2} \right]^{\frac{\gamma}{\gamma - 1}} \tag{2.101}$$

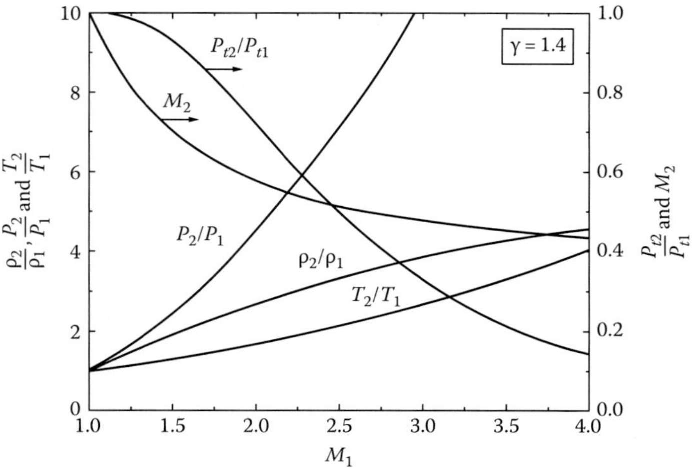

FIGURE 2.8 Variation of properties across a normal shock wave with upstream Mach number.

Combining Equations 2.89 and 2.94, we can express static pressure ratio in terms of inlet Mach number, M_1:

$$\frac{P_2}{P_1} = \frac{1+\gamma M_1^2}{1+\gamma M_2^2} = \frac{1+\gamma M_1^2}{1+\gamma\left(\dfrac{M_1^2+\dfrac{2}{\gamma-1}}{\dfrac{2\gamma}{\gamma-1}M_1^2-1}\right)} \quad (2.102)$$

By simplifying this equation, we can get density and temperature ratio as given in the following:

$$\frac{P_2}{P_1} = \frac{2\gamma}{\gamma+1}M_1^2 - \frac{\gamma-1}{\gamma+1} = \frac{2\gamma}{\gamma+1}\left(M_1^2-1\right)+1 \quad (2.103)$$

Similarly, we can get expression for

$$\frac{\rho_2}{\rho_1} = \frac{\left(\gamma+1\right)M_1^2}{\left(\gamma-1\right)M_1^2+2} \quad (2.104)$$

By using Equations 2.103 and 2.104, we can derive an expression for T_2/T_1 as

$$\frac{T_2}{T_1} = \frac{P_2}{P_1}\cdot\frac{\rho_1}{\rho_2} = \left[\left(\frac{2\gamma}{\gamma+1}\left(M_1^2-1\right)+1\right)\frac{2+\left(\gamma-1\right)M_1^2}{\left(\gamma+1\right)M_1^2}\right] \quad (2.105)$$

The entropy across the shock can be determined as

$$s_2 - s_1 = C_P \ln\left[\left(\frac{2\gamma}{\gamma-1}\left(M_1^2-1\right)+1\right)\frac{2+\left(\gamma-1\right)M_1^2}{\left(\gamma+1\right)M_1^2}\right] - R\ln\left[\frac{2\gamma}{\gamma+1}\left(M_1^2-1\right)+1\right] \quad (2.106)$$

The preceding expression indicates that entropy change across the normal shock wave is only a function of M_1 and γ. It can be noted that when flow is supersonic ($M_1 > 1$), then change in entropy across the normal shock ($s_2 - s_1$) becomes positive. But if the flow is subsonic, then change in entropy across the normal shock ($s_2 - s_1$) becomes negative, which violates the second law of thermodynamics. Hence, it is not possible to accelerate the flow across a shock wave. That means the shock can appear only when the flow is supersonic. The property ratios P_{t2}/P_{t1}, T_2/T_1, P_2/P_1, and V_2/V_1 are provided in table (Appendix B: Table B3-4) in terms of M_1 for a flow of an ideal gas through a normal shock ($\gamma = 1.4$ and 1.33), which can be used easily for solving the problem.

2.5.6 Oblique Shocks

We know that the normal shock wave is more likely to occur in one-dimensional flow, in which a compression wave can occur normal to the flow direction. However, in actual nozzle flow in rocket engine, the supersonic flow is two-/three-dimensional, and a compression wave is likely to occur inclined at an angle to the flow, which is known as oblique shock. This is a special case of an oblique shock wave that occurs in supersonic flow.

When the flow is passed through a wedge and deflected through an angle θ, an oblique shock wave is formed having a wave angle β with respect to upstream velocity V_1, as shown in Figure 2.9. The velocity behind the

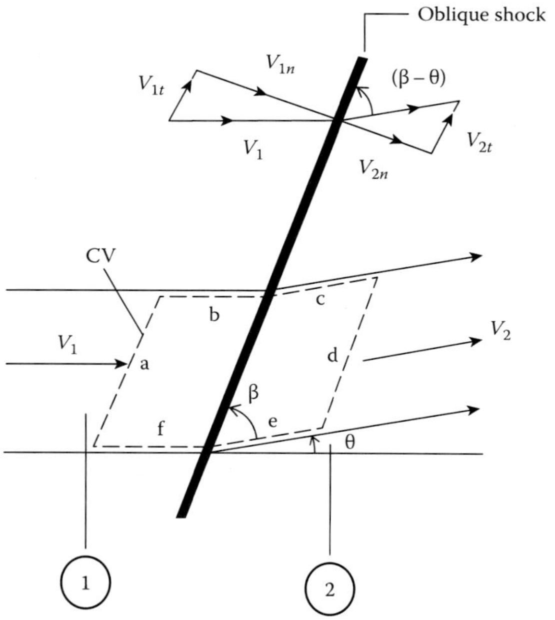

FIGURE 2.9 CV for an oblique shock wave.

shock wave is V_2, which is parallel to the wedge surface with an angle θ. Let us consider a CV as indicated by the dashed lines in Figure 2.9. The upper and lower sides of this CV coincide with the flow stream lines to the left-, and right-hand sides are parallel to the oblique shock. Note that the velocity component perpendicular and parallel to the shock wave are V_n and V_t, respectively, as shown in Figure 2.9. We will apply the conservation equations to this oblique shock wave. By integrating the mass conservation equation across the CV between stations (1) and (2), we can have

$$\rho_1 V_{1n} A_1 = \rho_2 V_{2n} A_2 \Rightarrow \rho_1 V_{1n} = \rho_2 V_{2n} \quad \text{for } A_1 = A_2 \tag{2.107}$$

Let us first apply momentum equation in normal and tangential directions. Invoking momentum balance along tangential direction, we can have

$$\int(\rho V \cdot dA)V = -\int PdA \tag{2.108}$$

For CV shown in Figure 2.9, this equation becomes

$$-\rho_1 V_{1n}^2 + \rho_2 V_{2n}^2 = -(-P_1 - P_2) \tag{2.109}$$

The tangential component of momentum equation would be

$$-\rho_1 V_{1n} V_{1t} + \rho_2 V_{2n} V_{2t} = 0 \tag{2.110}$$

Note that tangential component of PdA is zero as the momentum due to pressure component on "b" and "c" cancels that of on faces "f" and "e," respectively. By using the preceding equation and continuity equation, we can have

$$V_{1t} = V_{2t} \tag{2.111}$$

Note that the tangential velocity component remains constant across an oblique shock. Hence, an oblique shock can be easily considered as a normal shock relative to the coordinate system moving with velocity V_{1t}. Hence, the normal shock relations in terms of normal component of upstream Mach number M_{1n} can be applied easily for this analysis. However, the normal Mach number is given by

$$M_{1n} = M_1 \sin\beta \tag{2.112}$$

By determining the angle between wedge centerline and shock β, the properties' ratios across oblique shock waves are obtained easily by using the following normal shock relationships:

$$\frac{P_2}{P_1} = 1 + \frac{2\gamma}{\gamma+1}\left(M_{1n}^2 - 1\right) \tag{2.113}$$

$$\frac{\rho_2}{\rho_1} = \frac{\left(\rho+1\right)M_{1n}^2}{\left(\gamma-1\right)M_{1n}^2 + 2} \tag{2.114}$$

$$\frac{T_2}{T_1} = \frac{P_2}{P_1} \cdot \frac{\rho_1}{\rho_2} \tag{2.115}$$

$$M_{2n}^2 = \frac{M_{1n}^2 + \left[2/\left(\gamma-1\right)\right]}{\left[2\gamma/\left(\gamma-1\right)\right]M_{1n}^2 - 1} \tag{2.116}$$

The Mach number at the downstream of the shock can be easily evaluated by using geometry, shown in Figure 2.9, as

$$M_2 = \frac{M_{2n}}{\sin\left(\beta-\theta\right)} \tag{2.117}$$

Note that to find out M_2, the flow deflection angle θ has to be evaluated. However, the flow deflection angle θ is uniquely related to M_1 and β. The relation between θ, β, and M_1 can be derived by considering the velocity triangle, shown in Figure 2.9, as follows:

$$\tan\beta = \frac{V_{1n}}{V_{1t}} \tag{2.118}$$

$$\tan\left(\beta-\theta\right) = \frac{V_{2n}}{V_{2t}} \tag{2.119}$$

By using these two equations and noting that $V_{1t} = V_{2t}$, we can have

$$\frac{\tan\left(\beta-\theta\right)}{\tan\beta} = \frac{V_{2n}}{V_{1n}} \tag{2.120}$$

But from the continuity, we know that

$$\frac{V_{2n}}{V_{1n}} = \frac{\rho_1}{\rho_2} \tag{2.121}$$

Now, by using Equations 2.117, 2.114, and 2.119 and simplifying it further, these equations can be expressed in terms of inlet Mach number as

$$\tan\theta = 2\cot\beta\left[\frac{M_1^2\sin^2\beta - 1}{M_1^2(\gamma + \cos 2\beta) + 2}\right] \tag{2.122}$$

Note that this is an explicit relation for θ in terms of M_1 and β that is often termed as the β–θ–*M* relation, which is very important for the analysis of oblique shock. The variations of β with θ for a wide range of Mach numbers are plotted in Figure 2.10. Note that for a given upstream Mach number M_1, the deflection angle has a maximum value, beyond which ($\theta > \theta_{max}$) no solution exists for a straight shock wave. Beyond this condition, the shock gets detached and curved. On the other hand, for a given upstream Mach

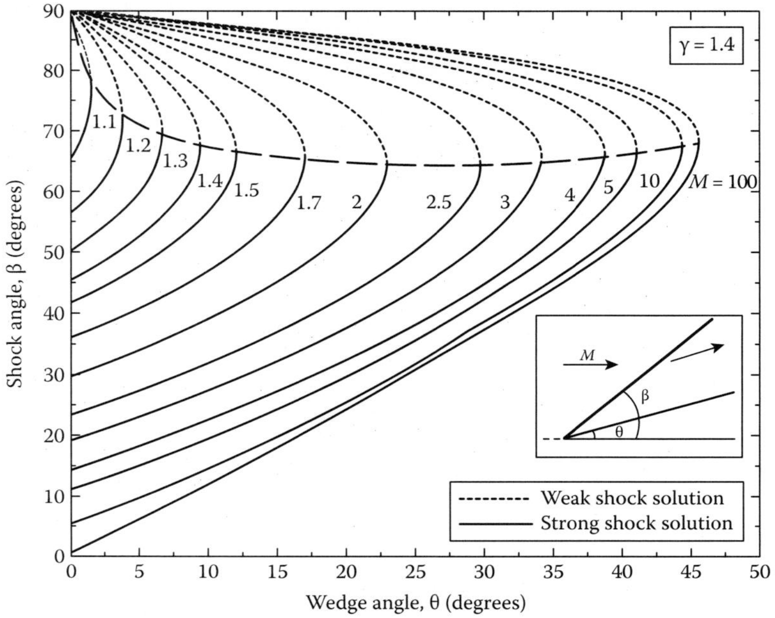

FIGURE 2.10 β–θ–*M* diagram for air (γ = 1.4) for an oblique shock wave.

number M, if the deflection angle is less than θ_{max} ($\theta < \theta_{max}$), two values of β can be obtained from the β–θ–M relation (see Figure 2.10). For example, for given $\theta = 20°$, $M_1 = 3.0$, β can either be 37.76° (weak solution) or 82.15° (strong solution). Note that the strong shocks do not occur very often except for special situations, where the back pressure is increased by some independent mechanisms. When the deflection angle becomes zero, then oblique shock becomes a normal shock wave. Note that for a particular deflection angle θ, the wave angle β increases for the weak solution with a decrease in free stream Mach number from high to low supersonic value.

Example 2.6

Air with $M_1 = 2.5$, $P_1 = 85$ kPa, and $T_1 = 228$ K approaches a wedge with an included angle of 30°. Determine (1) shock angle β; (2) exit Mach number M_2.

Solution

From β–θ–M (Figure 2.10), $\theta = 15°$, we can get shock angle $\beta = 37°$. Then we can estimate the normal component of inlet Mach number as

$$M_{1n} = M_1 \sin\beta = 2.5 \sin 37° = 1.37$$

From normal shock (Table B.2) for $M_{1n} = 1.37$, we can get

$$\frac{T_2}{T_1} = 1.235; \quad \frac{P_2}{P_1} = 2.023; \quad M_{2n} = 0.753 \quad \text{and} \quad \frac{P_{t2}}{P_{t1}} = 0.965$$

Hence, we can easily evaluate properties' ratio across oblique shock wave as follows:

$$P_2 = \frac{P_2}{P_1} \times P_1 = 2.023 \times 85 = 171.95 \text{ kPa}$$

$$T_2 = \frac{T_2}{T_1} \times T_1 = 1.235 \times 228 = 281.58 \text{ K}$$

$$M_2 = \frac{M_{2n}}{\sin(\beta - \theta)} = \frac{0.753}{\sin 22°} = 2.01$$

It is interesting to note that the Mach number at the downstream of oblique shock is less than inlet Mach number M_1, but remains in supersonic flow regime.

REVIEW QUESTIONS

1. What do you mean by calorically perfect gas? Explain it invoking kinetic theory of gas. Can it be employed for the analysis of components of rocket engine?
2. Draw the variation of specific heats of monatomic, diatomic, and triatomic species with temperature and explain the difference among the three types of species.
3. Derive an expression of the first law of thermodynamics for a control volume system by stating the assumptions.
4. Why is the second law of thermodynamics important for the analysis of various components of rocket engine?
5. What do you mean by equivalence ratio? Why is it used very often while dealing with gas turbine engine?
6. What do you mean by adiabatic flame temperature? How can it be estimated? Explain it by an example.
7. Draw the variation of adiabatic flame temperature with equivalence ratio for C_2H_5OH–air mixture at ambient temperature and pressure.
8. How does the adiabatic flame temperature vary with an increase in initial temperature? Why is it so? Explain it by providing proper argument.
9. Enumerate the procedure to estimate adiabatic flame temperature under a constant volume condition.
10. What do you mean by incompressible flow? How does it differ from compressible flow?
11. Explain why it is desirable to locate normal shock at the throat of the CD nozzle.
12. What is the difference between compression and expansion wave in a compressible flow?
13. Under what circumstances does a compression wave change into a shock wave?
14. What do you mean by shock strength? Explain it physically.
15. What are the differences between normal and oblique shock?

PROBLEMS

2.1 The combustion products with specific heat ratio of 1.25 from a rocket engine at T_1 = 3200 K and P_1 = 2.5 MPa enter into a nozzle. Assuming the flow is isentropic, estimate density and temperature at a point when pressure attains 0.02 MPa.

2.2 The combustion product (C_p = 1.145 kJ/kg K, and MW = 25) is expanded from initial state (P_1 = 3.5 MPa, T_1 = 3050 K) to final state (P_2 = 0.1 MPa, T_2 = 1300 K). Determine the specific enthalpy change between these two states. Assuming ideal gas behavior, evaluate and show that entropy change between two given states is same for both isochoric and isobaric processes.

2.3 The compositions of natural gas at 0.1 MPa are CH_4 = 95.5%, C_2H_6 = 1.2%, C_3H_8 = 1.5%, CO_2 = 1.8%. If this gas enters into storage tank at 400 K, find out the mass fraction of each component, the absolute enthalpy of the mixture in both a mole (kJ/kmol) and a mass basis (kJ/kg).

2.4 The natural gas from Mahanadi field is used in a combustor that operates with an oxygen concentration of 5% in the flue gas. If the composition of this natural gas is H_2 = 2%, CH_4 = 95%, CO = 1.9%, N_2 = 1.1%, determine the operating air–fuel ratio and the equivalence ratio.

2.5 Ethyl alcohol is burnt with oxygen in a combustion. The volumetric analysis of products on dry basis is CO_2 = 15.5%, O_2 = 6.76%, CO = 0.93%. Determine (1) A/F ratio, (2) equivalence ratio, and (3) percentage of stoichiometric air used.

2.6 If 1.5 kg of hydrogen is burnt with certain amount of oxygen leading to complete combustion in a burner, determine the amount of heat released and change in mass due to the heat release during this combustion process.

2.7 In a burner, propane gas is burnt with oxygen with stoichiometric proportion. Estimate the adiabatic temperature at 1 atm when the initial reactant is at 298 K. Take C_{p,CO_2} = 56.21 kJ/kmol K, C_{p,H_2O} = 43.87 kJ/kmol K. If the reactant temperature becomes 1000 K, determine its adiabatic temperature and comment on it while comparing with previous value.

2.8 Determine the adiabatic flame temperature of the stoichiometric iso-butane–oxygen mixture at 298 K, 0.1 MPa, assuming no dissociation of the products, for the following two cases: (1) evaluating C_P of each species at 298 K, and (2) evaluating C_P of each species at 2000 K.

2.9 In a vessel, the equilibrium reaction $O_2 \Leftrightarrow 2O$ occurs. Estimate the mole fraction of O_2 and O for (1) $T = 2300$ K, $P = 0.1$, and (2) $T = 3000$ K, $P = 10$ MPa. Assume that there is no dissociation during this reaction.

2.10 In a vessel, 3 mol of hydrogen and 2 mol of oxygen are allowed to react at $T = 1800$ K and $P = 10$ MPa. Determine equilibrium composition assuming their action to be $H_2 + 0.5\ O_2 \Leftrightarrow 2\ H_2O$. If the pressure is reduced by 10 times from 10 to 0.1 MPa, what will be its equilibrium composition. Comment on your results. You may have to use trial-and-error method for solving this problem. Please indicate all steps of this method.

2.11 Hydrogen gas tank at pressure of 100 MPa and temperature 300 K has a hole of 1 mm. Determine the velocity and the mass flow rate of hydrogen through this tiny hole.

2.12 The combustion products from thrust chamber of rocket engine at 4.5 MPa and 3100 K is expanded through a CD nozzle. If the back pressure is maintained at 50 kPa, determine (1) A_e/A_{th}, (2) V_e. Take, $\gamma = 1.33$.

2.13 A spacecraft moving through the atmosphere (288 K) with velocity of 1000 m/s has a detached bow shock formed ahead of it. Determine the pressure ratio across the bow shock, assuming the central portion of this bow shock to be a normal shock $\gamma = 1.33$.

2.14 A spacecraft moving through the atmosphere (288 K) with velocity of 1020 m/s has an attached oblique shock with a full wedge angle of 30°. Determine the static pressure ratio, temperature, and total pressure ratio across the oblique shock.

REFERENCES AND SUGGESTED READINGS

1. Mishra, D.P., *Engineering Thermodynamics*, Cengage Learning India Pvt. Ltd., New Delhi, India, 2011.
2. Mishra, D.P., *Fundamentals of Combustion*, PHI Learning Pvt. Ltd., New Delhi, India, 2011.

3. Mishra, D.P., *Experimental Combustion*, CRC Press, New York, 2014.
4. Mishra, D.P., *Gas Turbine Propulsion*, MV Learning, London, U.K., 2015.
5. Anderson, J.D., Jr., *Modern Compressible Flow*, McGraw-Hill, New York, 1982.
6. Rathakrishnan, E., *Gas Dynamics*, 5th edn., PHI Learning Pvt. Ltd., New Delhi, India, 2013.

CHAPTER 3

Elements of Rocket Propulsion

It all looked so easy when you did it on paper—where valves never froze, gyros never drifted, and rocket motors did not blow up in your face.

MILTON W. ROSEN, ROCKET ENGINEER

3.1 INTRODUCTION

We learnt in Chapter 1 that both air-breathing and non-air-breathing (rocket) engines work on the principle of jet propulsion, but the air-breathing engine is different from the rocket engine in the sense that it carries both fuel and oxidizer during its flight. As a result, it can fly beyond the earth's atmosphere even to the deep-space region. In order to understand the basic principles of rocket propulsion, the fundamentals of thermodynamics, chemistry, and gas dynamics are reviewed in Chapter 2. In this chapter, the elements of rocket propulsion are discussed in detail. As the processes involved in rocket engine are quite complex, certain assumptions are made for an ideal engine. Subsequently, thrust equation for rocket engine is derived. The performance parameters, namely, specific impulse, impulse to weight ratio, specific propellant flow rate, mass flow coefficient, thrust coefficient, characteristic velocity, and propulsive efficiencies are defined and discussed, which are useful in characterizing rocket engines.

3.2 IDEAL ROCKET ENGINE

We know that the processes involved during the operations of the chemical rocket engine are quite complex in nature. The flow is inherently three-dimensional in nature. Besides, flow is likely to be unsteady and highly turbulent, but the fluctuations in propellant supply line may vary from 1% to 4% of average value. Of course, if combustion instability occurs in the combustion chamber, there might be higher levels of fluctuation in the rocket engine. Generally, the combustion instability should be avoided at any cost as it may lead to failure of the entire rocket engine system itself. As a large amount of heat is released during the combustion of solid/liquid at a very fast rate, heat transfer does take place through the walls of the combustion chamber and nozzle. The total loss of heat from a typical rocket engine varies only between 1% and 2%. As solid/liquid propellants are used in the chemical rocket engine, it is more likely that two-/three-phase flow can occur in the combustion chamber and nozzle itself. Besides, shock and expansion waves are likely to occur during expansion of gas in the exhaust nozzle. There will be interactions between shock/expansion wave and boundary layers during its operation. Hence, it can be concluded that it is quite complex to deal with such a complex flow conditions in the rocket engine. In order to make these complex problems tractable, certain simplifying assumptions can be made for obtaining a general understanding of the main features in the chemical rocket engine. Although the theory developed under the following assumptions for the ideal rocket engine may not depict the complex features in an actual chemical rocket engine, it is good enough to arrive at certain solutions that can handle the majority of chemical rocket engine systems. This is because the difference between the performance parameters obtained by the idealized model and actual measurements lies only between 1% and 5%. In this idealized model, rocket flow is essentially considered as steady quasi-one-dimensional and isentropic in nature with the following assumptions [1,2]:

1. Steady one-dimensional inviscid flow.
2. Details of combustion are ignored.
3. The working fluid is homogeneous in nature. The amount of mass due to condensed phase (solid/liquid) is negligible as compared to gaseous fluid.
4. Ideal gas law with constant specific heat can be applied.

5. No heat transfer from the rocket engine.
6. Flow is isentropic except across the shock.
7. Uniform chamber conditions at the nozzle entrance; flow properties, namely, pressure, temperature, and density remain constant.
8. Velocity at the nozzle entrance is negligibly small as compared to its exit.
9. Gas composition across the nozzle remains constant. Frozen flow in the nozzle.
10. Gas leaves nozzle exit along axial direction only.

These simplifying assumptions are quite useful in deriving simplified performance parameters that are helpful in characterizing the chemical rocket engine. Subsequently, we will discuss the correction factors by which this idealized model can be improved further.

3.3 THRUST EQUATION OF ROCKET ENGINES

An expression for the thrust developed by a rocket engine under static condition can be obtained by applying the momentum equation. For this, let us consider a control volume (CV), as shown in Figure 3.1. The propulsive thrust "*F*" acts in a direction opposite to V_e. The reaction to the thrust "*F*" on the CV is opposite to it. The momentum equation for such CV is given by [4]

$$\frac{d}{dt}\int_{cv} \rho V_x \, dV + \int_{cs} V_x \left(\rho V_x \cdot n\right) dA = \Sigma F_x \tag{3.1}$$

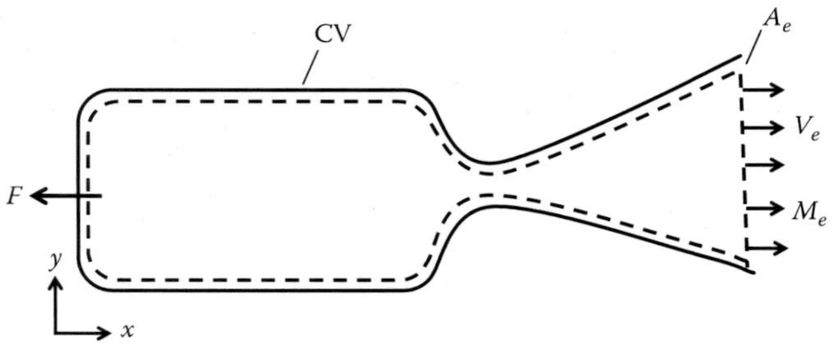

FIGURE 3.1 Control volume of rocket engine.

As the flow is steady in nature, we can neglect the unsteady term in Equation 3.1:

$$\frac{d}{dt}\int_{cv} \rho V_x \, dV = 0$$

Let us now evaluate the momentum flux term and sum of forces acting on CV of Equation 3.1, as given in the following:

$$\int_{cs} V_x \left(\rho V_x \cdot n\right) dA = \int V_x \, dm = \dot{m} V_e \tag{3.2}$$

$$\sum F_x = F + P_a A_e - P_e A_e \tag{3.3}$$

where

F is the thrust
V_x is the velocity component in x-direction
V_e is the velocity component at exit of nozzle
$\dot{m}$ is the mass flow rate of propellant
P_e is the pressure at exit plane of nozzle
P_a is the ambient pressure

By combining Equations 3.3, 3.2 and 3.1 for steady flow, we can have,

$$F = \dot{m} V_e + \left(P_e - P_a\right) A_e \tag{3.4}$$

where the first and second terms represent the momentum contribution and pressure components of thrust, respectively.

3.3.1 Effective Exhaust Velocity

We know that velocity profile at the exit of nozzle need not be one-dimensional in nature in the practical situation. It is quite cumbersome to determine the velocity profile at the exit of a rocket nozzle. In order to tackle this problem, we can define effective exhaust velocity by using Equation 3.4 as given below:

$$V_{eq} = \frac{F}{\dot{m}} = V_e + \frac{\left(P_e - P_a\right)}{\dot{m}} A_e \tag{3.5}$$

where V_{eq} is the effective exhaust velocity that would be produced equivalent thrust which could have produced due to both momentum and pressure components of thrust. Note that when the exit pressure is same as that of the ambient pressure, the effective exhaust velocity V_{eq} becomes equal to nozzle exit velocity V_e. Otherwise, the effective exhaust velocity V_{eq} is less than the nozzle exit velocity V_e. Note that the second term in Equation 3.5 is smaller than the nozzle exit velocity V_e. Hence, the effective exhaust velocity V_{eq} is quite close to the nozzle exit velocity V_e. Note that the effective exhaust velocity V_{eq} is often used to compare the effectiveness of different propellants in producing thrust in rocket engine. It can be applied to all mass expulsion systems that undergo expansion through a nozzle. It can be evaluated experimentally by measuring the thrust and total propellant flow rate. Keep in mind that it can be easily related to specific impulse and characteristic velocity, which will be discussed in a subsequent section. In recent times, some groups are advocating the use of effective exhaust velocity V_{eq} in place of specific impulse for the purpose of comparison.

3.3.2 Maximum Thrust

We need to evaluate the condition under which maximum thrust can be achieved for a given chamber pressure and mass flow rate with a fixed throat area of nozzle. We can obtain this condition by differentiating the thrust expression, Equation 3.4, given in the following:

$$dF = \dot{m} dV_e + \left(P_e - P_a\right) dA_e + A_e dP_e \tag{3.6}$$

Note that mass flow rate remains constant ($\dot{m}$). By invoking momentum equation for one-dimensional steady inviscid flow, we can have

$$\dot{m} dV_e = -A_e dP_e \tag{3.7}$$

By clubbing Equations 3.6 and 3.7, we can get

$$dF = \left(P_e - P_a\right) dA_e \tag{3.8}$$

The condition $P_e = P_a$ is called optimum expansion because it corresponds to maximum thrust for the given chamber conditions. The maximum thrust can be obtained only when the nozzle exhaust pressure is equal to the ambient pressure ($P_e = P_a$), as given in the following:

$$\frac{dF}{dA_e} = \left(P_e - P_a\right) = 0; \quad \text{when } P_e = P_a \tag{3.9}$$

Then, thrust expression for maximum thrust can be expressed as

$$F = \dot{m} V_e \tag{3.10}$$

This condition provides a maximum thrust for a given chamber pressure and propellant mass flow rate. The rocket nozzle in which this condition is achieved is known as optimum expansion ratio nozzle.

3.3.3 Variation of Thrust with Altitude

We can also note from Equation 3.4 that the thrust of a rocket engine is independent of the flight velocity, unlike the gas turbine engine. However, the thrust varies with altitude particularly while operating within the earth's atmosphere, as there will be changes of ambient atmospheric pressure with altitude. When the rocket propels, the thrust increases with increase in altitude because the atmospheric pressure decreases with increase in altitude. It can be noted that around 10%–30% of overall thrust changes may occur due to change in altitude. Let us see the variation of the thrust and the specific impulse I_{sp} with altitude, shown in Figure 3.2, for a typical rocket engine. It can be observed that the thrust and specific impulse increase with altitude to respective asymptotic value at high altitude, indicating that thrust remains invariant beyond the earth's atmosphere envelope.

3.3.4 Effect of Divergence Angle on Thrust

Recall that we have derived the thrust expression (Equation 3.4) assuming the exhaust velocity at nozzle exit to be parallel to its axis. It would not be true particularly for conical nozzle as divergence angle in the range of 10°–20° [3,4] is used in rocket nozzle. In order to determine the effects of divergence angle on the ideal thrust, we can consider flow in divergent portion of nozzle as shown in Figure 3.3, in which all streamlines

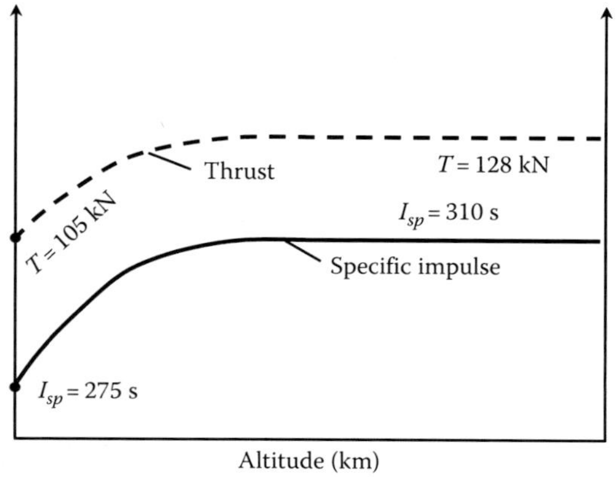

FIGURE 3.2 Variation of thrust of a typical rocket engine with altitude.

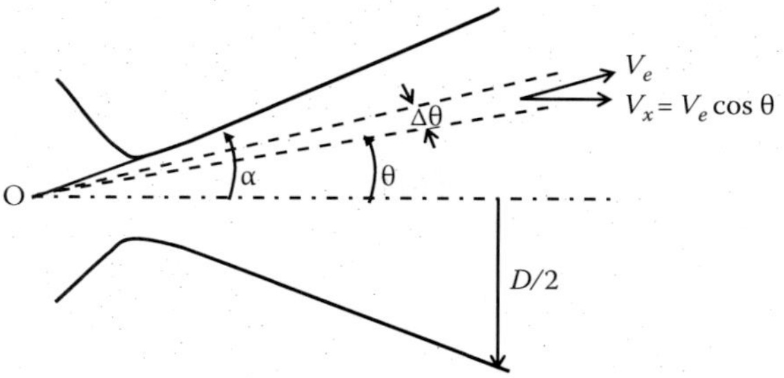

FIGURE 3.3 Schematic of divergent flow in a conical nozzle.

are considered to be straight that intersect at the point O, in the upstream region of nozzle exit. Note that mass flow crosses only through the spherical exit segment of nozzle, as shown in Figure 3.3. The cross-sectional area segment at nozzle exit can be evaluated as

$$dA = 2\pi R \sin\theta R d\theta \tag{3.11}$$

The velocity at nozzle exit in the axial direction would be $V_e \cos\theta$. By using Equations 3.1 and 3.10, we can have expression for thrust under steady-state condition:

$$F = \int_0^{\alpha} \dot{m} V_e \cos\theta \; 2\pi R \sin\theta \; R d\theta + \left(P_e - P_a\right) A_e \tag{3.12}$$

By integrating this equation, we get

$$F = \frac{\left(1 + \cos\alpha\right)}{2} \dot{m} V_e + \left(P_e - P_a\right) A_e = \lambda_d \dot{m} V_e + \left(P_e - P_a\right) A_e \tag{3.13}$$

where λ_d is the divergence correction factor whose values are closer to unity. For example, for nozzle divergence angle of 15°, λ_d happens to be around 0.983. Hence, for preliminary calculation of the rocket engine, the divergence correction factor can be taken to be unity.

Example 3.1

A booster rocket engine with nozzle exit diameter of 225 mm is designed to propel a satellite to altitude of 20 km. The chamber pressure at 12 MPa is expanded to exit pressure and temperature of

105 kPa and 1400 K, respectively. If the mass flow rate happens to be 15 kg/s, determine the exit jet velocity, effective jet velocity, and thrust at $Al = 20$ km. $MW = 25$ kg/kmol.

Solution

At 20 km altitude, $P_a = 5.53$ kPa.

Assuming one-dimensional flow at the exit of nozzle, and assuming ideal gas law, we can evaluate exhaust velocity from mass flow rate at exit as

$$V_e = \frac{\dot{m}}{\rho_e A_e} = \frac{4R_e T_e \dot{m}}{MWP_e \pi D_e^2} = \frac{4 \times 8.314 \times 1400 \times 15}{25 \times 105 \times 3.14 \times (0.225)^2} = 1673.65 \text{ m/s}$$

By using Equation 3.5, we can evaluate effective velocity as

$$\begin{aligned} V_{eq} &= V_e + \frac{(P_e - P_a)}{\dot{m}} A_e \\ &= 1673.65 + \frac{(105 - 5.53) \times 10^3}{15} \frac{3.14}{4} (0.225)^2 \\ &= 1673.65 + 263.5 = 1937.15 \text{ m/s} \end{aligned}$$

Note that the equivalent velocity is almost the same as the exit velocity, indicating that thrust due to momentum will be predominant. We can evaluate total thrust as

$$F = \dot{m} V_{eq} = 29057.25 = 29.06 \text{ kN}.$$

3.4 ROCKET PERFORMANCE PARAMETERS

In order to characterize the performance of the rocket engine, several parameters are being used by engineers. Some of the important performance parameters are specific impulse, propellant consumption, thrust coefficient, and characteristic velocity. The importance of volumetric specific impulse and impulse efficiency is also defined and discussed. Besides these parameters, several efficiencies, namely, propulsive, thermal, overall efficiencies are being devised to characterize the performance of the rocket engine. All these parameters are defined and discussed in the following [1–4].

3.4.1 Total Impulse and Specific Impulse

We know that the thrust force by the rocket engine imparts motion to the spacecraft due to momentum change caused by expansion of propellants in the exhaust nozzle. Impulse is imparted due to this change in momentum over a certain period of time. In other words, total impulse I, imparted to the vehicle during its acceleration can be obtained by integrating over the burning time t_b:

$$I = \int F\,dt = \int_0^{t_b} \dot{m}_p V_{eq}\,dt = m_p V_{eq} \tag{3.14}$$

Note that we have obtained the equation by assuming V_{eq} to be constant with time.

The specific impulse is an important performance variable. The specific impulse I_{sp} can be defined as the impulse per unit weight of the propellant:

$$I_{sp} = \frac{I}{m_p g} \tag{3.15}$$

where

I is the total impulse imparted to the vehicle during acceleration
m_p is the total mass of expelled propellant
g is the acceleration due to gravity at earth's surface

The denominator $m_p g$ in Equation 3.15 represents the total effective propellant weight, which is generally evaluated with gravitational acceleration at sea level. In SI units, the unit specific impulse I_{sp} happens to be in seconds. That does not mean that it indicates a measure of elapsed time. Rather, it represents the time during which the thrust delivered by the rocket engine is equal to the propellant weight. It also indicates how much impulse can be generated per unit weight of propellant. Note that specific impulse I_{sp} represents the time average specific impulse over entire period of operation of the rocket engine. The value of specific impulse I_{sp} does vary during transient operations, namely, start, shutdown period.

Then,

$$I_{sp} = \frac{m_p V_{eq}}{m_p g} = \frac{V_{eq}}{g} = \frac{F}{\dot{m}_p g} \tag{3.16}$$

It must be noted that I_{sp} does not depend on the flight velocity. The presence of g in the definition of I_{sp} is arbitrary. But it makes the unit of I_{sp} uniform in all common systems of units. Note that the specific impulse depends on both the type of propellant and the rocket engine configuration. Besides this, the specific impulse I_{sp} can be stated as the amount of impulse imparted to vehicle per kilogram of effective propellant. Note that I_{sp} can also be defined as thrust per unit weight flow rate consumption of propellant. Hence, lower $\dot{m}_p$ implies higher I_{sp}, which is desirable for chemical rockets. A higher specific impulse I_{sp} is directly related to long flight range, which indicates superior range capability of air-breathing engine over chemical rockets at relatively low speeds. Note that in liquid-propellant engine, propellant flow rate and thrust can be measured easily and thus specific impulse I_{sp} can be easily estimated experimentally. But in case of solid-propellant rocket engine, it is quite cumbersome to measure the propellant flow rate accurately and hence specific impulse I_{sp} can be determined from the total impulse and propellant weight using Equation 3.15. Typical values of specific impulse I_{sp} for various types of rocket engines are given in Table 3.1. It can be noted that specific impulse I_{sp} of solid-propellant rocket engine varies between 200 and 300 s, while liquid-propellant rocket exhibits its value between 300 and 400 s. The values of I_{sp} for hybrid rocket engine are higher than that of solid-propellant rocket engine. Of course, nonchemical rocket engines have higher values of I_{sp}. But these engines have capability of producing very less thrust. Therefore, it is only preferred for deep-space applications.

3.4.2 Specific Impulse Efficiency

We have learnt that specific impulse can be determined from the experimental data. It can also be determined from aerothermodynamical calculation theoretically. Note that software SP273 developed by NASA is routinely used for theoretical calculation of I_{sp}. The measured value of

TABLE 3.1 Typical Values of Specific Impulse I_{sp} for Various Kinds of Rocket Engines

Rocket Engines	I_{sp} **(s)**
Solid	200–310
Liquid	300–460
Hybrid	300–500
Solar	400–700
Nuclear	600–1000
Electrical (arc heating)	400–2000

specific impulse I_{sp} will be lower than the theoretical value as several kinds of losses, namely, friction, gas dynamic, incomplete combustion, and so on, can occur during actual operation of the rocket engine. In order to ascertain the extent of difference between theoretical and measured values of specific impulse I_{sp}, we can define specific impulse efficiency, which is the ratio between measured specific impulse I_{sp} and theoretical specific impulse I_{sp}, as given in the following:

$$\eta_{I_{sp}} = \frac{I_{sp}}{I_{sp,theory}} \tag{3.17}$$

where $I_{sp,theory}$ is the theoretical specific impulse. Note that specific impulse efficiency, $\eta_{I_{sp}}$ for well-designed rocket engine varies from 0.95 to 0.99, indicating that theoretical calculation can be used easily for design calculation.

3.4.3 Volumetric Specific Impulse

In order to ascertain the effect of engine size on the performance of rocket engine, it is important to consider volumetric specific impulse I_V in place of specific impulse. Generally, it is defined as the total impulse per unit propellant volume, which can be expressed mathematically as

$$I_V = \frac{Ft}{V_p} = \frac{m_p g I_{sp}}{V_p} = \rho_p g I_{sp} \tag{3.18}$$

where

V_p is the propellant volume
ρ_p is the average density of propellant

It is desirable to have higher volumetric specific impulse as volume required for propellant storage in the spacecraft will be smaller. In other words, for propellant with particular density, higher volumetric specific impulse calls for higher specific impulse. Hence, it is essential to have higher propellant density to have higher volumetric specific impulse for a constant specific impulse system. That is the reason why solid-propellant rocket engine is preferred over liquid-propellant engine even though it has lower specific impulse. We will discuss it in detail in Chapter 4, while dealing with the rocket equation.

3.4.4 Mass Flow Coefficient

In the case of the rocket engine, mass flow rate through the nozzle is governed by the chamber pressure and throat area of the nozzle. Hence, it is important to have a parameter that can represent all these variables. This parameter, known as the mass flow coefficient $C_{\dot{m}}$, is defined as the ratio between mass flow rate of propellant $\dot{m}_p$, and product of chamber pressure P_c and throat area A_t, given as follows:

$$C_{\dot{m}} = \frac{\dot{m}_p}{P_c A_t} \tag{3.19}$$

Note that mass flow coefficient $C_{\dot{m}}$ can be estimated easily from experimental data for parameters, namely, mass flow rate of propellant $\dot{m}_p$, chamber pressure P_c, and throat area A_t. The mass flow coefficient $C_{\dot{m}}$ can be determined from experimental data and documented in the form of design charts, which is quite useful for the design and development of rocket engine.

3.4.5 Thrust Coefficient

In Section 3.3, we have derived the expression for thrust (Equation 3.4) that indicates that thrust is dependent on exhaust velocity V_e, and exit pressure P_e and ambient pressure P_a, and mass flow rate of propellant $\dot{m}_p$. But we know that exhaust velocity V_e, exit pressure P_e, and mass flow rate of propellant $\dot{m}_p$ are affected by chamber pressure P_c and throat area A_t of the nozzle. Hence, in case of the rocket engine, thrust is dependent on $\dot{m}_p$, chamber pressure P_c, and throat area A_t of the nozzle. Therefore, let us relate the thrust F with $\dot{m}_p$, chamber pressure P_c, and throat area A_t of the nozzle through a parameter known as thrust coefficient C_F, which is defined as the ratio of thrust F, and product of chamber pressure P_c and throat area A_t of the nozzle, given in the following:

$$C_F = \frac{F}{P_c A_t} \tag{3.20}$$

Note that thrust coefficient C_F can be determined easily from the experimental data of chamber pressure P_c and throat area A_t. Generally, the design chart for thrust coefficient C_F in terms of pressure ratio P_c/P_o and area ratio A_e/A_t is generated, which can be used for design and development of the rocket engine. Besides this, thrust coefficient C_F as a function pressure ratio P_c/P_o and area ratio A_e/A_t can be determined theoretically by carrying out aerothermodynamic analysis, which will be discussed in Chapter 4. We will also prove in Chapter 4 using theoretical analysis that the thrust coefficient C_F represents the performance of the nozzle only.

Let us relate specific impulse I_{sp} to the thrust coefficient C_F and mass flow coefficient $C_{\dot{m}}$ by using Equations 3.19 and 3.20, as given in the following:

$$I_{sp} = \frac{F}{\dot{m}_p g} = \frac{C_F}{C_{\dot{m}} g} \tag{3.21}$$

It can be noted that specific impulse I_{sp} can be easily estimated from the thrust coefficient C_F and mass flow coefficient $C_{\dot{m}}$ data.

3.4.6 Specific Propellant Consumption

In modern times, it is important to know how much fuel is being burnt for unit power produced in heat engines. Similarly, it is important to know how much propellant is consumed per unit impulse. Hence, specific propellant consumption (SPC) for the rocket engine can be defined as the amount of propellant weight consumed per total impulse delivered. It is similar to the thrust-specific fuel consumption (TSFC) for air-breathing jet engine and brake-specific fuel consumption for automobile engines. Unfortunately, specific propellant consumption (SPC) is not commonly used in the rocket engine. It is basically reciprocal of specific impulse, as given by

$$SPC = \frac{1}{I_{sp}} = \frac{\dot{m}_p g}{F} \tag{3.22}$$

3.4.7 Characteristic Velocity

Another experimental performance parameter used routinely for rocket engine is the characteristic velocity C^*, which is defined as the ratio of equivalent exit velocity V_e and the thrust coefficient C_F, given as follows:

$$C^* = \frac{V_{eq}}{C_F} \tag{3.23}$$

Let us express this equation in terms of measurable parameters, namely, $\dot{m}_p$, chamber pressure P_c, and throat area A_t by using Equations 3.23 and 3.20, given as follows:

$$C^* = \frac{P_c A_t}{\dot{m}_p} = \frac{1}{C_{\dot{m}}} \tag{3.24}$$

It is interesting to note that this characteristic velocity is reciprocal of the mass flow coefficient as discussed in Section 3.4.8. Note that it does not depend on the thrust produced by the rocket engine. It represents the effectiveness with which combustion takes place in the combustion chamber of the rocket engine. The characteristic velocity C^* can be expressed in terms of thrust coefficient and specific impulse I_{sp} by using Equation 3.21, given as follows:

$$C^* = \frac{1}{C_{\dot{m}}} = \frac{I_{sp} g}{C_F} \tag{3.25}$$

Note that the characteristic velocity C^* is dependent on thrust coefficient and specific impulse. We can rewrite Equation 3.25 as

$$I_{sp} = \frac{C^* C_F}{g} \tag{3.26}$$

It can be observed from this equation that specific impulse is dependent on two independent parameters, namely, the characteristic velocity C^* and thrust coefficient C_F. Note that the characteristic velocity, C^* represents the combustion efficiency while thrust coefficient C_F indicates the effectiveness with which high pressure hot gases are expanded in the nozzle to produce requisite thrust. We will discuss further these two parameters in Chapter 4.

3.4.8 Impulse-to-Weight Ratio

In order to find out the efficacy of overall design of rocket engine, it is prudent to use impulse–weight ratio, which is defined as the total impulse to the initial vehicle weight as given by

$$\frac{I}{W} = \frac{I_{sp} \cdot m_p}{\left(m_p + m_s + m_l\right) g} \tag{3.27}$$

where

W is the total weight of the vehicle
m_p is the propellant mass
m_s is the structural mass of the vehicle
m_l is the payload mass

It is desirable to have higher value of impulse–weight ratio.

Example 3.2

A rocket engine with chamber pressure of 4.5 MPa and nozzle throat diameter of 110 mm produces thrust of 17 kN by consuming propellant flow rate of 7.5 kg/s with calorific value of 25 MJ/kg. If the flight velocity happens to be 850 m/s, determine the specific impulse, effective exhaust velocity, specific propellant consumption and thrust power, and thrust coefficient.

Solution

The specific impulse can be calculated by using Equation 3.21 as

$$I_{sp} = \frac{F}{\dot{m}_p g} = \frac{17\times10^3}{7.5\times9.81} = 231.06 \text{ s}$$

The effective velocity can be determined using Equation 3.16, given as follows:

$$V_{eq} = I_{sp} g = 231.06\times9.81 = 2266.7 \text{ m/s}$$

The specific propellant consumption can be determined using Equation 3.22, as

$$SPC = \frac{1}{I_{sp}} = \frac{1}{231.06} = 0.0043 \text{ kg/N}\cdot\text{s}$$

The thrust power can be determined as

$$P_F = FV = 17\times850 = 14,450\,\text{kW} = 14.45\,\text{MW}$$

The thrust coefficient can be evaluated using Equation 3.21, as

$$C_F = \frac{F}{P_c A_t} = \frac{4 \times 17}{2500 \times 3.14 \times (0.11)^2} = 0.72$$

The characteristic velocity can be evaluated using Equation 3.23, as

$$C^* = \frac{V_{eq}}{C_F} = \frac{2266.7}{0.72} = 3148.2 \text{ m/s}$$

3.4.9 Energy Balance and Efficiencies

We need to evaluate the various kinds of efficiencies, namely, propulsive, thermal, overall efficiencies, and others for the rocket engine to analyze the performance of energy conversion during its operation. These efficiencies are to be defined and used while carrying out aerothermodynamics analysis of components of propulsive devices. Note that although these efficiencies are used seldom for rocket engines, they are routinely used in air-breathing engines to evaluate their performance. For this purpose, let us consider the energy balance of the rocket engine as shown in Figure 3.2. It can be noted that $\dot{E}_{in}$ is the total rate of energy supplied to the propulsion system. Thus, this rate of energy input, $\dot{E}_{in}$ to the chemical rocket engine can be expressed as

$$\dot{E}_{in} = \dot{m}_p \Delta H_p \tag{3.28}$$

where

$\dot{m}_p$ is the mass flow rate of propellant
ΔH_p is the heat of combustion of propellant

As illustrated in Figure 3.2, a small fraction of this input energy during combustion process is lost due to incomplete combustion. Thus, the combustion efficiency η_C can be defined as the ratio of the rate of heat energy liberated during combustion $\dot{E}_t$ and the rate of chemical energy input from the propellant $\dot{m}_p \Delta H_p$.

$$\eta_C = \frac{\dot{E}_t}{\dot{m}_p \Delta H_p} \tag{3.29}$$

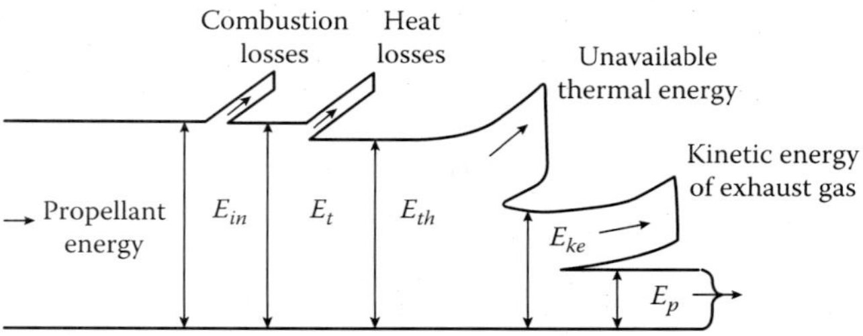

FIGURE 3.4 Energy balance of a typical rocket engine.

Basically, it is a measure of losses during the combustion of propellant. Generally, the combustion efficiency η_C of a typical rocket engine varies from 95% to 99%. Around 1%–3% of total energy input to the rocket engine can be lost through walls of the combustion chamber and nozzle, as indicated in Figure 3.4. Besides, 2%–3% of the total energy input to the rocket engine could not be available for exhaust nozzle to harness thrust power, which is due to mixing and gas dynamic losses. Around 92%–98% of the total energy input to the rocket engine will be available for the exhaust nozzle to produce the thrust. Unfortunately, not all energy per unit time available for the exhaust nozzle can be converted into thrust power. A significant amount of total energy input in the range of around 40%–50% can be lost as the residual kinetic energy from the exhaust energy. The remaining portion of energy of the total energy input to the rocket engine denoted as E_p can be converted into propulsive power. Based on this energy balance, we can define three efficiencies, namely, propulsive, thermal, and overall efficiencies, as defined and discussed in the following.

3.4.9.1 Propulsive Efficiency

We know that the thrust power is the product of thrust and flight velocity. In order to ascertain the efficacy of the rocket engine producing this propulsive power, we can take the help of propulsive efficiency, which is defined as the ratio of thrust power to the rate of kinetic energy available at the inlet of exhaust nozzle, given as follows:

$$\eta_p = \frac{FV}{FV + \frac{\dot{m}_p}{2}\left(V_e - V\right)^2} \approx \frac{\dot{m}_p V_e V}{\dot{m}_p V_e V + \frac{\dot{m}_p}{2}\left(V_e^2 + V^2 - 2V_e V\right)} = \frac{2\left(V/V_e\right)}{1 + \left(\frac{V}{V_e}\right)^2} \tag{3.30}$$

where

V is the flight velocity
V_e is the exit velocity of exhaust nozzle
$\dot{m}_p$ is the flow rate of propellant

In Equation 3.30, it is assumed that the nozzle is expanded fully and the pressure term in thrust equation becomes zero. It can be observed from Equation 3.30 that propulsive efficiency is dependent only on the ratio between flight velocity V and exhaust jet velocity V_e, (V/V_e). The variation of propulsive efficiency with (V/V_e) is shown in Figure 3.5. It can be noted that propulsive efficiency increases with (V/V_e) and attains a maximum value of 100% when the flight velocity is equal to jet velocity. Subsequently, it recedes to a lower value with further increase in (V/V_e) due to inefficient energy conservation. It must be noted that the rockets do operate at flight speed much higher than exhaust velocity unlike the air-breathing engine, because, the thrust does not depend on the flight velocity, unlike in the air-breathing engine.

3.4.9.2 Thermal Efficiency

Another useful performance parameter of the rocket engine is the thermal efficiency, which is defined as the ratio of the rate of kinetic energy

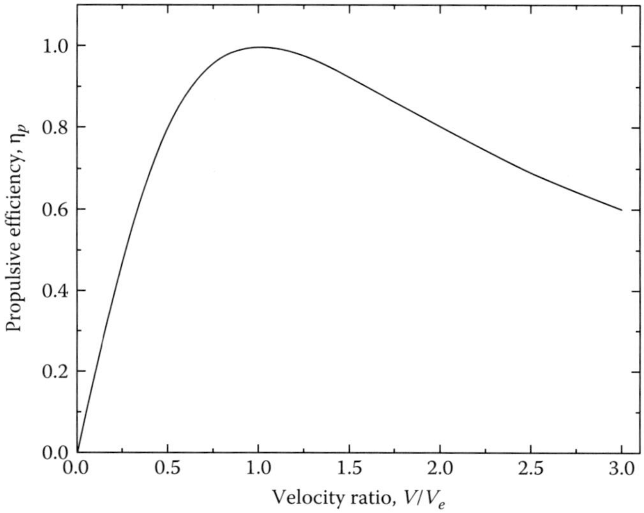

FIGURE 3.5 Variation of propulsive efficiency with (V/V_e).

available at the inlet of exhaust nozzle to the total chemical energy consumption rate from the propellant, as given by

$$\eta_{th} = \frac{FV + \frac{\dot{m}_p}{2}(V_e - V)^2}{\dot{m}_p \Delta H_P} \approx \frac{\frac{\dot{m}_p}{2}\left(V_e^2 + V^2\right)}{\dot{m}_p \Delta H_P} \tag{3.31}$$

where ΔH_p is heat of combustion per kilogram of propellant. It indicates the effectiveness of converting chemical energy in the propellant into kinetic energy for harnessing thrust power.

3.4.9.3 *Overall Efficiency*

The overall efficiency is the ratio of thrust power to the total chemical energy consumption rate from the propellant, as given by

$$\eta_o = \frac{FV}{\dot{m}_p \Delta H_P} = \frac{FV}{FV + \frac{\dot{m}_p}{2}(V_e - V)^2} \cdot \frac{FV + \frac{\dot{m}_p}{2}(V_e - V)^2}{\dot{m}_p \Delta H_P} = \eta_p \cdot \eta_{th} \tag{3.32}$$

It can be noted that thermal efficiency is basically the product of propulsive and thermal efficiency.

Example 3.3

A rocket engine with nozzle exit diameter of 125 mm produces 7.5 kN by consuming propellant flow rate of 3.5 kg/s with calorific value of 35 MJ/kg. The chamber pressure at 6.5 MPa is expanded to exit pressure of 85 kPa. If the flight velocity happens to be 1250 m/s at altitude of 20 km, determine nozzle exhaust velocity, propulsive efficiency, overall efficiency, and thermal efficiency.

Solution

At 20 km altitude, $P_a = 5.53$ kPa.

By using Equation 3.4, we can evaluate total thrust as

$$V_e = \frac{F - (P_e - P_a)A_e}{\dot{m}} = \frac{7500 - (85 - 5.53) \times 10^3 (3.14/4)(0.125)^2}{3.5}$$
$$= 1864.36 \text{ m/s}$$

By using Equation 3.30, we can determine propulsive efficiency as

$$\eta_p = \frac{FV}{FV + \frac{\dot{m}_p}{2}(V_e - V)^2} = \frac{7.5\times10^3\times1250}{7.5\times10^3\times1250 + \frac{3.5}{2}(1864.36 - 1250)^2}$$
$$= 0.934$$

By using Equation 3.32, we can determine overall efficiency as

$$\eta_o = \frac{FV}{\dot{m}_p \Delta H_P} = \frac{7.5\times10^3\times1250}{3.5\times35\times10^6} = 0.0765$$

By using Equation 3.32, we can determine thermal efficiency as

$$\eta_{th} = \frac{\eta_o}{\eta_p} = \frac{0.076}{0.934} = 0.081$$

Note that thermal and overall efficiency of the rocket engine are quite low as compared to gas turbine engines.

REVIEW QUESTIONS

1. Derive an expression for the static thrust of a rocket engine by stating assumptions.
2. What do you mean by thrust power? How is it different from propulsive power?
3. What is the condition under which maximum thrust can be obtained?
4. What do you mean by effective exhaust velocity? What is its utility?
5. What do you mean by propulsive efficiency? How is it different from thermal efficiency?
6. How is the specific impulse varied with altitude for a typical rocket engine?
7. What do you mean by total impulse? How is it different from specific impulse?

8. What do you mean by volumetric impulse? How is it different from specific impulse?
9. Define thrust coefficient. How is it related to specific impulse?
10. What is the difference between thrust coefficient and mass flow coefficient?
11. Define specific fuel consumption. What is its significance?
12. What do you mean by characteristic velocity? Can it be related to thrust?
13. Why is impulse to weight important for a rocket engine?
14. What do you mean by propulsive efficiency? Derive its expression for rocket engine and compare it with that for an air-breathing engine.
15. What is thermal efficiency? How is it related to overall efficiency?

PROBLEMS

3.1 A rocket engine produces a thrust of 800 kN at sea level with a propellant flow rate of 300 kg/s. Calculate the specific impulse.

3.2 A rocket having an effective jet exhaust velocity of 1100 m/s and flying at 8500 km/h has propellant flow rate of 6 kg/s. If the heat of reaction of propellants is 43 MJ. Calculate the propulsive efficiency, thermal efficiency, and overall efficiency.

3.3 If the thermal efficiency of a rocket engine is 0.5, fuel–air ratio is 0.3, effective jet velocity is 1200 m/s, and flight to jet speed ratio is 0.8, calculate the heat of reaction per kilogram of the exhaust gases.

3.4 A rocket vehicle has the following data:

Initial mass = 300 kg

Final mass = 180 kg

Payload mass = 130 kg

Burn duration = 5 s

Specific impulse = 280 s

Determine the mass ratio, propellant mass fraction, propellant flow rate, thrust-to-weight ratio, and impulse-to-weight ratio of the vehicle.

3.5 A certain rocket has an effective exhaust velocity of 1550 m/s; it consumes 5 kg/s of propellant mass, each of which liberates 7000 kJ/kg. The rated flight velocity equals 2600 m/s. Calculate (1) propulsion power (2) engine output.

3.6 A spacecraft's engine ejects mass at a rate of 32 kg/s with an exhaust velocity of 3200 m/s. The pressure at the nozzle exit is 6 kPa and the exit area is 0.7 m^2. What is the thrust of the engine in a vacuum?

3.7 In a rocket engine with nozzle exit diameter of 105 mm, hot gas at 2.5 MPa is expanded to exit pressure and temperature of 85 kPa and 1200 K, respectively. If the mass flow rate happens to be 75 kg/s, determine the exit jet velocity, effective jet velocity, and thrust at an altitude of 25 km. Take calorific value of propellant as 22 MJ/kg.

3.8 The hot propellant gas at chamber pressure of 3.5 MPa with a flow rate of 5.5 kg/s is expanded fully through a CD nozzle with throat diameter of 80 mm to produce thrust of 12 kN. If the flight velocity happens to be 750 m/s, determine the specific impulse, effective exhaust velocity, specific propellant consumption and thrust power, and thrust coefficient. Take calorific value of propellant as 22 MJ/kg.

3.9 A rocket engine nozzle (D_e = 105 mm) produces 8.2 kN by expanding gas from 7.5 MPa to 95 kPa by consuming propellant flow rate of 3.75 kg/s. If the flight velocity happens to be 1650 m/s at an altitude of 20 km, determine the nozzle exhaust velocity, propulsive efficiency, overall efficiency, and thermal efficiency. Take calorific value of propellant as 30 MJ/kg.

REFERENCES AND SUGGESTED READING

1. Sutton, G.P. and Ross, D.M., *Rocket Propulsion Elements*, John Wiley & Sons, New York, 1975.
2. Barrere, M., Jaumotte, A., de Veubeke, B.F., and Vandenkerckhove, J., *Rocket Propulsion*, Elsevier Publishing Company, New York, 1960.
3. Turner, M.J.L., *Rocket and Spacecraft Propulsion*, Springer Verlag, Heidelberg, Germany, 2001.
4. Hill, P.G. and Peterson, C.R., *Mechanics and Thermodynamics of Propulsion*, Addison Wesley Publishing Company, Reading, MA, 1965.

CHAPTER 4

Rocket Nozzle

> Great spirits have always encountered violent opposition from mediocre minds.
>
> ALBERT EINSTEIN

4.1 INTRODUCTION

The main purpose of the nozzle in a rocket engine is to expand the high-pressure hot gases generated by burning the propellant to a higher jet velocity for producing the requisite thrust. In order to have a large value of specific thrust, the kinetic energy of the exhaust gas must be large enough to produce a higher exhaust velocity. Keep in mind that the pressure ratio across the nozzle controls the expansion process. We can achieve maximum thrust in an engine only when the exit pressure P_e is the same as the ambient pressure, P_a. Generally, two types of nozzles, namely, (1) convergent nozzle and (2) convergent–divergent (CD) nozzle, are used to produce thrust in the jet engine [1,2,4–6]. However, CD nozzles are mostly used for producing thrust in rocket engines.

4.2 BASICS OF CD NOZZLE FLOW

We know that when the pressure ratio across the nozzle becomes greater than its critical value, the maximum gas velocity attained is equal to the speed of sound. In this situation, a divergent nozzle is added to enhance its velocity further because, in the convergent section, the gas velocities cannot be increased above the speed of sound. This CD nozzle is basically a convergent duct followed by a divergent duct as shown in Figure 4.1. Note that

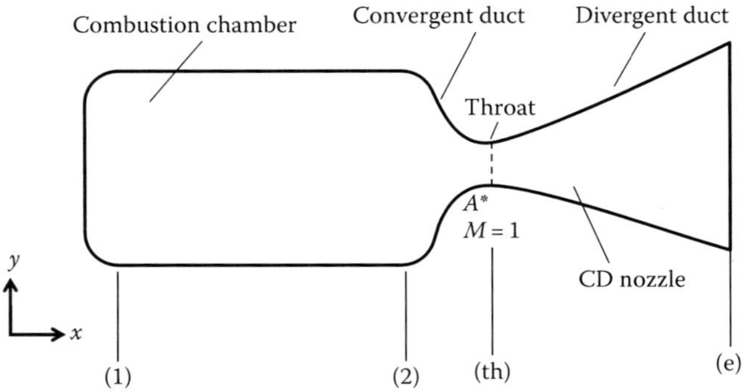

FIGURE 4.1 Flow through convergent–divergent nozzle.

there is a minimum cross-sectional area between the convergent and divergent portions which is known as throat. The gas attains a sonic speed at the throat under supersonic flow conditions. This is also known as "de Laval" or sometimes simply "Laval" nozzle as it was devised for the first time by the Swedish engineer de Laval. Hence, this CD nozzle is employed in a rocket engine to achieve a higher exhaust velocity beyond sonic speed.

The flow through a CD nozzle as shown in Figure 4.2 is governed by the ratio of total pressure to back pressure. For a particular total pressure P_{t2} at the upstream chamber, when the back pressure is reduced below the upstream P_{t2}, the pressure decreases in the nozzle as shown by the curve "A" in Figure 4.2 and the flow remains subsonic throughout the nozzle. If the back pressure is reduced further as shown by the curve "B," sonic flow will be attained at the throat section for a certain value of P_b. This condition at the throat is called a choked condition as the mass flow rate through the nozzle attains a maximum value. In other words, the mass flow cannot be increased by any further decrease in P_b for a constant P_{t2} and T_{t2}. If the back pressure P_b decreases further, say to P_{bC}, then the flow cannot attain isentropic conditions since the throat is already choked. Rather, the pressure decreases further from the throat until a normal shock is formed in the divergent portion of the CD nozzle as shown by the curve "C" in Figure 4.2. Note that once normal shock is formed, the flow decelerates in the divergent duct. When P_b is decreased further, the normal shock moves downstream and at a certain P_{bD}, it gets located exactly at the nozzle exit plane as shown in Figure 4.2. Note that isentropic expansion takes place (see curve E in Figure 4.2) only when the back pressure matches with the correct one for which the nozzle is designed. This situation corresponds to the nozzle being expanded fully. Besides this, the nozzle can be overexpanded and

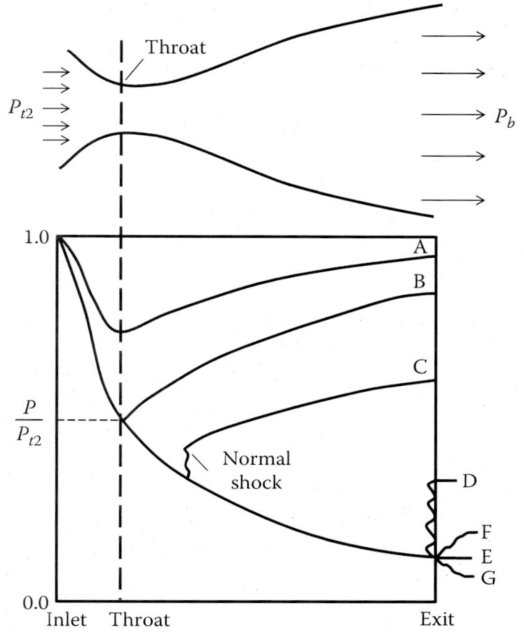

FIGURE 4.2 Flow features in a CD nozzle.

underexpanded under certain back pressure conditions, which will be discussed in Section 4.4 in detail.

4.2.1 Exhaust Velocity

Generally, the flow in the CD nozzle is three-dimensional in nature and the temperature, pressure, and composition of gas vary along the length of the nozzle. Besides this, the flow becomes more complex due to the presence of shock and flow separation in the nozzle. Hence, the shape of the nozzle plays a very important role in dictating the flow and heat transfer from its wall, which affects the thrust produced by the rocket engine. In order to analyze the flow in the nozzle, we may consider the flow to be ideal with the following assumptions:

1. Steady one-dimensional isentropic flow.
2. The working fluid is homogeneous in nature. The amount of mass due to condensed phase (solid/liquid) is negligible compared to gaseous fluid.
3. Ideal gas law with constant specific heat can be applied.

4. No heat transfer from the rocket engine.
5. Uniform flow properties, namely, pressure, temperature, and density, at the nozzle entrance.
6. Velocity at the nozzle entrance is negligible compared to its value at the exit.
7. Gas compositions across the nozzle remain constant.
8. Gas leaves the nozzle exit along the axial direction only.

We need to analyze the rocket nozzle by simplifying the complex flow with these assumptions. By applying the energy equation between station (2) and exit (e) as shown in Figure 4.1, we have

$$h_{t2} = C_p T_{t2} + \frac{V_2^2}{2} = h_{te} = C_p T_{te} + \frac{V_e^2}{2} \tag{4.1}$$

Note that this relationship indicates that total temperature remains invariant although static temperature and local velocity vary along the length of the adiabatic nozzle. Recall that total/stagnation temperature is defined as the temperature when the fluid element is brought to rest isentropically. We have assumed that velocity in the reservoir at station (2) is almost zero. Then the exit velocity can be expressed with Equation 4.1 as follows:

$$V_e = \sqrt{2C_p\left(T_{t2} - T_e\right)} \tag{4.2}$$

Assuming the flow to be isentropic and invoking the isentropic relation (Equation 2.14), we can express Equation 4.2 in terms of the pressure ratio across the nozzle as follows:

$$V_e = \sqrt{\frac{2\gamma}{\left(\gamma - 1\right)} \frac{R_u T_{t2}}{MW} \left[1 - \left(\frac{P_e}{P_{t2}}\right)^{\frac{\gamma-1}{\gamma}}\right]} \tag{4.3}$$

where

P_{t2} and P_e are the reservoir and nozzle exit pressures, respectively
R_u is the universal gas constant
MW is the molecular weight of gas
γ is the specific heat ratio

It may be noted from Equation 4.3 that the nozzle exit velocity is dependent on the reservoir (combustion chamber) temperature, pressure ratio (P_e/P_{t2}), MW, and γ of the flowing gas. The nozzle exit velocity gets enhanced by increasing the combustion chamber temperature. Hence, it is desirable to choose a propellant that has a higher value of adiabatic flame temperature. However, the designer faces two limitations. One has to choose a proper material with higher thermal stress at high pressure for a sustained period of time. Besides this, cooling of the nozzles and combustion chamber is to be considered for a longer period of time, which makes the rocket engine bulky and complex. However, if the duration of the flight is small, one can avoid using a complex cooling system by employing a certain high-temperature coating of heated parts of a rocket engine. Another limitation is the dissociation of combustion products, due to which desired high temperature would not be attained as a large amount of heat is absorbed during the dissociation process. For general propellants, this limitation due to dissociation lies between 2800 and 3500 K. However, for a fluorine-hydrogen system, one can easily achieve a higher temperature of 5000 K. Note that the exhaust velocity increases with the pressure ratio across the nozzle. The maximum nozzle exit velocity can be attained only when the pressure ratio (P_e/P_{t2}) is tending toward zero, corresponding to the vacuum condition ($P_e = 0$). Thus, the expression for the maximum nozzle exit velocity $V_{e,\max}$ can be obtained from Equation 4.3 as follows:

$$V_{e,\max} = \sqrt{2\frac{\gamma}{(\gamma-1)}\frac{R_u T_{t2}}{MW}} \tag{4.4}$$

Let us derive an expression for the velocity ratio $V_e/V_{e,\max}$ by dividing Equation 4.3 with Equation 4.4 as follows:

$$\frac{V_e}{V_{e,\max}} = \sqrt{\left[1-\left(\frac{P_e}{P_{t2}}\right)^{\frac{\gamma-1}{\gamma}}\right]} \tag{4.5}$$

The variation in the velocity ratio $V_e/V_{e,\max}$ for three specific ratios is plotted in Figure 4.3 with the pressure ratio across the nozzle. It may be observed that the exhaust velocity ratio increases initially at a faster rate till it reaches a pressure ratio of 25. Subsequently, it also increases beyond a pressure ratio of 25 but at a slower rate and reaches almost an asymptotic value at a pressure ratio of 1000, indicating its diminishing effect.

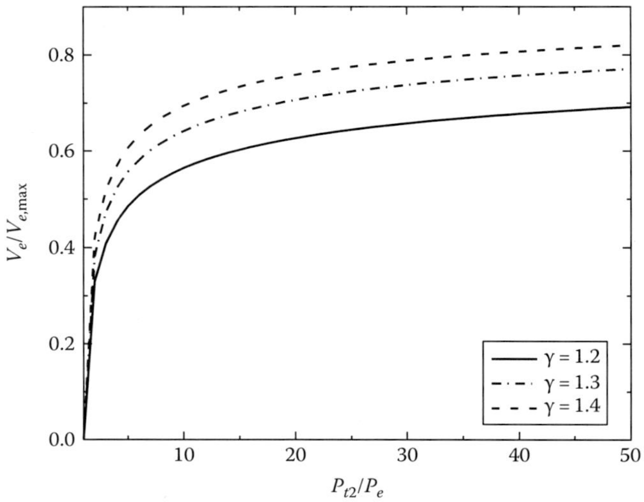

FIGURE 4.3 Effect of pressure ratio on $V_e/V_{e,\max}$.

Of course, with an increase in chamber pressure, the dissociation of chemical species decreases, leading to a higher combustion temperature. In spite of this advantage, a higher chamber pressure cannot be favored from a practical point of view as this would increase the dead weight, leading to lower performance. Hence, a moderate range of pressure ratios across the nozzle of 25–50 is preferred, particularly when the rocket is designed for medium altitude. However, for deep-space applications, higher values of pressure ratio across the nozzle are preferred for achieving a higher performance level. Exhaust gases with a smaller molecular weight MW produce a larger exhaust velocity for the same pressure ratio. Hence, a higher percentage of hydrogen in the exhaust products is preferred in order to have an overall lower molecular weight MW so that a higher nozzle exhaust velocity can be obtained. Note that the specific heat ratio γ can affect the exhaust velocity in two ways. First, it can affect the maximum nozzle exit velocity $V_{e,\max}$ significantly as the initial enthalpy C_pT_{t2} is highly dependent on γ, as depicted in Figure 4.4. It may be noted that the maximum nozzle exit velocity $V_{e,\max}$ decreases with an increase in specific heat ratio γ of the exhaust gas. The second factor is the expansion governed by the pressure ratio across the nozzle (see Equation 4.5), which is significantly influenced by the specific heat ratio γ. Both these effects get nullified and thus the nozzle exit velocity V_e varies insignificantly with the specific heat

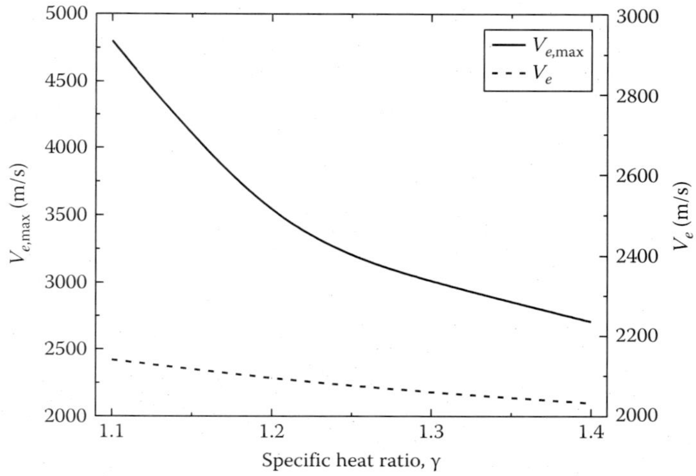

FIGURE 4.4 Effect of specific heat ratio γ on V_e and $V_{e,max}$.

ratio γ as depicted in Figure 4.4. Note that these inferences are drawn and discussed assuming that the flow is frozen. In other words, the gas compositions remain almost invariant in the nozzle. But in an actual system, the gas compositions undergo changes which may lie between frozen and equilibrium flow conditions. Although the expression for the nozzle exit velocity V_e is derived under ideal conditions, its predicted value differs from experimental values by 5%–6%. Hence, it can only be used for preliminary design calculation.

4.2.2 Mass Flow Rate and Characteristics of Velocity

The mass flow rate through the CD nozzle under steady-state condition is dependent on the pressure, temperature of combustion chamber, and cross-sectional area. By considering one-dimensional and steady isentropic flow, we can invoke the continuity equation as follows:

$$\dot{m} = \rho A V = \text{constant} \tag{4.6}$$

where

- ρ is the density
- A is the cross-sectional area
- V is the flow velocity at any location in the nozzle

As the flow is isentropic in nature, we can use the isentropic relation for density in terms of pressure as follows:

$$\rho = \rho_{t2}\left(\frac{P}{P_{t2}}\right)^{1/\gamma} \tag{4.7}$$

By using Equations 4.7 and 4.3 along with the ideal gas law, we obtain an expression for mass flux through the CD nozzle at any location as follows:

$$\frac{\dot{m}}{A} = P_{t2}\sqrt{\frac{2\gamma}{(\gamma-1)}\frac{1}{RT_{t2}}\left(\frac{P}{P_{t2}}\right)^{\frac{2}{\gamma}}\left[1-\left(\frac{P}{P_{t2}}\right)^{\frac{\gamma-1}{\gamma}}\right]} \tag{4.8}$$

Note that the mass flux through the CD nozzle will be dependent on the combustion chamber conditions, pressure ratio P/P_{t2}, and specific heat ratio γ. The nondimensional mass flux can be derived from Equation 4.8 as follows:

$$\frac{\dot{m}\sqrt{RT_{t2}}}{AP_{t2}} = \sqrt{\frac{2\gamma}{(\gamma-1)}\left(\frac{P}{P_{t2}}\right)^{\frac{2}{\gamma}}\left[1-\left(\frac{P}{P_{t2}}\right)^{\frac{\gamma-1}{\gamma}}\right]} \tag{4.9}$$

The variation in the nondimensional mass flux through the nozzle is plotted in Figure 4.5. It may be observed that mass flux increases from zero

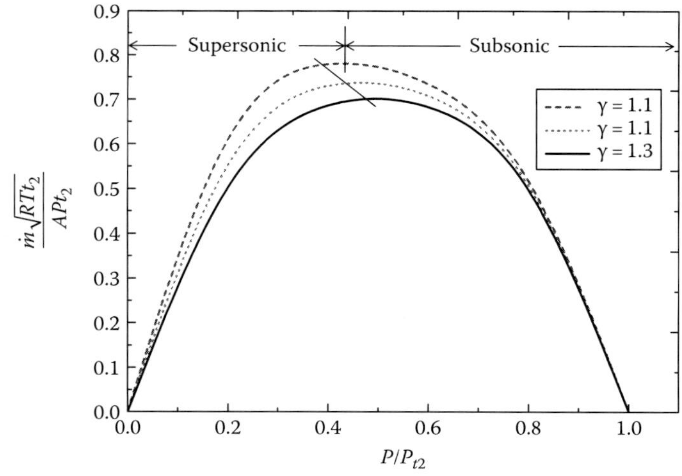

FIGURE 4.5 Effect of pressure ratio P/P_{t2} and specific heat ratio γ on mass flux rate.

value at pressure ratio P/P_{t2} to a maximum value at critical pressure ratio and subsequently decreases to zero value again when the pressure ratio P/P_{t2} becomes unity. As there is no pressure gradient across the nozzle at unity pressure ratio P/P_{t2}, no flow occurs in the nozzle, thus making the mass flux zero. However, when the pressure ratio P/P_{t2} is zero, mass flux can be zero due to infinite area, as gas has to be expanded to vacuum pressure. Note that maximum mass flux is attained at minimum cross-sectional area of the nozzle. The pressure ratio corresponding to maximum mass flux can be obtained by differentiating Equation 4.9 and equating it to zero as follows:

$$\frac{P}{P_{t2}} = \left(\frac{2}{\gamma+1}\right)^{\gamma/(\gamma-1)} \tag{4.10}$$

We know that this relation corresponds to the critical condition under which flow is considered to be aerodynamically choked. The relation for choked mass flux can be easily obtained by substituting Equation 4.10 in Equation 4.8 as follows:

$$\frac{\dot{m}}{A_t} = \frac{\Gamma P_{t2}}{\sqrt{RT_{t2}}} \quad \text{where } \Gamma = \sqrt{\gamma\left(\frac{2}{(\gamma+1)}\right)^{\frac{\gamma+1}{\gamma-1}}} \tag{4.11}$$

where A_t is the throat area of the CD nozzle. It may be noted that critical mass flux is dependent on A_t, chamber pressure, temperature, and specific heat ratio. Generally, in such cases, the rocket engine nozzle is choked for the major portion of its operation. Hence, the chamber pressure P_{t2} can be expressed easily in terms of mass flux and $\sqrt{RT_{t2}}/\Gamma$ by using Equation 4.11 as follows:

$$P_{t2} = \frac{\sqrt{RT_{t2}}}{\Gamma}\frac{\dot{m}}{A_t} \tag{4.12}$$

where the term $\sqrt{RT_{t2}}/\Gamma$ represents the relationship between mass flux and chamber pressure P_{t2} and is measured in m/s. This indicates the capacity of the propellant to generate a certain pressure in the combustion chamber for a given mass flow rate per unit throat area, and hence it is often

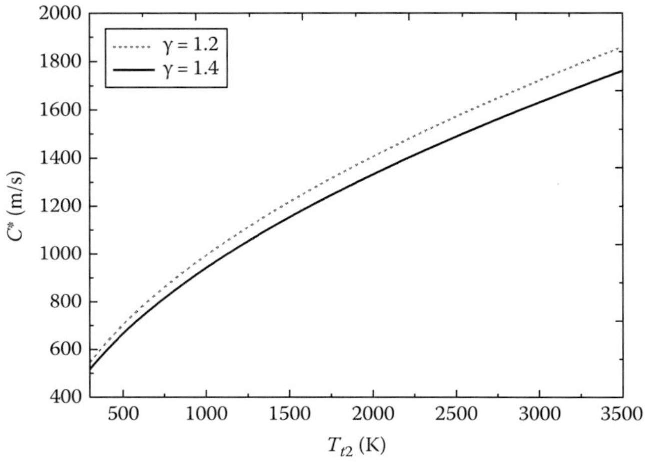

FIGURE 4.6 Variation of characteristic velocity C^* with adiabatic flame temperature T_{t2}.

known as the characteristic velocity C^*. Then the mass flow rate through the nozzle can be expressed by using Equation 4.11 as follows:

$$\dot{m} = \frac{P_{t2}A_t}{C^*} \quad \text{where } C^* = \frac{\sqrt{RT_{t2}}}{\Gamma} \tag{4.13}$$

Note that the characteristic velocity C^* increases with adiabatic flame temperature and decreases with molecular weight of gases. It increases with decreasing value of specific heat ratio γ. The effect of adiabatic flame temperature T_{t2} on the characteristic velocity C^* is depicted in Figure 4.6 for three different cases of specific heat ratio γ and a gas molecular weight of 21 kg/kmol. It may be observed that the characteristic velocity C^* increases with temperature. The effect of specific heat ratio γ on C^* is more prominent at a higher temperature range compared to a lower temperature range.

Example 4.1

In a convergent–divergent nozzle with a throat area of 0.5 m^2 of a rocket engine, gas at 5.75 MPa and temperature of 2800 K is expanded fully to 5.5 kPa for producing thrust. Assuming flow to be isentropic, determine (1) the exit Mach number, (2) the maximum exit mass flow rate passing through this nozzle, and (3) the exit area. Assume $\gamma = 1.25$ and $MW = 25$ kg/kmol.

Solution

Assuming the flow to be isentropic, the expression for the exit Mach number can be expressed as follows:

$$M_e = \sqrt{\frac{2}{\gamma-1}\left[\left(\frac{P_{t2}}{P_e}\right)^{\frac{\gamma-1}{\gamma}} - 1\right]} = \sqrt{\frac{2}{0.25}\left[\left(\frac{5750}{5.5}\right)^{0.2} - 1\right]} = 4.9$$

Note that we have assumed that the total pressure in the chamber is the same as the nozzle exit total pressure. We know that the nozzle exit velocity can be obtained from M_e as follows:

$$V_e = M_e\sqrt{\gamma R T_e}$$

We need to evaluate T_e by using the following isentropic relation:

$$T_e = T_{te}\left(\frac{P_e}{P_{te}}\right)^{\frac{\gamma-1}{\gamma}} = 2800\times\left(\frac{5.5}{5750}\right)^{0.2} = 683.9\ \text{K}$$

Now we can determine V_e as follows:

$$V_e = M_e\sqrt{\gamma R T_e} = 4.9\times\sqrt{(1.25\times 8314\times 683.9)/25} = 2612.66\ \text{m/s}$$

For maximum flow to occur in the nozzle, the throat of the nozzle must be choked. Let us evaluate the choked mass flow rate through the nozzle's throat:

$$\dot{m} = \rho^* A^* V^* = \rho^* A^* \sqrt{\gamma R T^*}$$

But the density at the throat for a choked condition can be estimated as follows:

$$\rho^* = \left(\frac{2}{\gamma+1}\right)^{\frac{1}{\gamma+1}} \frac{P_{te}}{RT_{te}} = \left(\frac{2}{1.25+1}\right)^{\frac{1}{1.25-1}} \frac{5750\times 1000\times 25}{8314\times 2800} = 3.85\ \text{kg/m}^3$$

Then, the temperature T^* for a choked condition can be determined as follows:

$$T^* = \left(\frac{2}{\gamma+1}\right) T_{te} = 2488.9 \text{ K}$$

Then the velocity at the throat can be estimated as follows:

$$V^* = a^* = \sqrt{\gamma R T^*} = \sqrt{(1.25 \times 8314 \times 683.9)/25} = 533.2 \text{ m/s}$$

By using the ideal gas law, we can evaluate the density at the exit of the nozzle:

$$\rho_e = \frac{5.5 \times 10^3 \times 25}{8314 \times 683.9} = 0.024$$

Then the nozzle exit area is determined as follows:

$$\dot{m}_e = \rho^* A^* V^* = \rho_e A_e V_e$$

$$\Rightarrow A_e = \frac{\rho^* A^* V^*}{\rho_e V_e} = \frac{3.85 \times 0.5 \times 533.2}{0.024 \times 2612.66} = 16.4 \text{ m}^2$$

4.2.3 Expansion Area Ratio

By using Equations 4.11 and 4.8, we can obtain an expression for the expansion area ratio A/A_t as follows:

$$\frac{A}{A_t} = \frac{\Gamma}{\sqrt{\frac{2\gamma}{(\gamma-1)}\left(\frac{P}{P_{t2}}\right)^{\frac{2}{\gamma}}\left[1-\left(\frac{P}{P_{t2}}\right)^{\frac{\gamma-1}{\gamma}}\right]}} \tag{4.14}$$

This equation depicts the relation between the expansion area ratio and the local pressure ratio. We can evaluate the area ratio directly for a given pressure ratio and the specific heat ratio γ. Note that for a given pressure ratio, two solutions of Equation 4.14 exist for subsonic and supersonic

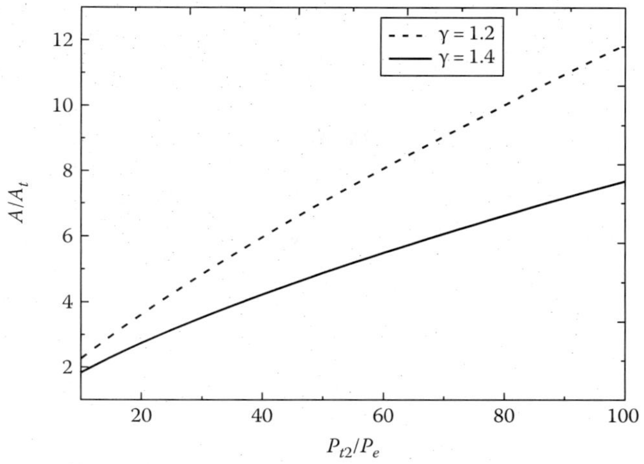

FIGURE 4.7 Effect of pressure ratio and specific heat ratio γ on expansion area ratio A/A_t.

flow. However, for a given area ratio, the pressure ratio can be evaluated only by using an iterative method. The variation in the local pressure ratio with the expansion area ratio A/A_t is plotted in Figure 4.7 for three specific heat ratio cases for a supersonic flow in the CD nozzle. The relationship for the area ratio between the throat and the exit for the CD nozzle can be obtained from Equation 4.14 by setting $A = A_e$ and $P = P_e$ as follows:

$$\frac{A_e}{A_t} = \frac{\Gamma}{\sqrt{\frac{2\gamma}{(\gamma-1)}\left(\frac{P_e}{P_{t2}}\right)^{\frac{2}{\gamma}}\left[1-\left(\frac{P_e}{P_{t2}}\right)^{\frac{\gamma-1}{\gamma}}\right]}} \tag{4.15}$$

We can easily determine the flow variables, namely, pressure, temperature, velocity, and density, along the nozzle length provided we know the local expansion ratio by using the theory developed earlier. The variations in pressure, temperature, velocity, and Mach number along with the length are shown in Figure 4.8. It may be noted that pressure and temperature decrease with length while velocity and Mach number increase as depicted in Figure 4.8.

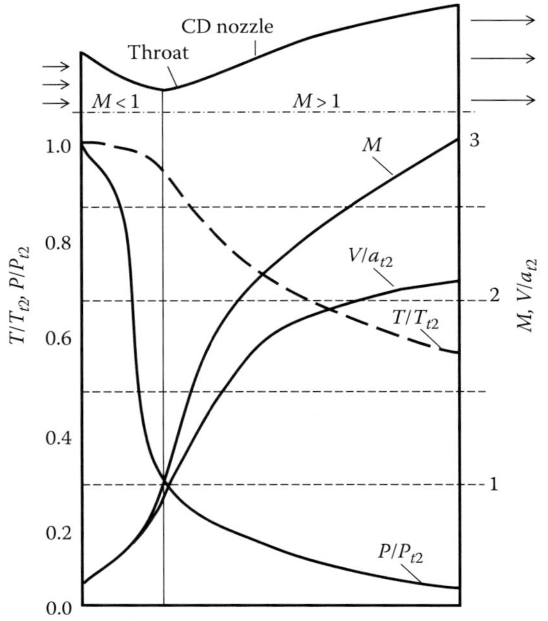

FIGURE 4.8 Variations in pressure, temperature, density, and Mach number along with the length of the CD nozzle.

4.3 CD NOZZLE GEOMETRY

Generally, both combustion chamber and nozzle with circular cross section are used in rocket engine applications although several other configurations have been evolved over the years which will be discussed briefly in the subsequent sections. In this section, we mostly discuss the conical nozzle. Note that a typical CD nozzle consists of a converging section emanating from the combustion chamber, a throat section, and a diverging section as shown in Figure 4.1. Generally, the converging section between the combustion chamber and the nozzle throat does not affect the performance of the rocket nozzle. As the subsonic flow prevails in this converging section, it can incur low pressure losses although a higher apex angle of the converging section in the range of 60°–90° is preferred in the nozzle design. Of course the pressure gradient will be very high for a converging section with a higher apex angle. Flow is less likely to get separated within this range of converging angle as flow accelerates in this zone with a favorable pressure gradient. The inlet cross-sectional area of a CD nozzle in the rocket engine is the same as that of the combustion chamber. The throat area can be calculated by using Equation 4.11 for a particular chamber pressure and mass flow rate. Note

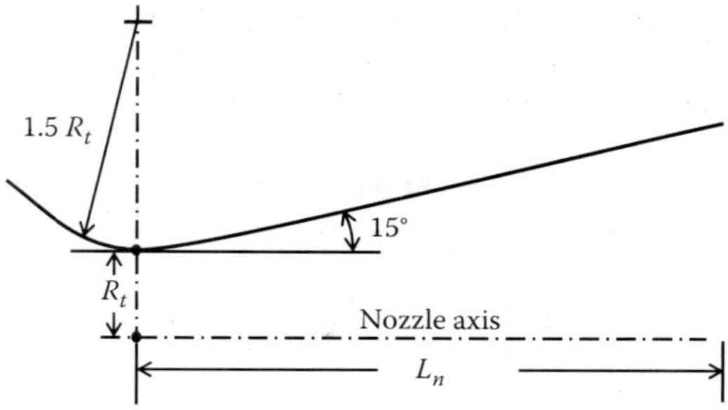

FIGURE 4.9 Schematic of a conical CD nozzle.

that the shape of the throat does not affect the performance of the nozzle significantly. But the radius of curvature of the profile for the throat section must be chosen such that a smooth decrease of static pressure along with an increase in velocity is ensured. In practice, the profile of the convergent section is joined with the throat section with the help of the radius of curvature, which is equal to 1.5 times the throat radius R_t, as shown in Figure 4.9. In some cases, a small cylindrical tube with a [4] constant diameter is used as the throat. Subsequently, with a smooth increase of the cross-sectional area, the divergent section emanates from the throat for generating a supersonic flow in the rocket nozzle. The exit area A_e and the exit diameter D_e of the CD nozzle can be obtained for a particular pressure ratio across the nozzle by using Equation 4.15. The length of the nozzle can be determined by using a suitable divergence angle. Generally, the divergence angle of a CD nozzle is restricted to 30° to avoid flow separation in the divergent section. Unfortunately, the flow in this kind of conical nozzle will no longer be one-dimensional in nature. Rather the flow becomes two-dimensional in nature due to flow divergence at its exit, which incurs thrust losses. Besides this, a larger area nozzle is used for high-altitude applications, which results in a longer and heavier nozzle, leading to a lower performance of the rocket engine.

In order to enhance the thrust, the exhaust gas in the nozzle can be expanded fully to ambient pressure with parallel and uniform exit flow. This kind of nozzle can be designed using the method of characteristics [1,2], which happens to be quite lengthy and heavy in nature. In order to reduce the length L_n of the diverging section in the CD nozzle, the flow can be expanded at a faster rate compared to an ideal nozzle by providing a higher divergence angle from the sonic throat. Subsequently, the divergent flow can be straightened along the length of the divergent section into an approximately axial flow at its exit, as shown in Figure 4.10. Such a kind of CD nozzle is known as

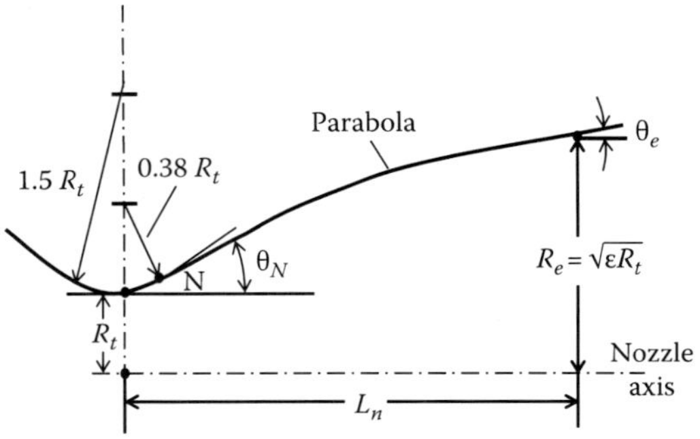

FIGURE 4.10 Schematic of a bell-shaped CD nozzle.

a bell-shaped nozzle, which was designed by Rao using semiempirical parabolic approximation methods [3] for obtaining maximum thrust. The optimum contour depends on ambient pressure, nozzle curvature in the throat region, length, and gas composition. Generally, a circular arc with a radius of 1.5 R_t is used for the nozzle contour immediately upstream of the throat while a circular arc of 0.382 R_t is used for the divergent section of the contour from the throat and point N, as shown in Figure 4.10. From this point N, a parabola is formed till the nozzle exit that matches the expansion ratio. It may be noted that the angle downstream of the throat of a bell-shaped nozzle is much larger than 30°. The length of the bell-shaped nozzle is reduced considerably by around 20% compared to a conical nozzle with the same divergent angle.

4.4 EFFECT OF AMBIENT PRESSURE

In Section 4.2, we initiated a discussion on the effect of ambient pressure, which is also known as back pressure P_b that affects the nozzle flow significantly. During the flight of a rocket engine, ambient pressure is the same as the back pressure acting on the exhaust nozzle. Recall that for constant chamber pressure and fixed nozzle area ratio, gas flow is said to be fully expanded only when the nozzle exit pressure P_e is equal to the ambient pressure at a particular altitude for which it is designed. When the nozzle is operated at a higher altitude, the ambient pressure is less than the nozzle exit pressure P_e, and the gas is expanded from P_e to P_a beyond the nozzle exit as a free jet. This fixed nozzle area operating under constant chamber pressure is said to undergo underexpansion. In contrast, when the nozzle operates at a lower altitude than the designed one, the exit pressure P_e will be less than the ambient pressure P_a, and the nozzle is said to undergo overexpansion. We will discuss further about underexpansion and overexpansion in the subsequent sections.

4.4.1 Underexpansion in CD Nozzle

As discussed, for a fully expanded nozzle, isentropic flow in the divergent section of the nozzle will be supersonic in nature. As a result, when the nozzle exit pressure P_e is higher than the ambient pressure P_a, at a higher altitude during its flight, there will not be any change in the upstream pressure as the pressure change due to the change in altitude cannot propagate upstream because of its sonic velocity. As a result, the nozzle exit pressure remains the same as that of the design value for isentropic flow and the supersonic flow expands further from P_{ed} to P_a, through a fan of Prandtl–Meyer-type expansion downstream of the nozzle exit in the free jet wave as shown in Figure 4.11a. In order to adjust this phenomenon, these expansion waves get reflected from the free jet boundary downstream of the nozzle

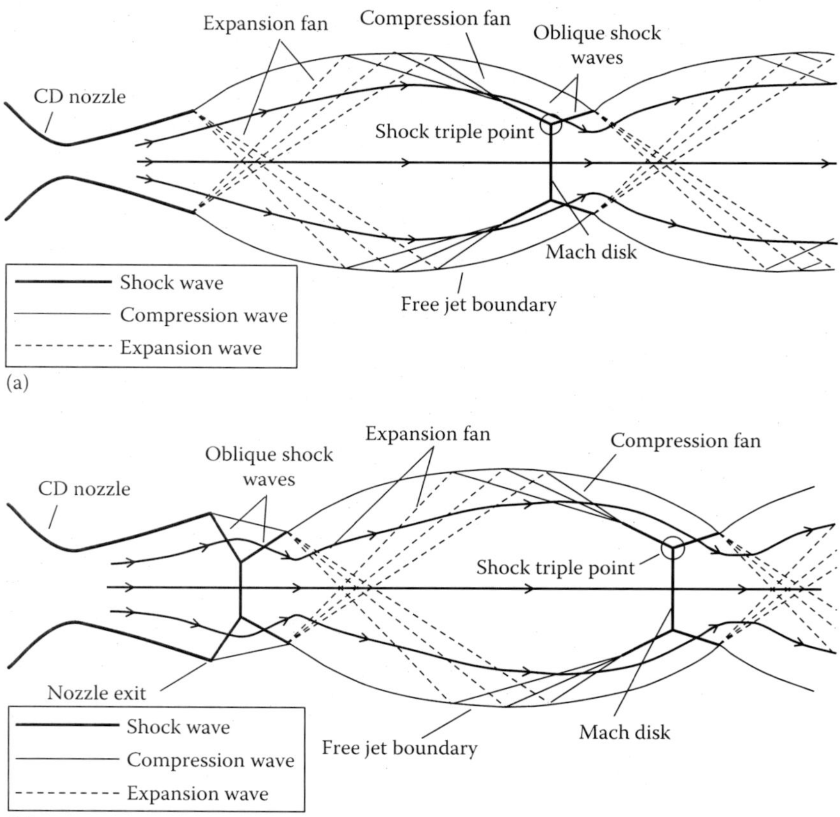

FIGURE 4.11 Flow pattern downstream of the nozzle exit under (a) underexpansion and (b) overexpansion.

exit and get converted into compression waves as shown in Figure 4.11a. Subsequently, these compression waves get reflected, unlike expansion waves from the free jet boundary. This kind of flow pattern of the expansion wave followed by compression waves is repeated downstream of the nozzle exit till its effect reduces. In such a situation, the gas is not expanded completely inside the nozzle and the kinetic energy of the jet flow at the nozzle exit is lower compared to that of the complete isentropic expansion of gas from P_{t2} to P_a. As a result, the thrust produced by the nozzle under this condition will be lower than the one corresponding to complete expansion.

4.4.2 Overexpansion in CD Nozzle

We know that overexpansion in a CD nozzle takes place when the nozzle is operated at a lower altitude than the designed one. In other words, at a lower altitude, the gas is expanded in the nozzle such that its exit pressure P_e is less than the ambient pressure P_a. This increase in back (ambient) pressure will not cause any change in the upstream pressure as the nozzle flow moves at supersonic speed. As a result, the nozzle exit pressure remains the same as that of the design value for isentropic flow. However, the change in back pressure can propagate upstream with sonic velocity through the boundary layer as subsonic flow prevails within it. On the other hand, at the exit plane, the nozzle exit pressure gets adjusted to the ambient pressure with the formation of a normal shock. Region "C" in Figure 4.2 discussed in Section 4.2 indicates the formation of a normal shock with abrupt changes in the flow properties in the divergent portion of the CD nozzle. Downstream of this normal shock, the flow is subsonic and hence decelerates with an increase in static pressure at the nozzle exit to match the ambient pressure. With a further increase in altitude, the ambient pressure decreases, which results in relocation of the normal shock toward its exit. Note that the strength and location of a normal shock are dependent on the magnitude of back pressure. Region "D" in Figure 4.2 indicates the formation of a normal shock at the exit of the CD nozzle in which isentropic supersonic flow prevails in the divergent section of the nozzle, during which the nozzle exit pressure before the shock is almost equal to the design pressure. But the pressure downstream of the shock happens to be subsonic and almost equal to the ambient pressure. When the back (ambient) pressure decreases slightly due to an increase in altitude, oblique shock waves are formed at the lip of the nozzle exit. In order to adjust to this phenomenon, these oblique shock waves get reflected from the free jet boundary downstream of the nozzle exit and are converted into expansion waves as shown in Figure 4.11b. Subsequently, these expansion waves get reflected from the free jet boundary and coalesce to form a compression wave. This kind of flow pattern of expansion waves followed by compression waves is repeated downstream of the nozzle exit till its effect becomes reduced. In this situation, the gas is not expanded completely inside the nozzle and the kinetic energy of the jet flow at the nozzle exit is lower compared to that of the complete isentropic expansion of gas from P_{t2} to P_a. As a result, the thrust produced by the nozzle under this condition is lower than that corresponding to the complete expansion case.

Note that sudden compression by the formation of shocks may lead to the separation of flow from the wall of the nozzle as shown in Figure 4.12. The extent of flow separation is dependent on the rate of acceleration nozzle divergence angle and local pressure. But the location of flow separation will be dependent on the contours of divergent section and oblique shock. It has been observed that flow separation point moves back and forth with time at nozzle exit particularly for contour nozzle. Hence certain finite angle at nozzle exit is provided even for a bell nozzle. When the flow separation occurs in the presence of an oblique shock at the nozzle exit, the shock may move upstream into the divergent section, as shown in Figure 4.12. Downstream of an oblique shock, slight recompression of gas can occur, which may move further upstream with an increase in ambient pressure P_a. The compilation of several experimental data by Summerfield [4] for conical CD nozzle with divergence angle of 30° indicates that flow separation is likely to occur when exit pressure is less than or equal to 0.4 times the ambient pressure, provided P_{t2}/P_a is greater than 16. This is known as the Summerfield criterion for flow separation in a conical nozzle. For higher-divergence-angled nozzle, flow separation is likely to occur at a lower value than the Summerfield criterion ($P_e/P_a < 0.4$). Besides, it has also been found that flow separation is influenced by local conditions,

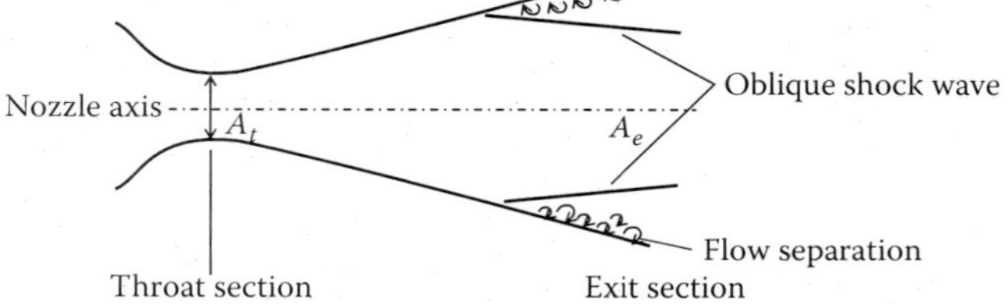

FIGURE 4.12 Effect of shock–boundary layer interaction in a CD nozzle under overexpanded condition.

namely, wall pressure, Mach number, wall roughness, and so on. Another criterion for flow separation is the following semiempirical relationship relating P_e, P_a, and M_e [5]:

$$\frac{P_e}{P_a} = \left(1.88M_e - 1\right)^{-0.64} \tag{4.16}$$

where M_e is the Mach number at the nozzle exit. In order to avoid flow separation, the nozzle exit area is chosen such that P_e must be greater than $P_a(1.88M_e - 1)^{-0.64}$ for a particular nozzle exit Mach number. In other words, an adverse pressure gradient is less likely to occur in the divergent section of the nozzle to avoid flow separation. With separation of flow, the nozzle may adapt to any altitude change as it provides higher pressure at the nozzle exit. Hence, it is expected to have a higher thrust under the flow separation condition for the same nozzle area ratio. However, flow separation does not occur in a symmetrical manner, as shown in Figure 4.12. Rather flow separation in a practical system is found to occur in an asymmetrical manner, giving rise to side thrust. Hence, flow separation must be avoided for rocket propulsion applications.

4.5 ADVANCED ROCKET NOZZLE

We have already discussed two conventional rocket nozzles, namely, (1) the conical nozzle and (2) the bell-shaped nozzle, in Section 4.3. Recall that the bell-shaped nozzle is preferred over the conical nozzle due to its divergence losses, and reduced length and weight. However, both nozzles do suffer from losses due to their nonadaptability with altitude as discussed in Section 4.4. In order to enhance a nozzle's adaptability with altitude for augmenting its performance over an entire range of flight, several nonconventional nozzles have been developed over the last few decades. Note that most of these advanced nozzles are intended to reduce nozzle length and weight which can adapt well to under/overexpansion conditions encountered along with altitude. We will restrict our discussions to the following types of nozzle: (1) extendible nozzle, (2) dual bell–shaped nozzle, (3) expansion–deflection nozzle, and (4) plug nozzle.

4.5.1 Extendible Nozzle

The area ratios of the nozzle need to be changed continuously to adapt the nozzle performance with altitude. As it is quite difficult to change the exit area of the nozzle, the extendible nozzle consisting of two or three segments

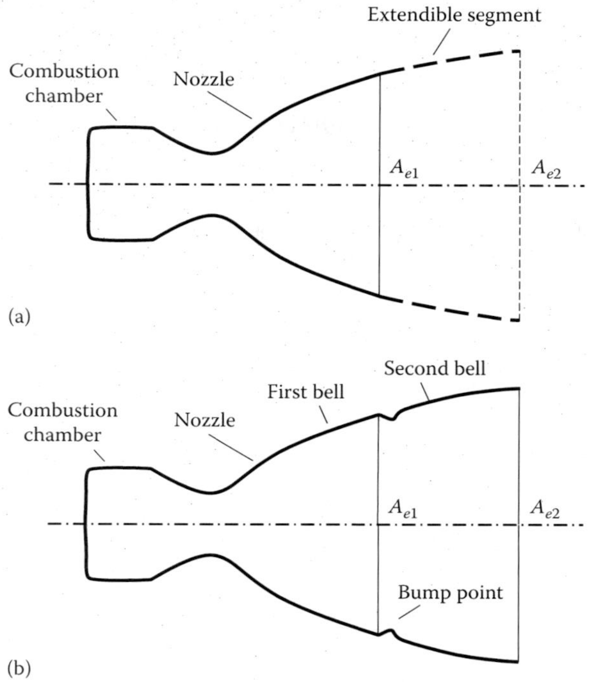

FIGURE 4.13 Schematic of (a) an extendible nozzle and (b) a dual bell–shaped nozzle.

has been designed and developed. A two-step extendible bell nozzle consisting of two segments is shown in Figure 4.13a. Initially, the extendible nozzle with a smaller exit area A_{e1} is used for a certain range of low altitude. Subsequently, a second segment is actuated using a special mechanism such that the nozzle exit area is increased to A_{e2} at higher altitude for better performance. This kind of extendible nozzle has been employed successfully in several solid rocket engines and few liquid-propellant engines, particularly before initiation of ignition. A three-segment extendible nozzle has been devised but has not yet been used during flight due to additional complexities. Some of the pertinent problems with the actuation mechanism, the seal between the nozzle segments, and the added weight are to be solved to make the extendible nozzle viable for future application in rocket engines.

4.5.2 Dual Bell–Shaped Nozzle

In order to overcome the problem of actuation mechanism of the extendible nozzle, a dual nozzle has been designed and developed by combining the two segments of the nozzle with a hump between them as shown in Figure 4.13b. During low-altitude flight, the first bell nozzle segment with the smaller exit area A_{e1} is operated for the expansion of gas as the flow separation occurs at the hump point and does not adhere to the larger exit area A_{e2}. In other words, the thrust is developed by the expansion of gas up to the exit area A_{e1}. In contrast, at higher altitude, due to lower ambient pressure, gas expands further for a particular chamber pressure even beyond the hump point by getting attached up to the larger exit area A_{e2}. As a result, the nozzle operates at a higher area ratio and enhances performance even at higher altitude. However, there will be marginal decrease in performance compared to that of the extendible nozzle. This type of nozzle has been used to improve the performance of cryogenic engines on the space vehicle Araine 5.

4.5.3 Expansion–Deflection Nozzle

Both extendible and dual bell–shaped nozzles cannot adapt with changes in altitude in a continuous manner, but rather in a discrete manner during the flight of a rocket engine, and hence will have lower performance level. In order to adapt the nozzle to the changes in altitude in a continuous manner, two different types of nozzles have been developed, namely, (1) the expansion–deflection nozzle and (2) the aerospike nozzle, in which the free jet boundary helps in adjusting the expansion process automatically and in a continuous manner with changes in ambient pressure along with the altitude. A schematic of a typical expansion–deflection nozzle with annual cross section with the central body/plug is shown in Figure 4.14a. In this case, gas flow from the combustion chamber is turned around the curved contour of the plug and moves outward away from its central axis along the curved and diverging surface of the bell nozzle. The purpose of the plug is to force the flow to remain attached to, or to stick to, the nozzle walls. An aerodynamic interface is formed between the inner gas layer along the curved diverging section and the ambient air. When ambient pressure changes with altitude, hot gases fill the larger portion of the diverging section of the nozzle. In other words, this nozzle adapts well to altitude in a continuous manner.

4.5.4 Aerospike Nozzle

In the case of an aerospike nozzle, an aerospike/aerodynamic plug is placed in the center of the nozzle as shown in Figure 4.14b. Note that the flow moves inward toward its axis guided by the centrally placed aerospike. The outer gas

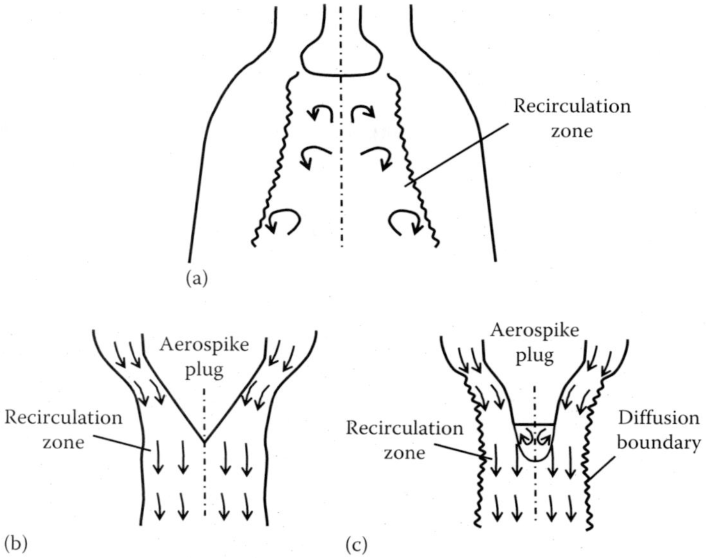

FIGURE 4.14 Schematic of (a) an expansion–deflection nozzle, (b) an aerospike nozzle, and (c) a truncated aerospike nozzle.

boundary is interfaced directly with ambient air. The expansion of gas flow around the aerospike gets adjusted directly with the changes in ambient pressure due to changes in altitude. At the designed pressure, the boundary will be almost parallel to the axis with similar performance as that of a conventional bell nozzle. But for higher ambient pressure, the boundary moves inward raising the exhaust pressure. On the other hand, if the boundary moves outward, it allows the gas to expand to a lower pressure. As a result, this kind of nozzle can perform in a far superior way than the designed condition compared to the conventional bell nozzle. The full parabolic contoured nozzle is much longer in order to achieve minimum flow losses when fluid is turned off around the aerospike. Hence, the nozzle with a truncated aerospike, as shown in Figure 4.14b, is designed to overcome this problem. Generally, the aerospike in the nozzle is truncated to such an extent that it provides least performance loss. Of course, with the use of a truncated aerospike, the nozzle becomes shorter and sturdier with minimum problems due to heat transfer.

4.6 THRUST-VECTORING NOZZLES

We know that the rocket nozzle produces thrust by expanding hot gases only in a certain direction. But we need to change the direction of the thrust at the time of maneuver for which proper provisions have to be made in the nozzle design. This is commonly known as thrust vectoring. Several methods of thrust vectoring have been designed and developed over the years. We will now discuss five kinds of thrust vectoring:

1. *Gimballing system*: In the gimballing system, the entire engine is rotated using a universal joint as shown in Figure 4.15a. A hinge is permitted to rotate around each axis only using proper bearing such that thrust vectoring can be achieved. This calls for smaller control forces to execute the thrust vector, particularly for small changes in angle. Hence, it incurs negligible losses in its performances in terms of

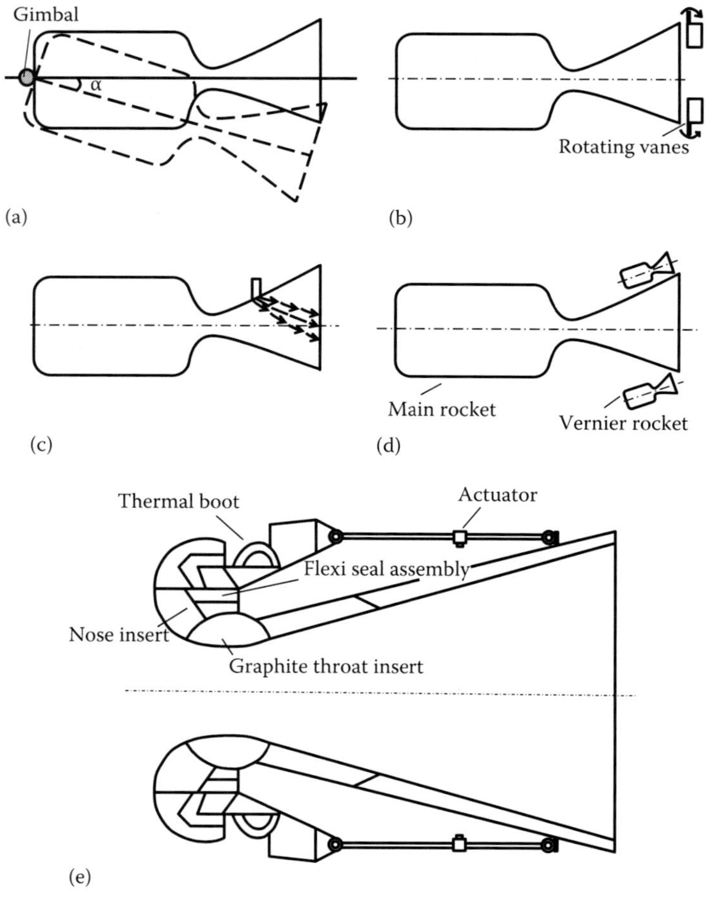

FIGURE 4.15 Schematic of (a) a gimballing system, (b) jet vanes and jetavator, (c) a side liquid injection system, (d) a Vernier rocket nozzle, and (e) a flexible nozzle.

thrust and specific impulse, and is being used profusely. It is preferred in liquid rocket engines, particularly when flexible supply lines are being used. It is less preferred in solid rocket engines as it is heavier and larger in size. Most U.S. spacecraft (e.g., space shuttle) use the gimballing system while Russian spacecraft prefer the hinge system.

2. *Jet vanes and jetavator*: In this method, the exhaust jet from the nozzle can be deflected by using a pair of heat-resistant vanes as shown in Figure 4.15b. These aerodynamic-shaped vanes are placed at the exit of a fixed nozzle, which causes a certain drag that reduces the specific impulse from 2% to 5% with severe vane erosion. In order to reduce the vane erosion, graphite is used for the fabrication of jet vanes. Germany's V2 missiles and Iraq's Scud missiles use graphite jet vanes. But these can crack and break easily. Recently, glass-phenolic molded vanes have been used for this purpose. These can provide maximum side thrust as large as 20% of the axial thrust with higher reliability. Besides, jet tabs are used for thrust vectoring due to their low actuating power and light weight. But these are also subject to erosion and thrust loss. Nozzles with an extendible portion are known as jetavators, which are used for thrust vectoring purpose on demand. In other words, these devices can allow jets to be deflected whenever thrust vectoring is required. As a result, losses due to thrust vectoring are reduced drastically compared to jet vanes.

3. *Side liquid injection*: In this method, a secondary liquid is injected through the side wall into the main gas stream which forms an oblique shock in the diverging portion of the nozzle, as shown in Figure 4.15c. The liquid jet must have sufficient jet momentum compared to the gas momentum in the diverging portion of the nozzle. This bow shock upstream of the cross jet causes the flow to separate from its wall due to a large pressure gradient. A recirculation is formed upstream due to this large pressure gradient. Note that the liquid jet gets broken and forms a region of vapor and droplets downstream of the bow shock. As a result, asymmetrical distribution along the azimuthal direction will occur and hence a side force is produced effecting thrust vectoring of the rocket engine. Axial force is reduced marginally, particularly for small flow rate of injectant, because any decrease in axial thrust due to the formation of oblique shock is compensated by gain due to liquid injection. This method has been used for a few large solid-propellant rocket engines (e.g., Titan IIIC). Generally, high-density liquid is preferred to reduce the storage tank

volume. Strontium perchlorate with water solution (30:70) is preferred as a liquid injectant. It is mostly preferred for small deflection. It is not favored for large side forces as large amounts of injectants have to be used. Note that liquid requirement for the worst-case scenario has to be carried for a successful mission.

4. *Vernier rocket*: Small auxiliary nozzles known as Vernier rockets are used to provide roll control of the rocket engine. These Vernier nozzles are small and can be actuated easily to deflect the exhaust nozzle flow. Note that these nozzles are supplied from the same feed system of the main rocket engine. A typical Vernier rocket system for thrust vectoring is shown in Figure 4.15d in which four small rocket nozzles can be used to impart vectoring of thrust produced by the engine. Vernier rockets were used routinely in early Atlas missiles. The space shuttle has six Vernier rockets for its reaction control. In recent times, it is not much used because of its weight and complex feeding system.
5. *Flexible nozzle*: Among the recent types of nozzles, the flexible nozzle has found wider application for thrust vectoring, particularly for solid-propellant rocket engines as it does not reduce thrust and specific impulse significantly compared to other methods. A typical flexible nozzle submerged in solid propellant is shown in Figure 4.15e. It consists of an actuator, a thermal boot, a throat hosing, a flex seal an assembly, a throat insert, and a divergent liner. In this submerged flexible nozzle, a number of high-temperature composite sheathing joints are used which can move when the nozzle is actuated by a mechanical/hydraulic actuator by reorienting its angle. As a result, the nozzle is rotated by an angle from 4° to 7°, thus making the nozzle flexible. In recent times, flexible nozzles have become popular in India, Japan, the United States, and Europe.

4.7 LOSSES IN ROCKET NOZZLE

We have already discussed at length about rocket nozzle losses under ideal conditions. The effects of ambient pressure (altitude) on rocket performance have also been discussed in detail. Besides this, several losses are incurred during the operation of the nozzle. Some of the major losses are enumerated here:

1. Flow divergence in a conical nozzle causes losses in the axial momentum, leading to lower specific impulse. Losses can be minimized by using a bell-shaped nozzle.

2. Generally, the area ratio between the combustion chamber and the throat must be greater than 4 to have low velocity in the combustion chamber. But a smaller combustion chamber area is preferred due to constraints of rocket engine design. As a result, the velocity in the combustion chamber will be considered to be almost zero under ideal conditions and will thus incur more pressure losses due to acceleration caused by heat addition during the combustion process, leading to lower thrust.
3. Extended combustion in the nozzle can change the flow properties, which can alter the production of the ideal nozzle.
4. Nonuniform gas composition can cause losses due to incomplete mixing and combustion.
5. Performance of the rocket nozzle becomes slack during takeoff and burnout period due to transient pressure operation.
6. A boundary layer develops on the nozzle wall, leading to slower exhaust velocity (0.5%–1.5%), which can result in less thrust.
7. Solid particles/droplets in the nozzle flow can cause losses.
8. Non-optimum nozzle expansion can cause losses in exhaust velocity, leading to lowering of specific impulse. As overexpansion and underexpansion of nozzle flow do take place during its flight, this can incur losses as high as 15% compared to an ideal nozzle. Hence, advanced nozzles are being developed to compensate their performance with altitude, as discussed in Section 4.5.
9. Unsteady combustion and pressure fluctuation can lower nozzle performance.

4.8 PERFORMANCE OF EXHAUST NOZZLE

The performance of an exhaust nozzle can be characterized by several performance parameters reported in the literature. However, we will restrict our discussion to three parameters, which are used profusely by designers:

1. Isentropic nozzle efficiency η_{in}
2. Discharge coefficient C_D
3. Thrust coefficient C_F

4.8.1 Isentropic Efficiency

The expansion process in a nozzle is shown in an T-s diagram in Figure 4.16. It may be noted that the velocity at the nozzle exit under isentropic expansion condition is higher than that in the non-isentropic condition due to increased entropy caused by friction, and turbulence effects. In order to ascertain the performance of the nozzle, efficiency is defined as the ratio of actual to isentropic enthalpy drop across the nozzle. By using the station numbers of the rocket engine, the expression for isentropic efficiency can be derived as follows:

$$\eta_{in} = \frac{h_{t2} - h_e}{h_{t2} - h_{es}} \approx \frac{T_{t2} - T_e}{T_{t2} - T_{es}} \tag{4.17}$$

This equation can be expressed in terms of pressure ratio across nozzle and actual exit velocity as follows:

$$\eta_{in} = \frac{V_e^2/2}{h_{t2} - h_{es}} = \frac{V_e^2/2}{C_P T_{t2}\left(1 - \frac{T_{es}}{T_{t2}}\right)} = \frac{V_e^2/2}{\frac{\gamma R T_{t2}}{\gamma - 1}\left(1 - \left(\frac{P_e}{P_{t2}}\right)^{\frac{\gamma-1}{\gamma}}\right)} \tag{4.18}$$

Note that the isentropic efficiency can be evaluated knowing the pressure ratio, chamber temperature, and actual nozzle velocity. It takes care of various kinds of losses that make the flow non-isentropic. Its value lies from 90% to 95% depending on the flow condition, nozzle size and shape, and so on.

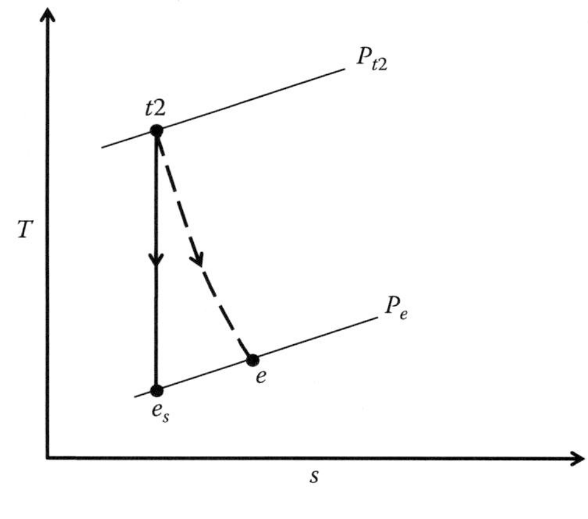

FIGURE 4.16 The expansion processes in a T-s diagram.

4.8.2 Discharge Coefficient

We know that the actual effective flow area of the nozzle is always slightly less than the geometrical area due to the presence of the boundary layer on the wall of the nozzle. Hence, we must define a discharge coefficient as follows:

$$C_D = \frac{\text{Actual mass flow rate}}{\text{Ideal mass flow rate}} = \frac{\dot{m}_e}{\dot{m}_{es}} = \frac{\rho_e A_e V_e}{\rho_{es} A_e V_{es}} = \frac{T_{es} V_e}{T_e V_{es}} = \frac{T_{es}}{T_e}\sqrt{\eta_{is}} \quad (4.19)$$

where C_D is assumed to be constant if the nozzle is fully expanded. Note that the discharge coefficient C_D is dependent on Reynolds number of the flow, which eventually is dependent on the pressure ratio P_{t2}/P_e across the nozzle. The variation of C_D with the pressure ratio P_{t2}/P_e for a typical CD nozzle is shown in Figure 4.17. Interestingly, it may be observed from this figure that the C_D does not vary with P_{t2}/P_e due to venturi effects during initial subsonic flow. With an increase in P_{t2}/P_e the C_D decreases due to increasing losses and attains a minimum value when the nozzle is choked.

4.8.3 Mass Flow Coefficient

In the case of a rocket engine, the mass flow rate through the nozzle is governed by the chamber pressure and throat area of the nozzle. Hence, it is important to have a parameter which can represent all these variables. This parameter, known as the mass flow coefficient $C_{\dot{m}}$, is defined as the ratio of

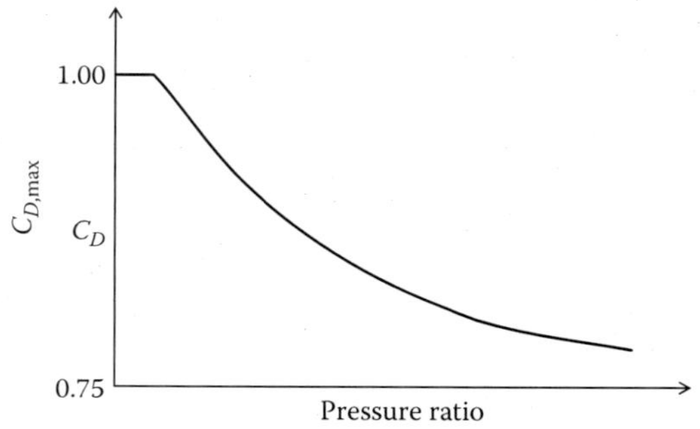

FIGURE 4.17 Variation in C_D versus pressure ratio P_{t7}/P_9 across a CD nozzle.

mass flow rate of propellant $\dot{m}_p$ to the product of chamber pressure P_c and throat area A_t as follows:

$$C_{\dot{m}} = \frac{\dot{m}_p}{P_c A_t} \tag{4.20}$$

Note that the mass flow coefficient $C_{\dot{m}}$ can be estimated easily from experimental data for the following parameters: mass flow rate of propellant $\dot{m}_p$, chamber pressure P_c, and throat area A_t. The mass flow coefficient $C_{\dot{m}}$ can be determined from experimental data and documented in the form of a design chart, which is quite useful for the design and development of rocket engines.

4.9 THRUST COEFFICIENT

We know that in an actual nozzle, the flow streamlines would not be parallel to the axis of the nozzle. In other words, there will be a velocity vector angularity in the flow, leading to lowering of the actual thrust from the ideal one. In addition, there will be a reduction in thrust due to a decrease in velocity caused by the presence of boundary layers over its surface. Due to all these reasons, the actual thrust will be different than the ideal one. In order to ascertain this effect, the thrust coefficient C_F was defined in Section 3.4.5 as the ratio of thrust F to the product of chamber pressure P_c and throat area A_t of the nozzle as follows:

$$C_F = \frac{F}{P_c A_t} \tag{4.21}$$

Recall that in Section 3.3, we had derived the expression for thrust F (Equation 3.4), which indicates that thrust is dependent on exhaust velocity V_e, exit pressure P_e, ambient pressure P_a, and mass flow rate of propellant $\dot{m}_p$. From Equation 4.3, we know that exhaust velocity V_e for an ideal nozzle is expressed in terms of exit pressure P_e, chamber pressure P_{t2}, specific heat ratio γ, and C_P of exhaust gas. Moreover, Equation 4.8 indicates that mass flow rate $\dot{m}$ is related to chamber pressure P_{t2}, chamber temperature T_{t2}, throat area A_t of the nozzle, and specific heat ratio γ. By using Equations 4.3, 4.8, and 4.21, we can derive an expression for the thrust coefficient C_F as follows:

$$C_F = \frac{F}{P_c A_t} = \sqrt{\frac{2\gamma^2}{(\gamma-1)}\left(\frac{2}{\gamma+1}\right)^{\frac{\gamma+1}{\gamma-1}}\left[1-\left(\frac{P_e}{P_c}\right)^{\frac{\gamma-1}{\gamma}}\right]} + \frac{A_e}{A_t}\left(\frac{P_e - P_a}{P_c}\right) \tag{4.21}$$

Note that thrust coefficient C_F is dependent on nozzle geometry, namely, nozzle exit area A_e and throat area A_t, apart from operating pressure ratios and specific heat ratio γ. Note that the term $(P_e - P_a)/P_c$ indicates how well the nozzle is suited to expansion due to the actual pressure ratio. It is interesting to note that the thrust coefficient does not depend on flame temperature T_c and gas constant R. Note that expansion area ratio A_e/A_t is dependent on P_e/P_c. Hence, let us plot the variation of C_F against A_e/A_t in Figure 4.18 for various P_{t2}/P_a and representative specific heat ratio of 1.2. It is interesting to note that the value of C_F varies from 0.85 at $P_{t2}/P_a = 2.5$ to 1.8 at $P_{t2}/P_a = 500$ in this plot. Let us consider the case of $P_{t2}/P_a = 8$ in Figure 4.18 for the value of C_F which increases to a peak value at an area ratio of 2 and drops subsequently. The peak value of C_F occurs when the nozzle is completely expanded ($P_e = P_a$). But for both underexpansion ($P_e > P_a$) and overexpansion ($P_e < P_a$) for a particular chamber pressure P_{t2}, C_F is lower than the peak value. Recall that over-expansion in the nozzle must be avoided to achieve higher performance

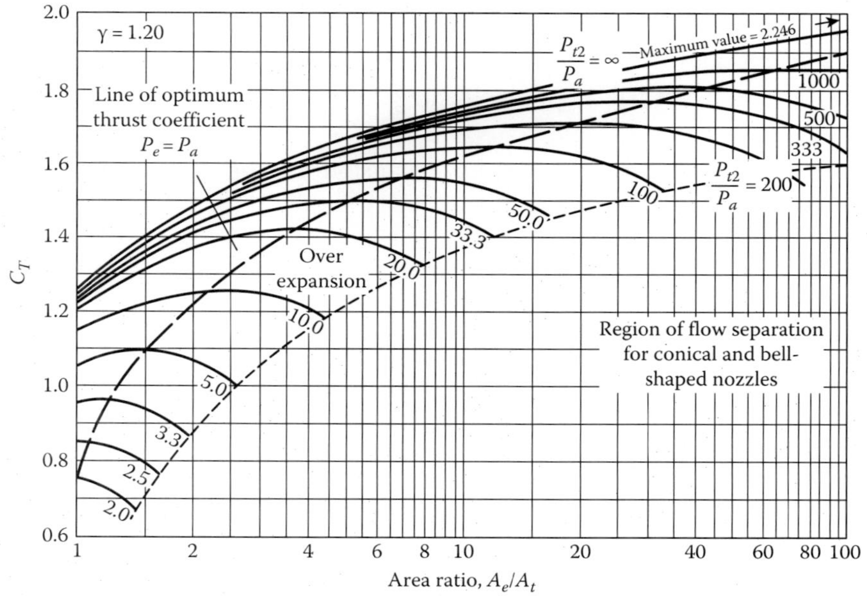

FIGURE 4.18 The variation of C_T against A_e/A_t. (From Sutton, G.P. and Biblarz, O., *Rocket Propulsion Elements*, 7th edn., John Wiley & Sons Inc., New York, 2001.)

in the nozzle, because strong oblique shocks can form inside the divergent portion of the nozzle as shown in Figure 4.12. If the strength of the oblique shock is high, then the boundary layer is likely to separate, resulting in exorbitant losses in thrust. Hence, the area ratio A_e/A_t for a CD nozzle has to be properly selected so that the flow separation in the nozzle can be avoided during overexpansion. Note that when the nozzle is expanded to near vacuum corresponding to a large area ratio, the asymptotic value of C_F is around 2.25 (not shown in Figure 4.18). Note that such an area ratio is not used for practical applications as the nozzle size and weight will be quite large. Generally, a nozzle pressure ratio from 5 to 100 is used for rocket engines. Note that the design chart for thrust coefficient C_F in terms of pressure ratio P_c/P_a and area ratio A_e/A_t is generated which can be used for the design and development of rocket engines.

Example 4.2

In a rocket engine, a CD nozzle is designed to expand gas fully from a chamber pressure of 5.4 MPa and a temperature of 3000 K to ambient pressure 15 kPa. Assuming flow to be isentropic, determine (1) the area ratio and (2) the thrust coefficient. If this engine is operated at sea level, what will be the percentage of change in the thrust coefficient. Assume $\gamma = 1.2$.

The area ratio for optimum expansion can be evaluated using Equation 4.15:

$$\frac{A_e}{A_t} = \frac{\left(2/[\gamma+1]\right)^{\frac{\gamma+1}{2(\gamma-1)}}\sqrt{\gamma}}{\sqrt{\dfrac{2\gamma}{(\gamma-1)}\left(\dfrac{P_e}{P_c}\right)^{\frac{2}{\gamma}}\left[1-\left(\dfrac{P_e}{P_c}\right)^{\frac{\gamma-1}{\gamma}}\right]}}$$

$$= \frac{\left(2/[1.2+1]\right)^{\frac{1.2+1}{2(1.2-1)}}\sqrt{1.2}}{\sqrt{\dfrac{2\times1.2}{(1.2-1)}\left(\dfrac{15}{5400}\right)^{\frac{2}{1.2}}\left[1-\left(\dfrac{15}{5400}\right)^{\frac{1.2-1}{1.2}}\right]}} = 31.94$$

The thrust coefficient can be determined for a fully expanded nozzle ($P_e = P_a$) by using Equation 4.21:

$$C_F^0 = \sqrt{\frac{2\gamma^2}{(\gamma-1)}\left(\frac{2}{\gamma+1}\right)^{\frac{\gamma+1}{\gamma-1}}\left[1-\left(\frac{P_e}{P_c}\right)^{\frac{\gamma-1}{\gamma}}\right]}$$

$$= \sqrt{\frac{2\times1.2^2}{(1.2-1)}\left(\frac{2}{1.2+1}\right)^{\frac{1.2+1}{1.2-1}}\left[1-\left(\frac{15}{5400}\right)^{\frac{1.2-1}{1.2}}\right]} = 1.78$$

When the engine is operated at sea level (P_a = 101.325 kPa), the thrust coefficient can be determined by using Equation 4.21:

$$C_F = \sqrt{\frac{2\gamma^2}{(\gamma-1)}\left(\frac{2}{\gamma+1}\right)^{\frac{\gamma+1}{\gamma-1}}\left[1-\left(\frac{P_e}{P_{t2}}\right)^{\frac{\gamma-1}{\gamma}}\right]} + \frac{A_e}{A_t}\left(\frac{P_e-P_a}{P_{t2}}\right)$$

$$= \sqrt{\frac{2\times1.2^2}{(1.2-1)}\left(\frac{2}{1.2+1}\right)^{\frac{1.2+1}{1.2-1}}\left[1-\left(\frac{15}{5400}\right)^{\frac{1.2-1}{1.2}}\right]}$$

$$-31.94\left(\frac{15-101}{5400}\right) = 1.78-0.51 = 1.27$$

The percentage of change in the thrust coefficient can be determined as 28.65.

REVIEW QUESTIONS

1. What are the types of exhaust nozzles used in rocket engines?
2. Show schematically the variations in total enthalpy, pressure, and temperature across a convergent nozzle. How are these comparable with their respective static quantities?
3. What is meant by back pressure control? Explain it.
4. What is meant by thrust vectoring. Describe various methods used in practical systems.
5. When is the CD nozzle preferred? Can you suggest methods to have a variable throat area in a CD nozzle?

6. What is meant by underexpansion in a CD nozzle? When is it likely to occur?
7. Describe the flow processes involved during the overexpansion of a CD nozzle.
8. Is the variable-geometry nozzle preferred in a rocket engine? Justify your answer.
9. Is it possible to have an arrangement of variable-geometry CD? Describe its operating principle.
10. What are the advanced nozzles used for rocket engines. Which is preferred among all? Why is it so?
11. What is meant by thrust vectoring? What are the types of thrust vectoring used currently for rocket engines?
12. What are the performance parameters used for nozzles? Define isentropic efficiency. What are the parameters that affect isentropic efficiency?
13. Define discharge coefficient. How does it vary with pressure ratio across a CD nozzle?
14. What is meant by mass flow coefficient? What is its physical meaning?
15. What is thrust coefficient? What are the parameters that affect thrust coefficient?

PROBLEMS

4.1 In a CD nozzle with a throat area of 0.75 m^2 of a rocket engine, gas at 7.5 MPa and temperature of 3200 K is expanded fully to 10.5 kPa for producing thrust. Assuming flow to be isentropic, determine (1) the exit Mach number, (2) the maximum exit mass flow rate passing through this nozzle, and (3) the exit area. Assume $\gamma = 1.2$ and $MW =$ 24 kg/kmol.

4.2 In a rocket engine, a CD nozzle is designed to expand gas fully from a chamber pressure of 6.4 MPa and temperature of 3200 K to an ambient pressure of 2.5 kPa. Assuming flow to be isentropic, determine (1) the area ratio and (2) the thrust coefficient. If this engine is operated at sea level, what will be the percentage of change in the thrust coefficient. Assume $\gamma = 1.2$.

4.3 A nozzle is fully expanded in a CD nozzle from a chamber pressure of 7.5 MPa and temperature of 3200 K to sea-level pressure. If the mass flow rate happens to be 4.3 kg/s, determine the exit velocity, exit temperature, and thrust coefficient. Assume $\gamma = 1.25$ and $MW = 20$ kg/kmol.

4.4 A nozzle is fully expanded in a CD nozzle from a chamber pressure of 2.5 MPa and temperature of 3200 K to 10 kPa. If the throat area happens to be 0.0006 m^2, determine the throat velocity, throat density, mass flow rate, specific impulse, and thrust. Assume $\gamma = 1.3$ and $MW = 22$ kg/kmol.

4.5 For a rocket engine, a CD nozzle with an exit to throat area ratio of 10 is designed to expand the hot gas with $\gamma = 1.25$ and $MW = 20$ kg/kmol at 2950 K to ambient pressure at an altitude of 20 km. Determine the chamber pressure, exit Mach number, exit temperature, and exit pressure.

4.6 A rocket engine has the following data:

Mass flow rate of propellant	5 kg/s
Nozzle exit diameter	10 cm
Nozzle exit pressure	1.02 bar
Ambient pressure	1.013 bar
Chamber pressure	20 bar
Thrust	7 kN

Calculate the effective jet velocity, actual jet velocity, specific impulse, and specific propellant consumption.

4.7 A rocket nozzle has a throat area of 18 cm^2 and combustion chamber pressure of 25 bar. If the specific impulse is 127.42 s and weight flow rate is 44.145 N/s, determine the thrust coefficient, propellant weight flow coefficient, specific propellant consumption, and characteristic velocity.

4.8 The data for a rocket engine are as follows:

Combustion chamber pressure	38 bar
Combustion chamber temperature	3500 K
Oxidizer flow rate	41.67 kg/s
Mixture ratio	5.0

If the expansion in the rocket nozzle takes place at an ambient pressure of 533.59 N/m^2, calculate the nozzle throat area, thrust, thrust coefficient, characteristic velocity, exit velocity of exhaust gases, and maximum possible exhaust velocity. Take $\gamma = 1.3$ and $R = 287$ J/kg · K.

4.9 Calculate the thrust, effective jet velocity, and specific impulse of a rocket operating at an altitude of 20 km with the following data:

Propellant flow rate	1 kg/s
Thrust chamber pressure	27.5 bar
Chamber temperature	2400 K
Nozzle area ratio	10.2

Take $\gamma = 1.3$ and $R = 355$ J/kg · K.

4.10 A rocket engine burning liquid oxygen and kerosene operates at a mixture ratio of 2.26 and a combustion chamber pressure of 50 atmospheres. If the propellant flow rate is 500 kg/s, calculate the area of the exhaust nozzle throat.

4.11 A convergent–divergent nozzle has an area ratio of 4 and is designed to expand the hot gases at total pressure and temperature of 5.5 MPa and 3230 K, respectively. What is the location of the normal shock wave and exit Mach number if the back pressure becomes 0.1 MPa?

4.12 A convergent–divergent nozzle is designed to expand the hot gases at total pressure and temperature of 4.5 MPa and 2830 K, respectively. This nozzle has an exit area of 0.86 m^2. If the isentropic efficiency of the nozzle is 0.95, estimate (1) the exit Mach number and (2) the mass flow rate, and A_e/A_t when the nozzle is operated under designed conditions (50 kPa and 226 K).

4.13 A nozzle with an inlet diameter of 50 mm, semi-convergence angle of 25°, and semi-divergence angle of 7° is to be designed to deliver 0.2 kg/s mass flow rate with Mach number of 2.2 at its exit. If the chamber pressure and temperature happen to be 1.5 MPa and 30°C, respectively, determine the exit diameter, throat diameter, and length of the nozzle. If the back pressure is the same as the atmospheric pressure, determine the expansion/compression waves and their angles at the nozzle exit and the maximum theoretical Mach number that can be achieved at its exit. Take $\gamma = 1.33$.

REFERENCES AND SUGGESTED READINGS

1. Hill, P. and Peterson, C., *Mechanics and Thermodynamics of Propulsion*, 2nd edn., Addison-Wesley Publishing Company, Inc., 1992.
2. Rathakrishnan, E., *Gas Dynamics*, 5th edn., PHI Learning Pvt Ltd, New Delhi, 2013.
3. Rao, G.V.R., Exhaust Nozzle Contour for optimum flight, *Jet Propulsion*, Vol. 28, No. 6, pp. 377–382, June 1958.
4. Barrere, M., Jaumotte, A., de Veubeke B. F., and Vandenkerckhove, J., *Rocket Propulsion*, Elsevier Publishing Company, New York, 1960.
5. Mishra, D.P., *Gas Turbine Propulsion*, MV Learning, London, U.K., 2015.
6. Sutton, G.P. and Biblarz, O., *Rocket Propulsion Elements*, 7th edn., John Wiley & Sons Inc., New York, 2001.

CHAPTER 5

Spacecraft Flight Performance

> If there is a small rocket on top of a big one and if the big one is jettisoned and small one is ignited, their speeds are added.
>
> HERMAN OBERTH, ROCKET EXPERT

5.1 INTRODUCTION

In the last few chapters, we have learnt about the various fundamentals which are required for understanding rocket propulsion. By expanding the high pressure and high temperature in a convergent–divergent (CD) nozzle, thrust is produced which is imparted to the vehicle for achieving the requisite flight velocity. Hence, it is important to understand the effect of exhaust nozzle jet velocity on the flight performance of space vehicles. The thrust imparted by the rocket engine is to overcome the drag forces and accelerate/decelerate the vehicle and change its direction in a controlled manner. For space applications, several flight regimes, namely, (1) atmospheric flight and (2) near-space flight and deep-space flight, are being considered. Note that the atmospheric flight regime is to be considered for air–surface missile, sounding rocket applications while near-space flight is considered for satellites, space labs, and so on. But the deep-space regime is to be considered for lunar, mars, and other planetary flight missions. We will be restricting our discussions to the simplified flight performance of a rocket engine in this book. Interested readers are advised to refer to advanced books for detailed information on space flight performance [1,2].

5.2 FORCES ACTING ON A VEHICLE

Several external forces, namely, thrust, aerodynamic forces (drag), gravitational forces, centrifugal forces, and coriolis forces, act on flying objects like a missile. The centrifugal and coriolis force being small in comparison to other forces can be neglected in the present analysis. The thrust produced by the rocket motor usually acts in the opposite direction of the high-temperature, high-pressure gas that expands and exits from the rocket nozzle. We know that thrust can be expressed as a function of the propellant mass flow rate $\dot{m}_p$ and effective exhaust velocity of the propellant in the nozzle. For a fully expanded nozzle, thrust from Equation 3.4 can be expressed mathematically as

$$F = \dot{m}_p V_e \tag{5.1}$$

In most cases, the propellant mass flow rate remains almost constant and thus can be expressed in terms of initial mass of propellant in the spacecraft and the burning time of the propellant t_b, neglecting of course the starting and ending transients, as follows:

$$\dot{m}_p = \frac{m_p}{t_b} \tag{5.2}$$

The instantaneous mass of the vehicle m can be expressed as a function of the initial mass of the full vehicle m_0 and the propellant mass m_p at an instantaneous time t as follows:

$$m = m_o - \dot{m}_p t \tag{5.3}$$

5.2.1 Aerodynamic Forces

We are aware of two aerodynamic forces, namely, drag and lift. The drag force acts in the direction opposite to the flight of the vehicle due to the resistance of the vehicle to the flight motion through a fluid (air) medium, whereas lift is the aerodynamic force acting in the direction normal to the flight of the vehicle. The lift L and the drag D can be expressed as functions of flight speed V, density of fluid, and a typical surface area which can be expressed as follows:

$$L = \frac{1}{2}\rho V^2 A C_L \tag{5.4}$$

$$D = \frac{1}{2}\rho V^2 A C_D \tag{5.5}$$

where

C_D is the drag coefficient
C_L is the lift coefficient
A is the frontal cross-sectional area

Note that the frontal area must be chosen properly as it varies from system to system. For example, in a missile system, the frontal area A is the maximum cross-sectional area normal to the missile axis. The drag and lift coefficients are mainly functions of flight measured in Mach number and angle of attack. The variation of C_D and C_L with changes in flight Mach number is depicted in Figure 5.1 for a typical missile for two angles of attack ($\alpha = 3°$ and 10°). It may be noted that the effects of Mach number can be neglected

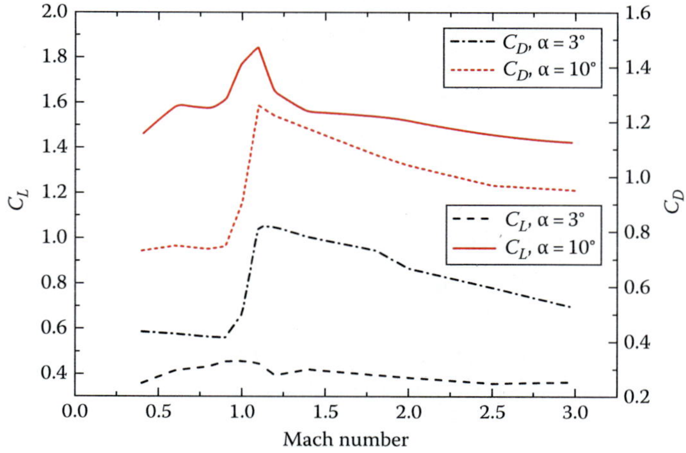

FIGURE 5.1 Variation of C_D and C_L with Mach number for a typical missile for two angles of attack $\alpha = 3°$ and 10°.

for low flight speed. The values of C_D and C_L reach a maximum around unity Mach number ($M \cong 1.1$).

5.2.2 Gravity

We know that gravitational attraction acts on a space vehicle by all terrestrial bodies which pull the vehicle toward their center of mass. But when the vehicle is in the immediate vicinity of the earth, the gravitational attraction of other planets and bodies is negligibly small compared to the earth's gravitational force. The variation of local gravitational acceleration with altitude can be estimated by the following expression:

$$g = g_e\left(\frac{R_e}{R}\right)^2 = g_e\left(\frac{R_e}{R+h}\right)^2 \tag{5.6}$$

where

g_e is the acceleration on the earth's surface
g is the local gravitational acceleration due to gravity
h is the altitude (km) from the earth's surface
R_e is the radius of the earth

Note that Equation 5.6 can be derived easily by using Newton's law of gravitation. Generally, the value of the radius of the earth R_e and the acceleration on the earth's surface g_e evaluated at the equator of the earth are 6378.5 km and 9.81 m/s^2, respectively, which can be used for calculating the local gravitational acceleration due to gravity. Note that the duration of flight in the earth's atmosphere is quite small and will not affect the effect of gravitational acceleration on the estimation of thrust and velocity increment. We will discuss this further in the subsequent section.

5.2.3 Atmospheric Density

We know that the density of atmosphere around the earth varies along with altitude. Hence, it has to be considered during a rocket engine's flight, particularly within earth's atmosphere. The approximate expression for atmospheric density ρ with altitude h is as follows [4]:

$$\rho = Ae^{-\left(Bh^{1.15}\right)} \tag{5.7}$$

where A and B are constants whose values are 1.2 and 2.9×10^{-5} when density and altitude are expressed in SI units. Besides this expression, density data with altitude can be used from the data table for international standard atmospheric properties as given in Appendix C1. Note that the air density in the atmosphere at an altitude of 30 km gets reduced to 1% of its value at sea level and can be neglected for this regime.

5.3 THE ROCKET EQUATION

We know that during a flight the mass of the vehicle changes a great deal due to consumption of the propellant. When the rocket vehicle flies nearer the earth's surface, the gravitational pull of other distant heavenly bodies is small enough to be neglected. During this motion of the rocket vehicle, several forces, namely, thrust, gravitational force, drag, lift, control forces, and lateral force, and all moments will be acting on it, which might cause the vehicle to turn in a certain direction. As a result, the motion of the vehicle is inherently three-dimensional in nature. However, for simplicity, we will assume a two-dimensional motion of the rocket vehicle. This is possible only when control forces, lateral forces that result in producing moments, are zero. In other words, this can happen only when the vehicle is having a rectilinear equilibrium flight. Let us assume the flight trajectory is two-dimensional and is constrained in the x–y plane as shown in Figure 5.2. However, we will consider the aerodynamic forces, namely, drag and lift, that are acting on the vehicle along with gravitational force from

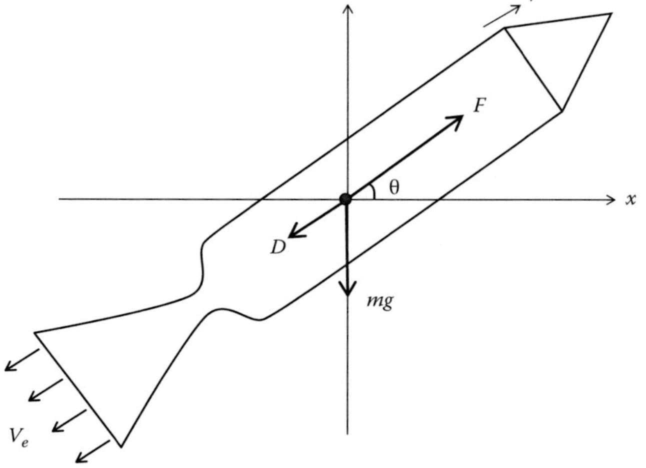

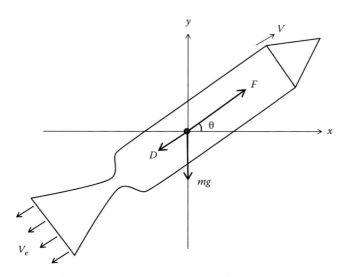

FIGURE 5.2 Free body force diagram in a two-dimensional plane for a flying rocket engine vehicle.

the earth. Note that we assume that the direction of flight is the same as that of the thrust.

In the direction of the flight path, we know that the product of mass and acceleration is equal to the sum of all acting forces:

$$m\frac{dV}{dt} = F - D - mg\cos\theta \tag{5.8}$$

where

m is the instantaneous mass of the vehicle
g is the acceleration due to gravity
D is the drag force acting on the vehicle
θ is the angle between the vehicle path and horizontal direction

We know that the vehicle mass is reduced as the propellant is ejected through the nozzle of the rocket engine for producing thrust. Hence, the instantaneous mass of the vehicle can be related to the propellant mass as follows:

$$m = m_o - \dot{m}_p t = m_o - \frac{m_p}{t_b}t = m_o\left(1 - PF\frac{t}{t_b}\right) \tag{5.9}$$

where

t is the instantaneous time
t_b is the total burning time
m_p is the initial propellant mass
m_o is the initial mass of the vehicle
$PF = m_p/m_o$ is the propellant fraction

By substituting Equation 5.9 for instantaneous mass in Equation 5.8, we obtain

$$m_o\left(1 - PF\frac{t}{t_b}\right)\frac{dV}{dt} = \frac{m_p}{t_b}V_{eq} - D - m_o\left(1 - PF\frac{t}{t_b}\right)g\cos\theta \tag{5.10}$$

By simplifying Equation 5.10, we obtain

$$\frac{dV}{dt} = \frac{m_p V_{eq}}{t_b\left[m_o\left(1 - PF\frac{t}{t_b}\right)\right]} - \frac{D}{\left[m_o\left(1 - PF\frac{t}{t_b}\right)\right]} - g\cos\theta \tag{5.11}$$

By expressing Equation 5.11 in terms of PF and substituting Equation 5.5, we obtain

$$\frac{dV}{dt} = \frac{(PF/t_b)V_{eq}}{\left[1 - PF\frac{t}{t_b}\right]} - \frac{\frac{1}{2}C_D \rho V^2 A}{\left[m_o\left(1 - PF\frac{t}{t_b}\right)\right]} - g\cos\theta \quad (5.12)$$

This equation is often known as the rocket equation. Note that the first term on the right-hand side representing acceleration due to applied thrust can be integrated easily. But the angle θ can vary continuously. The second term in Equation 5.12 that represents deceleration due to aerodynamic drag is difficult to integrate because we know that the drag coefficient C_D is a function of Reynolds number and Mach number. Besides this, C_D varies continuously with altitude, vehicle speed, and ambient density. Of course it can be integrated graphically or by numerical method. Its value can be designated as $B_1 C_D A/m_o$. The third term that represents the deceleration due to gravitational drag force of the earth is also equally difficult to evaluate as the value of g varies with altitude as per Equation 5.6. This term can be integrated with time provided we assume a mean value of cos θ. This assumption can be valid if the duration of thrust is quite short. This is possible as the thrust given to the vehicle is usually in the early part of its ascent, particularly for certain missile applications, and the rest of the journey is in ballistic (free flight) mode. Thus, we can take $\langle g\cos\theta\rangle$ as the mean value for calculation. Note that the drag term can contribute significantly only if the vehicle spends a considerable portion of its total time within the atmosphere. Therefore, we will neglect this term for simplicity for space flight. Then, on this simplification, Equation 5.13 can be expressed as follows:

$$\frac{dV}{dt} = \frac{(PF/t_b)V_{eq}}{\left[\left(1 - PF\frac{t}{t_b}\right)\right]} - \langle g\cos\theta\rangle \quad (5.13)$$

By integrating Equation 5.13 between interval $t = 0$ and $t = t_b$ ($V = V_0$ to $V = V_b$), we obtain

$$\int_{V_0}^{V_b} dV = (PF/t_b)V_{eq}\int_0^{t_b}\frac{dt}{\left[\left(1 - PF\frac{t}{t_b}\right)\right]} - \langle g\cos\theta\rangle\int_0^{t_b} dt \quad (5.14)$$

By integrating and simplifying Equation 5.14, we obtain

$$\Delta V = V_b - V_0 = -V_{eq} \ln\left[1 - PF\right] - \langle g \cos\theta \rangle t_b \qquad (5.15)$$

But we know that the propellant fraction PF can be expressed in terms of mass ratio MR as follows:

$$\frac{1}{1-PF} = \frac{1}{1-\dfrac{m_p}{m_0}} = \frac{m_0}{m_0 - m_p} = \frac{m_0}{m_b} = MR \qquad (5.16)$$

where

m_b is the burnout mass of the vehicle after all the propellant is consumed
MR is the mass ratio of the vehicle

We can express Equation 5.15 in terms of I_{sp} and MR as follows:

$$\Delta V = V_b - V_0 = V_{eq} \ln(MR) - \langle g \cos\theta \rangle t_b = I_{sp} g \ln(MR) - \langle g \cos\theta \rangle t_b \quad (5.17)$$

For larger ΔV, we need to have higher I_{sp} and MR and short burning duration when the vehicle is traveling through the gravitational field of any planet. As the time is too short to have a larger ΔV for the same I_{sp} and MR, the vehicle must be impulsive in nature. But if the vehicle is traveling in space, then the change in velocity due to gravitational force will be almost zero. Then Equation 5.17 for gravity-free flight becomes

$$\Delta V = V_{eq} \ln(MR) = I_{sp} g \ln(MR); \\ V_b = V_o + V_{eq} \ln(MR) = V_o + I_{sp} g \ln(MR) \qquad (5.18)$$

This equation was derived first in 1903 by the Russian scientist Konstantin E. Tsiolkovsky, who had advocated the use of rocket engines for space travel. This equation is known as Tsiolkovsky's rocket equation, which is the basis of rocket propulsion. It indicates that velocity change of the vehicle is dependent on the MR, which is the ratio of initial mass and burnout mass and the exhaust velocity. Note that it is not dependent on the applied thrust or the size of the rocket engine vehicle, or on the duration of propellant burning. Rather the velocity of the rocket vehicle increases as the propellant gets consumed for a particular value of I_{sp}. Besides, it also depends on the exhaust velocity for a particular value of mass ratio *MR*, indicating how fast a propellant can be ejected from the exhaust nozzle. In other words, to have higher rocket velocity, the mass ratio *MR* must have a higher value. For example, if the mass ratio happens to be 10, then the propellant mass would be around 90% of the total vehicle. In other words, the 10% total mass of the vehicle is used for structure, empty propulsion system mass, payload, guidance, and control and control surfaces. It calls for careful design acumen of the rocket engine designer. In order to appreciate the dependence of I_{sp} and *MR* better, the variation in velocity increment is plotted for different and shrewd values of I_{sp} and exhaust velocity. It may be observed that the vehicle velocity increases asymptotically for a certain value, particularly at higher *MR* for a particular value of I_{sp}. In other words, it would not be prudent to have higher *MR* beyond a certain limit as an increase in its value will have diminishing return. Of course, with further increase in the value of I_{sp}, vehicle velocity increases in a similar manner as that of lower I_{sp} values as shown in Figure 5.3 for $\langle g \cos \theta \rangle t_b = 825$ m/s. It may also be noted from this figure that a rocket vehicle can travel at a faster rate than its exhaust velocity. Of course, a rocket vehicle's velocity can be increased by increasing its exhaust velocity. Note that in earlier days the gun powder–propelled rocket engine had an exhaust velocity in the range of 2000 m/s, which could not be used for space application. But in modern

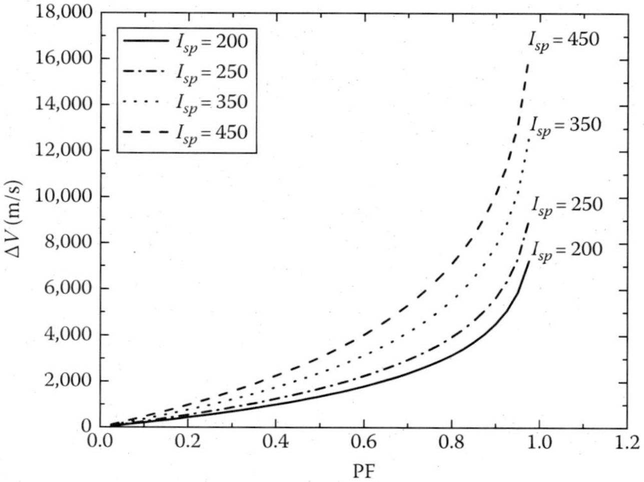

FIGURE 5.3 Variation of velocity increment with mass ratio MR.

cryogenic LPREs, an exhaust velocity of 4000 m/s can be achieved easily. As a result, the capability of a chemical rocket engine vehicle's velocity has increased to 11.98 km/s for an *MR* of 20 under ideal condition.

5.3.1 Burnout Distance

When the rocket is fired vertically, we can neglect both the drag force and gravitational force acting on the rocket engine. We can determine the distance covered during the burnout period of the propellant, known as the burnout distance h_b, by integrating Equation 5.18 into time duration t_b as follows:

$$h_b = \int_0^{t_b} V_b dt = \int_0^{t_b} \left[V_o + V_{eq} \ln\left(1 - PF\frac{t}{t_b}\right)\right] dt$$

$$= V_o t_b + V_{eq} t_b \left[\frac{1}{MR-1} \ln\left(\frac{1}{MR}\right) + 1 \right] \quad (5.19)$$

Note that the exhaust velocity is assumed to remain constant during this flight. It can be concluded from the relationship in Equation 5.19 that the burnout distance is dependent on the specific impulse I_{sp} and mass ratio *MR* and is independent of the size of the rocket engine. However, if the drag and gravitational forces are considered for this analysis, the burnout distance will depend on the size of the rocket engine in addition to I_{sp} and *MR*.

5.3.2 Coasting Height

Once the entire propellant of a rocket engine is burnt out, the rocket will travel further till all its kinetic energy is consumed by the vehicle. This vertical height achieved by the rocket engine is commonly known as the coasting height. Note that this height has to be determined routinely for the sounding rocket engine. By neglecting the drag and gravitational forces on the vehicle, the coasting height h_c can be determined by equating the kinetic energy possessed by the vehicle at burnout to the increase in potential energy due to the gain in vertical height as follows:

$$m_b \frac{V_b^2}{2} = \int_0^{h_c} m_b g dh \quad (5.20)$$

Assuming that acceleration due to the earth's gravity remains almost constant, we can integrate Equation 5.20 to get an expression for the coasting height h_c:

$$h_c = \frac{V_b^2}{2g} \quad (5.21)$$

The total vertical height h_t traveled by the vehicle can be determined as follows:

$$h_t = h_b + \frac{V_b^2}{2g} \quad (5.22)$$

5.3.3 Flight Trajectory

It is important to estimate the flight trajectory for guidance purpose. Of course, in an actual rocket vehicle, the flight trajectory can be three-dimensional in nature. However, for simplicity, we will consider the flight trajectory as two-dimensional in nature which will be on a single plane. This is possible only when the vehicle is not affected by the wind, thrust misalignment, solar attraction, and so on. In this case, the flight direction angle is the same as the thrust direction angle. The lift force can be assumed to be negligible as it is a wingless and symmetrical rocket engine vehicle. The simplified equation for velocity change in Equation 5.17 can be applied particularly when the vehicle is traveling through the earth's atmosphere as gravitational force will play a significant role in shaping the flight trajectory. However, Equation 5.11 can be invoked for determining the flight trajectory in a two-dimensional plane as it contains both drag and gravitational force terms. In order to integrate this equation, the calculation can be carried out over a small time interval compared to the burnout time. During this time interval, δt, g, θ, and D/m can be assumed to remain constant. Note that if the vehicle is launched vertically, it will not have a gravitational turn. However, if the vehicle has a certain initial angle of turn θ_1, the vehicle gradually takes a turn, known as the gravitational turn, and follows a curve path as shown in Figure 5.4. For a time interval δt, the overall vehicle increment ΔV is the vector sum of ΔV_F due to thrust, ΔV_g due to gravity, and ΔV_D due to drag, as shown in Figure 5.4a. Note that ΔV_D due to drag and *cosine* component of ΔV_g due to gravity are opposite in direction to ΔV_F due to thrust. In the subsequent time interval, the vehicle will

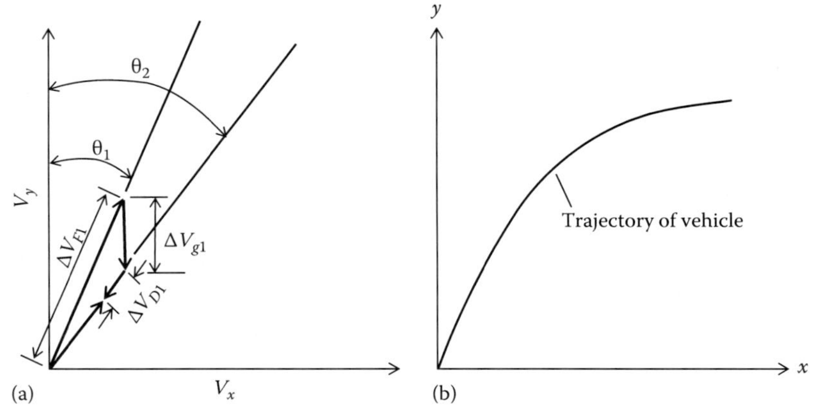

FIGURE 5.4 Flight trajectory calculation procedure for a vehicle with a gravity turn: (a) vector sum of all velocities for a small time interval, δt (b) flight trajectory of vehicle.

travel at a different turning angle. By obtaining the velocity as a function of time, we can easily evaluate its trajectory in the two-dimensional plane. Note that the accuracy of this procedure for the evaluation of flight trajectory will depend on the length of the time interval δt, which must be small compared to the burnout time.

5.4 SPACE FLIGHT AND ITS ORBIT

Rocket engines are employed to launch a spacecraft from the surface of the earth. The spacecraft moves around an orbit of the earth or of any other planet which is governed by the local gravitational field and momentum of the spacecraft. This orbit can be either circular or elliptic in shape. For the motion of the spacecraft in a circular orbit as shown in Figure 5.5, the gravitational force F_g holding the spacecraft can be determined by using Newton's law of gravitation:

$$F_g = \frac{GMm}{R^2} = m\omega^2 R \tag{5.23}$$

where

- M is the mass of the planet (earth)
- m is the mass of the space vehicle
- R is the distance between the two masses
- G is the universal gravity constant ($G = 6.67 \times 10^{-11}$ m^3/kg s^2)
- ω is the angular velocity of the mass m

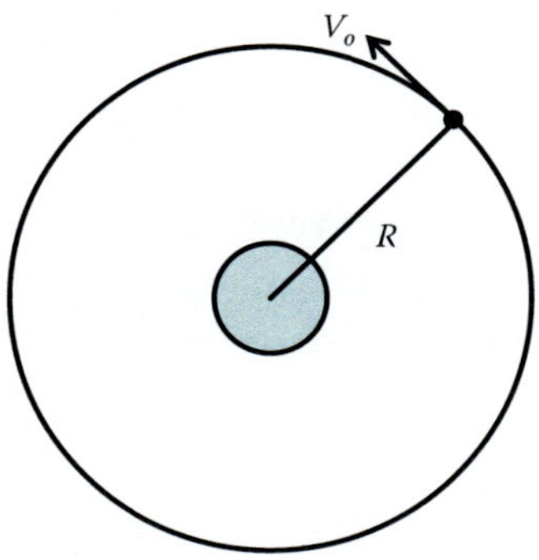

FIGURE 5.5 Flight trajectory around a circular orbit.

Note that the gravitational force F_g is balanced by the pseudo-centrifugal force $m\omega^2 R$. The angular velocity ω and orbital velocity V_o from Equation 5.23 can be evaluated easily as follows:

$$\omega = \sqrt{\frac{GM}{R^3}}; \quad V_o = \omega R = \sqrt{\frac{GM}{R}} \tag{5.24}$$

The orbital velocity decreases nonlinearly with the radius of the orbit. The time period required to revolve around this orbit can be determined as follows:

$$t_o = \frac{2\pi R}{V_o} = 2\pi\sqrt{\frac{R^3}{GM}} \tag{5.25}$$

The time period per revolution increases with the radius of the orbit at a higher rate compared to the orbital velocity.

Example 5.1

A satellite is placed at a height of 250 km from the surface of the earth in a circular orbit. Determine the orbital velocity V_o of this satellite.

Solution

The orbital velocity V_o can be determined from Equation 5.24 as

$$V_o = \sqrt{\frac{GM_E}{(R_E + h)}} = \sqrt{\frac{6.67\times10^{-11}\times5.97\times10^{24}}{(6380+250)10^3}}$$

$$= 7749.85 \text{ m/s} = 7.75 \text{ km/s}$$

5.4.1 Elliptic Orbit

We have discussed the circular orbit in which a spacecraft can orbit around a planet. But orbits need not to be circular always; rather they can be elliptic. When a spacecraft is orbiting in a circular orbit, if its velocity decreases, it can enter into an elliptic orbit. It may be noted that a planet is located at the focal point of the ellipse (Figure 5.6) while the satellite is orbiting at an orbital velocity of V_o. In a polar coordinate system, an elliptical orbit is described by the instantaneous radius R from the center of the planet, and the major axis a and minor axis b as shown in Figure 5.6. The motion of the satellite around the elliptical orbit is governed by Kepler's laws of planetary motion as follows:

1. *Law of ellipse*: The path of the satellite about the planet is elliptical in shape, with the center of the planet being located at one focus.
2. *Law of equal area*: The imaginary line joining the center of the planet shown in Figure 5.6 to the center of the satellite sweeps out an equal area of space in equal amounts of time.
3. *Law of harmonies*: The ratio of the squares of the periods of any two satellites is equal to the ratio of the cubes of their average distances from the planet.

By using these laws, the orbiting velocity of a satellite on an elliptic orbit in a polar coordinate system (see Figure 5.6) can be obtained:

$$V_o = \sqrt{GM\left(\frac{2}{R} - \frac{1}{a}\right)} \tag{5.26}$$

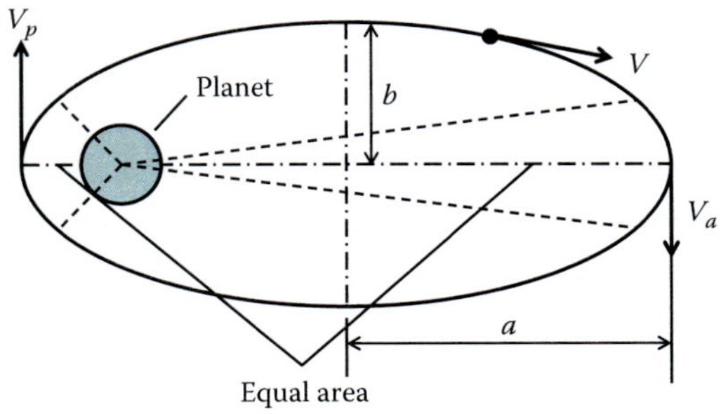

FIGURE 5.6 Flight trajectory around an elliptic orbit.

Note that the orbital velocity of the satellite (V_p) attains maximum value at perigee while it attains minimum value at apogee (V_a). We can define the shape of the ellipse, which is defined as the ratio of the distance between the center of the ellipse and each focus to the length of the semi-major axis. This is known as the eccentricity of the ellipse $e = \sqrt{(a^2 - b^2)}/a$. It is considered as a measure of how much the ellipse deviates from the circular orbit. Note that the circular orbit is a special case of the elliptical orbit, particularly when eccentricity is equal to zero. Note that the product of orbiting velocity and instantaneous radius remains constant for any location on the elliptical orbit ($V_aR_a = V_pR_p = VR$). Hence, the magnitude and direction of the injection velocity will dictate the path of the satellite around a planet.

Several types of orbits are used to place satellites around the earth. If a satellite is placed in an equatorial orbit along the line of the earth's equator, it is called an equatorial orbit (see Figure 5.7a). In order to have an equatorial orbit, a satellite must be launched from a location on the earth closer to the equator. This kind of orbit is used for satellites observing tropical weather patterns, as they can monitor cloud conditions around the earth. When the inclination angle between the equatorial plane and the orbital

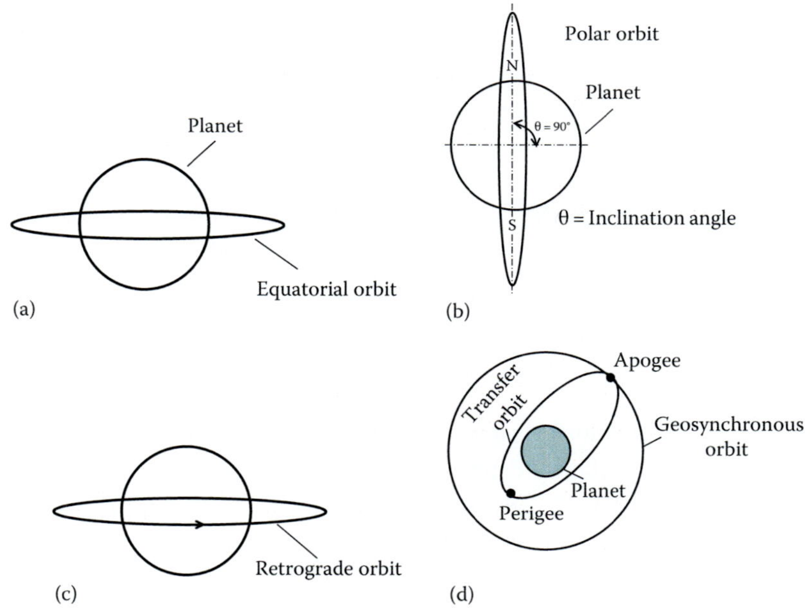

FIGURE 5.7 Types of orbits: (a) equatorial, (b) polar orbit, (c) retrograde orbit, and (d) geosynchronous orbit.

plane becomes 90° as shown in Figure 5.7b, it is known as a polar orbit as it passes through the northern pole to the southern pole. Polar orbits are preferred when mapping of an entire planet is required as the satellite has access to all the points on the surface of the entire planet. Interestingly, if the polar orbit around the earth can be precessed at the same period during which the earth precesses with the solar orbit, it is known as a polar sun-synchronous orbit. As a result, it can provide continuous illumination from the sun at a given point on the surface of the earth when viewed from a polar satellite. If the inclination angle of the orbit happens to be 180° with the equatorial plane, the satellite will be on the equatorial plane but in the opposite direction of rotation of the planet, which is known as a retrograde equatorial orbit (see Figure 5.7c).

Besides this, along the height of the altitude, the geocentric orbit can be classified into three types: low earth orbit (LEO: <2000 km), medium earth orbit (MEO: 2,000–37,786 km), and high earth orbit (HEO: 35,786 km). Note that for the orbit at an altitude of 37,786 km from the earth's surface, the period of the orbit happens to be equal to the period of the earth, as discussed earlier. Hence, this orbit is known as the geosynchronous orbit.

5.4.2 Geosynchronous Earth Orbit

It is important to place the satellite which can rotate around the earth in its equatorial plane at the same angular velocity as the earth's (see Figure 5.7d). As a result, the orbiting satellite will appear to be stationary to an observer on the earth's surface at all times. Hence, this is known as the geosynchronous earth orbit (GEO), which was first proposed by Arthur Clarke in 1945. The radius of the GEO R_G can be obtained by using Equation 5.24:

$$R_G = \sqrt[\frac{1}{3}]{\frac{GM_E}{\omega_E^2}} \tag{5.27}$$

By substituting the values of $\omega_E = 7.27 \times 10^{-5}$ rad/s, $G = 6.67 \times 10^{-11}$ N m²/kg², and $M_E = 5.97 \times 10^{24}$ kg, R_G can be calculated as 42,164 km. Hence, the GEO occurs at an altitude of 35,786 km. Note that synchronous orbits do occur for other planets and their satellites as well [3,4].

5.4.3 Requisite Velocity to Reach an Orbit

We need to place a satellite/space vehicle/space station on a particular orbit around a planet. For this purpose, the requisite velocity must be provided to the vehicle by using a rocket engine such that it can overcome the gravitational pull of the planet. For a particular circular orbit of radius R from the center of a planet with mass M, energy must be imparted to the vehicle mass m so that it can move from the surface of the planet with a radius of R_p to the desired orbit. The energy required to reach the orbit can be determined as follows:

$$E = \int_{R_p}^{R} \frac{GMm}{r^2} dr = GMm\left(\frac{1}{R_p} - \frac{1}{R}\right) \tag{5.28}$$

Besides this, energy must be provided to obtain the orbital velocity V_o. Then, the total energy E_t becomes

$$E_t = GMm\left(\frac{1}{R_p} - \frac{1}{R}\right) + \frac{mV_o^2}{2} = GMm\left(\frac{1}{R_p} - \frac{1}{2R}\right) \tag{5.29}$$

Note that we have used Equation 5.24 for the orbital velocity V_o to obtain Equation 5.29. But note that when $R = R_p + h$, where h is the altitude from the surface of the planet, this expression becomes

$$E_t = \frac{GMm}{2R_p}\left(\frac{R_p + 2h}{R_p + h}\right) \tag{5.30}$$

Considering that the total energy will be available to the vehicle as kinetic energy, we can determine the total velocity required for orbiting a body as follows:

$$V_t = \sqrt{\frac{GM}{R_p}\left(\frac{R_p + 2h}{R_p + h}\right)} \tag{5.31}$$

Note that the total velocity required for orbiting a body increases with altitude compared to the orbiting velocity as shown in Figure 5.8. It may be noted that both the total velocity and orbiting velocity remain almost constant and the same until an altitude of around 100 km. Beyond this, the orbital velocity decreases while the total velocity increases with altitude as shown in Figure 5.8.

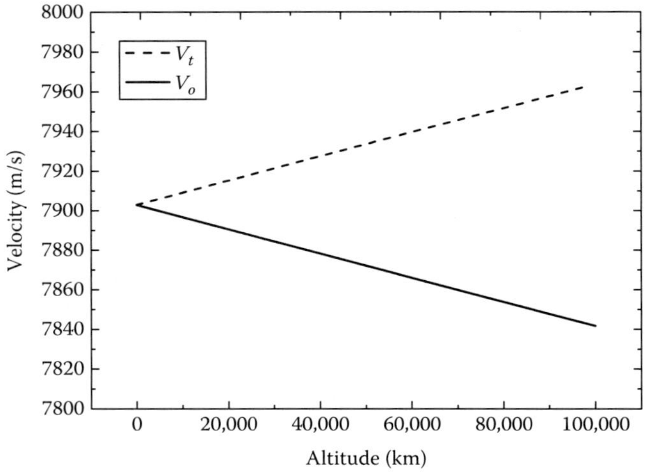

FIGURE 5.8 Variation with altitude of total velocity V_t and orbit velocity V_o around the earth.

5.4.4 Escape Velocity

In order to escape the gravitational pull of a planet, the rocket engine must provide sufficient amount of velocity to the vehicle which is known as escape velocity V_{es}. This escape velocity can be estimated by equating the kinetic energy of the moving body to the work needed to overcome the gravitational force of a planet:

$$\frac{1}{2}mV_{es}^2 = \int_{R_E}^{\infty} \frac{GMm}{r^2} dr \tag{5.32}$$

By integrating and simplifying this equation, we obtain

$$V_{es} = \sqrt{\frac{2GM}{R}} \tag{5.33}$$

Note that the escape velocity decreases with altitude. For the earth, the escape velocity is equal to 11.2 km/s. The escape velocity is equal to $\sqrt{2}$ time the orbital velocity V_o.

5.5 INTERPLANETARY TRANSFER PATH

In modern times, several agencies across the globe conduct interplanetary space flight, which is mostly confined between the planets of the solar system. Of course in the future man will be able to send space vehicles beyond the solar system. For this interplanetary space flight, the vehicle has to overcome the gravitational field of a planet, for example the earth, with the requisite escape velocity, which is known as circumnavigation. Subsequently, the space vehicle has to follow a low-thrust trajectory to enter into an orbit around another planet/satellite. This kind of flight must involve circumnavigation, landing, and return flight if required, like in a lunar mission. Generally, for an ideal transfer path, a simple transfer elliptical path based on minimum energy principle can be used. This is known as Hohmann's transfer orbit as shown in Figure 5.9, because Walter Hohmann, a German engineer, has demonstrated that an elliptical path is the minimum-energy path between two orbits. In this calculation, the pointless mass of two planets is considered. Besides an interplanetary mission, this kind of transfer path can be used for transferring the satellite into a geosynchronous orbit from the low earth orbit (LEO).

For this interplanetary transfer, the requisite velocity increment at perigee and apogee of the elliptical path between two circular orbits must be provided by a higher thrust of the rocket engine. Note that when the

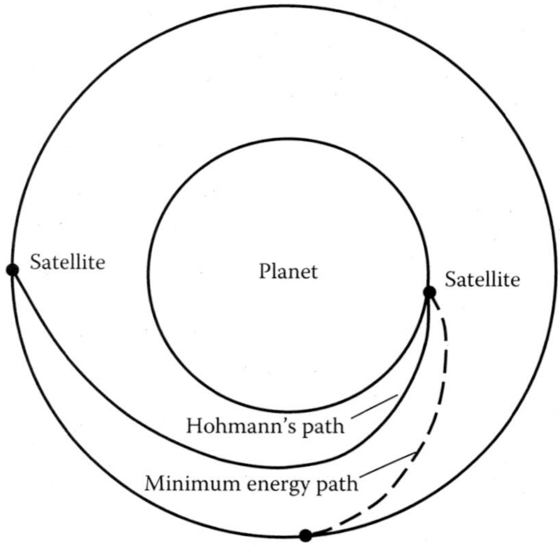

FIGURE 5.9 Schematic of Hohmann's transfer orbit path.

spacecraft arrives at the desired orbit of a target planet, the rocket engine must apply the requisite thrust to inject into the circular orbit with the necessary orbital velocity. This path is dictated by the relative position of the launch planet and target planet. Note that Hohmann's transfer orbit is reversible in nature and can be used to bring back the spacecraft from the target planet to the launching planet orbit easily by firing the rocket engine in the opposite direction [3,4]. Note that the space flight along Hohmann's transfer orbit takes around 259 days from the earth back to the earth. In order to reduce the transfer time, one can opt for a fast orbit by expending more energy as shown in Figure 5.9.

5.6 SINGLE-STAGE ROCKET ENGINES

Let us consider whether a single chemical rocket engine can provide sufficient velocity increment for various aerospace applications. Before initiating a meaningful discussion, we need to consider certain parameters in terms of mass of various components of the rocket engine which will be helpful in describing the performance of both single-stage and multistage rocket motors. As the main objective of the rocket engine is to place the payload, we need to consider the payload mass m_l. In a chemical rocket engine, the major portion of the mass is the propellant mass m_p. The structural mass m_s, which consists of engine structure, supporting structure, tankages, valves, guidance, and control, must be reduced to have higher velocity increment. Then, the total initial mass of the rocket engine m_0 is the sum of all the quantities:

$$m_0 = m_l + m_p + m_s \tag{5.34}$$

When the entire propellant is being burnt during its operation, the burnout mass m_b as defined earlier is equal to the sum of the structural and payload masses:

$$m_b = m_0 - m_p = m_l + m_s \tag{5.35}$$

Let us now define three mass ratios, namely, payload fraction *LF*, structural fraction *SF*, and propellant fraction *PF*, by dividing Equation 5.34 with the total initial mass m_0:

$$\frac{m_l}{m_0} + \frac{m_s}{m_0} + \frac{m_p}{m_0} = LF + SF + PF = 1 \tag{5.36}$$

As mentioned earlier, the mass ratio MR can be defined as follows:

$$MR = \frac{m_0}{m_b} = \frac{m_0}{m_l + m_s} = \frac{1}{LF + SF} \tag{5.37}$$

Similarly, this mass ratio MR can be expressed in terms of propellant fraction PF as follows:

$$MR = \frac{m_0}{m_b} = \frac{1}{1 - PF} \tag{5.38}$$

Let us define another important parameter known as payload coefficient:

$$\lambda = \frac{m_l}{m_0 - m_l} = \frac{m_l}{m_p + m_s} \tag{5.39}$$

It indicates the mass of the payload that can be carried compared to the propellant and structural masses. Unfortunately, this payload coefficient happens to be small for aerospace applications. Hence, engineers strive to obtain a higher payload fraction for meaningful space applications although it is very difficult to achieve in reality. Another important parameter known as structural coefficient ε is defined as the ratio of structural mass to the sum of structural and propellant masses as follows:

$$\varepsilon = \frac{m_s}{m_p + m_s} = \frac{m_b - m_l}{m_o - m_l} \tag{5.40}$$

Note that this expression is true only when the entire propellant is burnt out without any residual unburnt propellant mass. It is desirable to have a smaller value of structural coefficient for space applications because the smaller is its value, the lighter will be the vehicle. In other words, the structural coefficient indicates how far the designer can manage to reduce the structural mass. When the structural coefficient is small and the rocket engine is quite huge, then the total engine mass including its structural mass will be dictated by the initial propellant mass. Hence, under this condition, the structural coefficient can be considered to remain constant, indicating that it would not be dependent on the vehicle size and, in turn, velocity increment. By using the two mass ratios, namely, payload ratio λ and structural coefficient ε, we can express the mass ratio *MR* as follows:

$$MR = \frac{m_0}{m_b} = \frac{m_0}{m_o - m_p} = \frac{1+\lambda}{\varepsilon+\lambda} \tag{5.41}$$

For achieving higher attainable velocity increment as per Equation 5.17, with limited I_{sp}, higher MR must be used. In order to have a higher MR, lower m_b is to be used for a given initial engine mass. Thus, a very careful structural design is essential for a given payload. Generally, the payload capacity can be enhanced further for a given m_p and *MR*, by reducing structural mass (see Equation 5.37). Thus, it is advantageous to reduce the structural fraction SF as much as possible in consistence with the strength requirements of the vehicle. Let us understand further how payload mass can be related to velocity increment by considering Equations 5.17 and 5.37:

$$\Delta V = V_{eq} \ln\left(MR\right) = V_{eq} \ln\left(\frac{1}{LF + SF}\right) \tag{5.42}$$

The payload fraction can be expressed in terms of velocity increment and V_{eq} and *SF*:

$$LF = e^{-\left(\Delta V/V_{eq}\right)} - SF \tag{5.43}$$

Note that the payload fraction LF gets enhanced when higher V_{eq} is used for a lower value of velocity increment and structural fraction SF. In other words, when the velocity increment demand is high for a given value of structural mass fraction SF and V_{eq}, the payload fraction will be small. The variation in LF is plotted in Figure 5.10 with velocity increment ratio $\Delta V/V_{eq}$ for three values of SF. It may be noted that a higher payload fraction can be achieved only when the vehicle has a higher jet velocity and lower SF for a certain demand of velocity increment ΔV. Hence, the designer of a rocket engine has to enhance jet velocity while decreasing the payload fraction SF.

Example 5.2

A single-stage chemical rocket with I_{sp} = 250 is designed to escape with the following mass: m_l = payload mass = 200 kg; m_s = structural

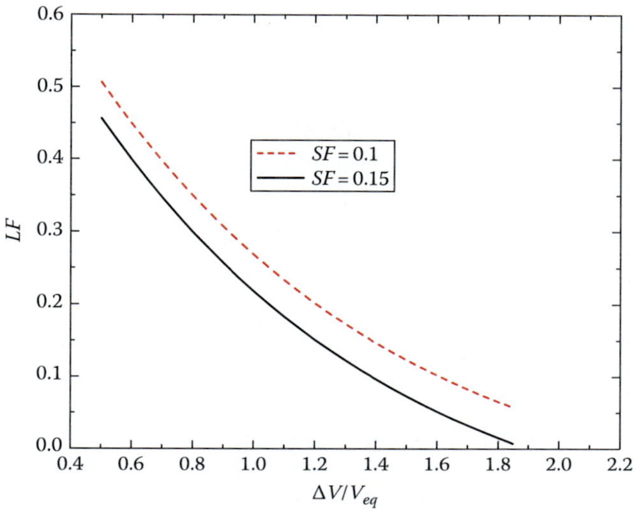

FIGURE 5.10 Variation in LF with velocity increment ratio $\Delta V/V_{eq}$ for two values of SF.

mass = 800 kg; m_0 = total mass = 30,000 kg. Determine the mass ratio, velocity increment, payload, and structural fraction for this rocket engine assuming there are no drag and gravity effects.

Solution

The burntout mass m_b can be determined as

$$m_b = 200 + 800 = 1000\,\text{kg}$$

By using Equation 5.16, the mass ratio MR can be determined as

$$MR = m_0/m_b = 30,000/1,000 = 30$$

Then the propellant fraction can be estimated using Equation 5.16:

$$PF = 1 - 1/MR = 0.97$$

This indicates that only 3% of the total mass can be used for structure and payload mass. In other words, a tiny payload can be carried into space by a monstrous rocket.

The payload fraction LF and structural fraction SF are determined as

$$LF = m_1/\mathrm{m}_0 = 200/30{,}000 = 0.0066$$

$$SF = m_s/m_0 = 800/30{,}000 = 0.0266$$

The velocity increment for an ideal case can be determined by using Equation 5.42:

$$\Delta V = 9.81 \times 250 \ln(30) = 8.34\,\mathrm{km/s}$$

The velocity can be attained ideally without taking into account drag and gravitational pull. In actual sense, it will be quite a low value. In order to escape the earth's gravity, velocity increment should be 11.2 km/s. Hence, it would not be possible to escape the earth's gravity with this rocket engine. One solution would be to use a multistage engine, which is discussed in the following section.

5.7 MULTISTAGE ROCKET ENGINES

We have already learnt from the rocket equation (Equation 5.17) and Example 5.1 that it would not be possible to have enough velocity increment with current chemical rocket engine technology (limited I_{sp}) for a rocket engine to place a satellite with a certain payload even if we neglect gravity and drag losses because the structural mass cannot be reduced beyond a certain minimum value. For example, the escape velocity required to overcome the earth's gravitational pull is around 11.2 km/s. For a rocket engine with a specific impulse of 350, this velocity can be achieved with a mass ratio of 24 and propellant fraction of 0.96 even without taking into account gravitational and drag losses, which are inevitable during its flight in the earth's atmosphere. In other words, 4% of the total weight will be used for payload, structural mass, guidance, and control, which is quite difficult to achieve. Besides this, we need to provide additional velocity increment for accommodating gravitational and drag losses apart from the velocity increment required for orbit injection, which is required for satellite application. In order to overcome this problem, we can use a multistage rocket engine as suggested by the Russian scientist K.E. Tsiolkovsky in 1924.

5.7.1 Multistaging

Instead of a single large rocket, a series of rocket motors each with its own structure, tanks, and engines are used to enhance the velocity increment for the entire vehicle for the same I_{sp}. As the propellant is consumed in each stage, its tank is dropped from the vehicle, at intervals. Thus, the propellant is not wasted in accelerating the unnecessary structural masses to attain a higher velocity. As a result, a launch vehicle can achieve higher velocity by using staging. Several kinds of staging have evolved over the years as shown in Figure 5.11 which will be discussed later. Now let us consider a tandem type shown in Figure 5.11a in which all the stages are placed in series on one another in the order of size. Generally, the first stage needs to impart higher thrust and total impulse as it is the largest mass to carry during launching operation. Hence, it is also known as the booster stage. Lower thrusts are to be provided in subsequent stages. Note that the payload for the first stage would be equal to the total mass of all upper stages along with the actual payload to be placed in the orbit. The second stage

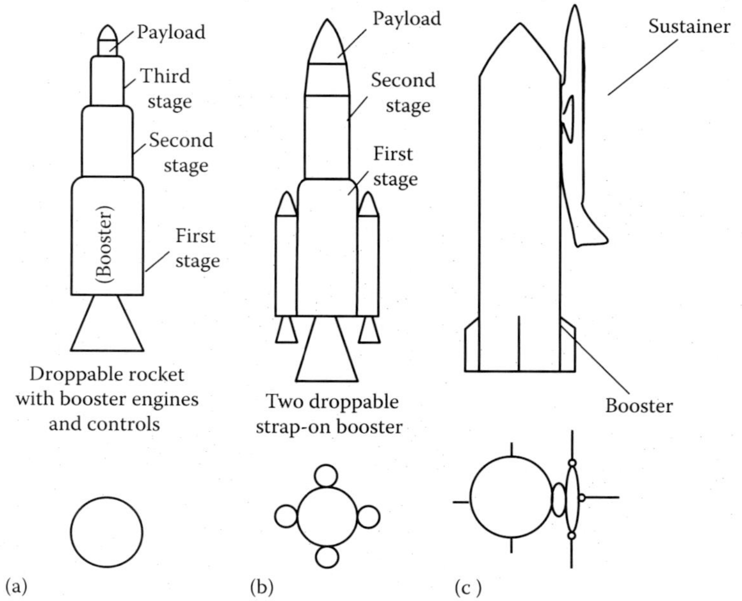

FIGURE 5.11 Three types of multistaging: (a) tandem, (b) parallel, and (c) piggyback.

gets started when its velocity is almost equal to the velocity increment provided by the first stage and releases the structural mass of the first stage, thus enhancing the mass ratio for getting higher velocity increment. If the velocity increment ΔV_1 is contributed by the first stage and ΔV_2 is contributed by the second stage, then the total velocity increment at the end of the second stage of the operation would be equal to $\Delta V_1 + \Delta V_2$. We can generalize the total velocity increment for n stages of a rocket engine as follows:

$$\Delta V_n = \Delta V_1 + \Delta V_2 + \cdots + \Delta V_i = \sum_{i=1}^{n} \Delta V_i \tag{5.44}$$

Let us consider n rocket stages without considering the drag and gravity effects. The net ideal velocity increment for n rocket stages from Equation 5.44 becomes

$$\Delta V_n = \sum_{i=1}^{n} \Delta V_i = V_{e1} \ln\left(MR_1\right) + V_{e2} \ln\left(MR_2\right) + \cdots + V_{en} \ln\left(MR_n\right) \tag{5.45}$$

By assuming the exit velocity of all stages to be the same, Equation 5.45 becomes

$$\Delta V_n = \sum_{i=1}^{n} \Delta V_i = V_{e1} \ln\left(MR_1 \cdot MR_2 \cdots MR_n\right) \tag{5.46}$$

By noting that the initial mass of the successive stage is equal to the burnt-out mass of the previous stage, as discussed earlier, Equation 5.46 becomes

$$\Delta V_n = \sum_{i=1}^{n} \Delta V_i = V_{e1} \ln\left(\frac{m_{o,1}}{m_{b,1}} \cdot \frac{m_{o,2}}{m_{b,2}} \cdots \frac{m_{o,n}}{m_{b,n}}\right) = V_{e1} \ln\left(\frac{m_{o,1}}{m_{b,n}}\right) \tag{5.47}$$

Note that $m_{o,1}/m_{b,n}$ is the ratio of the initial mass to final mass of the nth stage which represents the overall mass ratio of a multistage rocket engine. If the mass ratio of each stage is the same, Equation 5.46 becomes

$$\Delta V_n = \sum_{i=1}^{n} \Delta V_i = V_{e1} \ln\left(MR\right)^n = nV_{e1} \ln\left(MR\right) \tag{5.48}$$

This equation can be considered in terms of payload fraction and structural fraction by using Equation 5.37:

$$\Delta V_n = \sum_{i=1}^{n} \Delta V_i = nV_{e1} \ln\left(MR_i\right) = nV_{e1} \ln\left(\frac{1}{LF_i + SF_i}\right) \tag{5.49}$$

where SF_i and LF_i are the structure and payload fractions of the ith stage, respectively. The overall payload fraction LF_n is defined as the ratio of the final payload of the nth stage to the initial total mass as follows:

$$LF_n = \frac{m_{Ln}}{m_{01}} = \frac{m_{Ln}}{m_{0n}} \frac{m_{on}}{m_{0n-1}} \cdots \frac{m_{L2}}{m_{01}} = LF_i^n; \quad LF_i = \left(LF_n\right)^{\frac{1}{n}} \tag{5.50}$$

Note that for all stages, payload fraction and structural fraction remain almost the same.

$$\Delta V_n = \sum_{i=1}^{n} \Delta V_i = nV_{e1} \ln\left(MR_i\right) = nV_{e1} \ln\left(\frac{1}{\left(LF_n\right)^{\frac{1}{n}} + SF_i}\right) \tag{5.51}$$

The variation of change in velocity with exit velocity is plotted in Figure 5.12 for structural fraction for each stage and payload fraction for all stages.

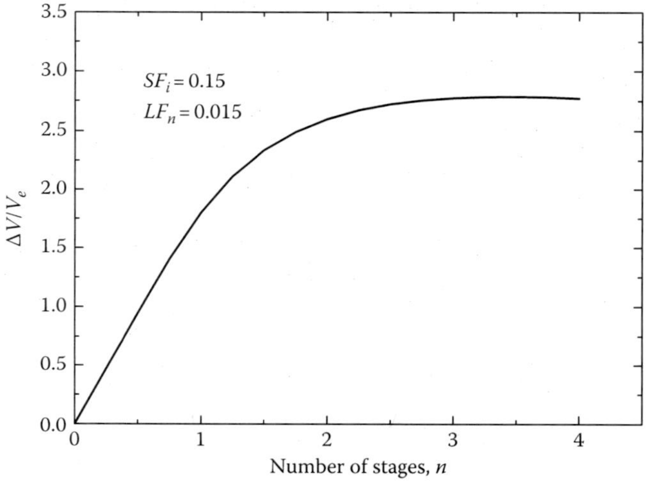

FIGURE 5.12 Variation of velocity increment with number staging.

It may be noted that for two stages, velocity increment increases by 50% compared to a single stage. As a result, the initial takeoff mass for the same payload decreases considerably. But with subsequent increase in stages, the velocity increment increases asymptotically. In other words, the gain in initial mass diminishes gradually with an increase in the number of stages. Hence, it is not prudent to go beyond four stages as the gain in initial mass for lifting the same payload would be substantial to compensate the enhanced complexities in the operation with an increase in number of stages. That is the reason why three stages are popular across the globe. For example, the space shuttle, GSLV, Titan V, Delta, and so on are some of the space vehicles which have adopted three stages of rocket engine.

Apart from the tandem type of multistaging, several other types have been designed and developed for space launch vehicles. Three of the most popular multistaging types are (1) tandem, (2) parallel, and (3) piggyback, which are depicted schematically in Figure 5.11. The first one is the tandem or series multistaging (see Figure 5.11a) in which stages are placed vertically on top of each other from larger mass to smaller mass as described earlier. This is the most commonly used method of multistaging as it is quite simple and very effective in achieving higher velocity. However, the ejection timing of a used stage and ignition timing of the next stage are critical for its successful operation. In a partial multistaging system, three or four strap-on booster rockets are used along with vertical rocket configurations which are dropped during the flight. Nowadays, both tandem and parallel stages are adopted together in the first stage of a space vehicle as this provides extra initial thrust to overcome the initial gravity losses and atmospheric drag. Of course, strap-on boosters are jettisoned to reduce the weight. The piggyback configuration has been used in the U.S. space shuttle in which two large solid boosters and expendable external propellant tanks are used to feed the shuttle's onboard main. The orbiter (aircraft) is carried as piggyback to make this vehicle reusable.

REVIEW QUESTIONS

1. What are the major forces acting on a rocket engine during its flight in the earth's atmosphere?

2. Derive a rocket equation mentioning assumptions made by you. What are the terms which play an important role during its flight in the earth's atmosphere?

3. What is the mass ratio for a class-one truck with a gross vehicle weight of 2225 kg which can carry 8 tons of materials?
4. What is meant by burnout distance? Derive an expression for it.
5. What is a sounding rocket? Derive a relationship for coasting height when a sounding rocket is fired.
6. How is the flight trajectory of a rocket estimated during its flight in the earth's atmosphere? Devise a procedure for it.
7. What is meant by escape velocity? What will be the escape velocity required to leave the moon?
8. Derive an expression for the time period required to revolve around a circular orbit.
9. Derive an expression for the total energy required to place a satellite on an orbit around the earth.
10. What is meant by a geosynchronous orbit? Derive an expression for the radius of a geosynchronous orbit.
11. How is an elliptical orbit different from a circular orbit? Is the elliptical orbit preferred over a circular orbit? Explain why.
12. What is meant by Hohmann's transfer orbit? How is it different from Obert's transfer orbit?
13. Derive an expression for the orbit velocity for an elliptical orbit.
14. What are the kinds of orbits one can use for launching a satellite around the earth?
15. Define payload fraction. How is it different from payload coefficient?
16. What is meant by structural coefficient? Derive an expression to show how the structural and payload coefficients can be related to mass fraction MR.
17. What is meant by multistaging? What are the advantages of multistaging over a single-stage rocket engine?
18. What is the optimum number of stages that can be used for a space vehicle? Justify your answer by deriving a mathematical expression for it.

19. What are the common types of multistaging used for launching space vehicles?

20. Discuss them by mentioning their pros and cons.

PROBLEMS

5.1 Determine the velocity and period of revolution of an artificial satellite orbiting the earth in a circular orbit at an altitude of 200 km above the earth's surface. Take the mass and radius of the earth as 5.97×10^{24} kg and 6380 km, respectively.

5.2 An artificial earth satellite is in an elliptical orbit which brings it in to an altitude of 200 km at perigee and out to an altitude of 650 km at apogee. Determine the velocity of the satellite at both perigee and apogee.

5.3 A satellite in the earth's orbit with an eccentricity of 0.055 has a semi-major axis of 3500 km. Determine the semi-minor axis and the altitude of this satellite at apogee if R_p is 7930 km.

5.4 A rocket engine with an initial mass of 5600 kg at takeoff is operated for 100 s with an effective jet velocity of 2100 m/s from the nozzle exit. If the maximum velocity of this rocket engine is equal to 1.2 times the effective jet velocity, determine the propellant consumption rate and the distance traveled by it.

5.5 A single-stage rocket engine during its vertical flight can withstand maximum acceleration of 10*g*. It can produce a thrust of 110 kN with an exhaust jet velocity of 2500 m/s for 75 s. Determine the maximum velocity attained and the initial mass of the vehicle neglecting the drag effects.

5.6 A 6000 kg spacecraft is in the earth's orbit traveling at a velocity of 7990 m/s. Its engine is burned to accelerate it to a velocity of 12,300 m/s for placing it on an escape trajectory. The engine expels mass at a rate of 12 kg/s and an effective exhaust velocity of 3100 m/s. Calculate the duration of the burn.

5.7 A single-stage rocket engine with a mass of 12,000 kg and specific impulse of 290 s during its vertical flight consumes 9,500 kg of propellant at a rate of 105 kg/s. Determine the acceleration at liftoff and 15 s after liftoff neglecting gravity and drag effects.

5.8 A sounding rocket with a payload of 250 kg and I_{sp} of 180 s is to be launched vertically from the earth's surface. It should not suffer acceleration greater than 6*g* during the burning period. The maximum propellant mass is 1500 kg with a structural coefficient of 0.12. Determine (1) the minimum allowable burning time and (2) the maximum height attainable. Derive the expression for burning time for maximum acceleration and maximum height neglecting drag forces.

5.9 The specific impulses of the first and second stages of a two-stage rocket engine are 300 and 350 s, respectively. The mass of each stage is as follows:

First-stage propellant mass = 150,000 kg

First-stage dry mass = 9,000 kg

Second-stage propellant mass = 22,000 kg

Second-stage dry mass = 2,300 kg

Payload mass = 2,200 kg.

Determine the total ΔV gained by this rocket engine. Assume negligible drag and gravitational effects on the vehicle.

5.10 A single-stage sounding rocket with I_{sp} of 225 s and constant thrust of 7.2 kN for 17 s is to be launched vertically from the earth's surface and can withstand maximum acceleration of 8*g*. If the empty mass is 21 kg, what is the maximum payload it can carry?

5.11 A three-stage rocket is to be designed to place 50 kg satellite in the low earth orbit of 400 km. The following data are to be used to determine (1) structure mass fraction of each stage, (2) payload mass fraction, and (3) total velocity increment.

Stage	m_p	m_s	V_j
First	10,000	1500	2300
Second	580	580	2500
Third	100	128	2900

5.12 A three-stage rocket is to be designed to place 750 kg satellite in low earth orbit of 400 km. The booster (first) stage is ignited along with four strap-on motors on the ground. The following data are to be used to determine (1) payload mass fraction and (2) total velocity increment.

Stage	m_p (kg) Propellant Mass	m_s (kg) Structure Mass	V_j (m/s) Jet Velocity
First	19,900	3500	2650
Second	7,800	2050	2820
Third	3,800	1000	3050

If the first stage is fired for 45 s, determine the rate of acceleration during takeoff, assuming the propellant mass flow rate to be constant.

REFERENCES AND SUGGESTED READINGS

1. Anderson, J.D., *Introduction to Flight*, 7th edn., McGraw-Hill Higher Education, New York, 2011.
2. Francis, J.H., *Introduction to Space Flight*, 1st edn., Prentice Hall, Englewood Cliffs, NJ, 1999.
3. Sutton, G.P. and Biblarz, O., *Rocket Propulsion Elements*, 8th edn., John Wiley & Sons, New York, 2011.
4. Hill, P.G. and Peterson, C.R., *Mechanics and Thermodynamics of Propulsion*, 2nd edn., 1992.

CHAPTER 6

Chemical Rocket Propellants

Rocket technology must not be used for mass destruction, rather for peaceful exploration.

D.P. MISHRA

6.1 INTRODUCTION

A propellant consists of all the chemical materials, including fuel and oxidizer, along with certain additives necessary for sustaining the combustion process to produce high-pressure hot gases, that which are expanded in a nozzle to produce thrust. Note that the principal ingredients of a propellant are the fuel and the oxidizer. Generally, a fuel is a chemical substance that reacts with an oxidizer while releasing thermal energy. Chemically, fuel can be defined as the chemical substance that can donate electrons during chemical reaction with oxidizer. In contrast, an oxidizer is that chemical substance that accepts electrons during chemical reaction with the fuel. The property of an atom by virtue of which it can either accept or donate electrons is known as electronegativity, which distinguishes a fuel and an oxidizer. Additives are used to enhance the properties of the propellant. We will see more about additives while discussing solid and liquid propellants in subsequent sections. Several kinds of solid, liquid, and hybrid chemical propellants have been devised over the years [2,5], which will be discussed in this chapter.

6.2 CLASSIFICATION OF CHEMICAL PROPELLANTS

Chemical propellants can be classified according to their respective physical states, namely, liquid, solid, gel, or hybrid, as shown in Figure 6.1. Generally, in solid propellants, fuel and oxidizer are stored and mixed together in a solid form. Note that a solid propellant needs to have a minimum ignition temperature for ignition to occur. Solid propellants can be broadly divided into two categories, namely, homogeneous and heterogeneous, based on the distribution of fuel and oxidizer in the propellant, as shown in Figure 6.1. In the case of homogeneous propellants, fuel and oxidizer are contained in the same molecule of the propellant. In the case of heterogeneous propellants, solid fuel and oxidizer retain their respective physical identities. Homogeneous propellants are further classified into two categories: (1) single base and (2) double base. Generally, a single-base propellant is a single compound that contains both fuel and oxidizer. An example of a single-base propellant is nitrocellulose, which has both oxidative and reduction capabilities, whereas, double-base propellants consist of two components. The most widely used double-base propellants are made of nitrocellulose (NC) and nitroglycerine (NG), of course, with a few additives. Heterogeneous propellants are also known as composite propellants, which consist of a fuel and oxidizer mixture. Generally, mineral salt is used as an oxidizer, while polymer and metal powders are used as fuel. For example, ammonium percolate (NH_4ClO_4) as an oxidizer and polybutadienes with metal powders as fuel are commonly used as composite propellant in solid-propellant rocket engines. Liquid propellants have several desirable characteristics over solid propellants and hence are used particularly when higher specific impulse is required. Based on the

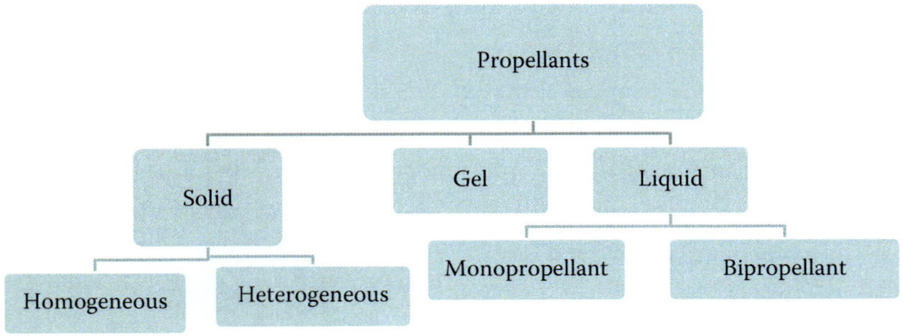

FIGURE 6.1 Classification of chemical propellants.

mode of ignition, liquid propellants can be classified into two categories: (1) hypergolic and (2) nonhypergolic. In hypergolic propellants, liquid fuel and oxidizer react spontaneously without external ignition energy. For example, hydrazine can be considered as a hypergolic propellant as it undergoes exothermic reactions and gets decomposed into water and oxygen spontaneously while coming in contact with suitable catalyst even in ambient temperature. But in the case of nonhypergolic propellants, suitable amount of ignition energy is provided to ignite the liquid fuel and oxidizer for combustion to take place. Both hypergolic and nonhypergolic propellants can be divided further into two categories, namely, monopropellants and bipropellants, as shown in Figure 6.1. Under the nonhypergolic propellant category, monopropellants can be further classified into two categories: (1) homogeneous and (2) composite, as shown in Figure 6.1. In the case of nonhypergolic monopropellants, fuel and oxidizer are contained in the same molecule. Methyl nitrate (CH_3NO_3) is an example of a simple nonhypergolic monopropellant. But in the case of composite nonhypergolic monopropellants, a mixture of fuel and oxidizer is used. Amyl acetate ($C_7H_{14}O_2$) and nitric acid (HNO_3) can be considered as composite nonhypergolic monopropellant. In order to enhance specific impulse with higher metal loading and overcome the problems of sloshing during flight of the liquid-propellant rocket engine, liquid fuel is converted into gel. Generally, gel propellant acts like solid and becomes liquid when subjected to a higher shear rate. A combination of solid and liquid is termed as hybrid propellant, which is being used in ram rocket engine in recent times.

6.3 GENERAL CHARACTERISTICS OF PROPELLANTS

All types of chemical propellants based on physical entity have their own desirable characteristics for high-level performance. Each propellant must have higher energetic properties that ensure higher heat release rate, accompanied by high combustion temperature, characteristic velocity, and specific impulse. Besides this, the propellant must have high ballistic properties, namely, low density, higher ignitability, performance reproducibility, and least combustion instability [1,4]. Using low-density propellants means that larger storage tanks will be required, thus increasing the mass of the launch vehicle. Other properties, namely, storage stability, less prone to explosion hazard, smokeless exhaust, low price, and easy processing are some of the desirable characteristics of propellants. The common desirable properties of chemical propellants are as follows [2,5]:

1. Propellant must have high chemical energy release so that it can have higher combustion temperature leading to high characteristic velocity C^*.
2. It can have low molecular weight of combustion product leading to high exhaust velocity V_e and thus can have high specific impulse I_{sp}.
3. It can have a high density such that large amount of chemical energy can be stored in the smallest volume and thus can have a compact design.
4. Easy to ignite even under low-pressure condition.
5. Physically and chemically stable with respect to time.
6. Smoke-free and nontoxic in nature.
7. Easy and expensive to manufacture and handle during operation.
8. Easily available and low price.
9. Less prone to explosion hazard.
10. Low emission level.

6.4 SOLID PROPELLANTS

Any solid propellant usually consists of fuel, oxidizer, and additives. Fuel and oxidizer are principal ingredients. Additives are used in very low percentage to enhance the burning rate, control fabrication process, minimize temperature sensitivity, to ensure chemical/physical stability during storage, to increase mechanical properties, and so on. Generally, solid propellants are designed for specific applications, namely, sounding rocket, missile, launch vehicle, gas generator, and so on. The desirable properties of solid propellants are enumerated as follows [2,5]:

1. Solid propellant must have higher heating value to have higher combustion temperature, leading to high characteristic velocity C^*.
2. Propellant combustion products should have low molecular weight to have high exhaust velocity V_e, leading to high specific impulse I_{sp}.
3. It should have a high density such that large amount of chemical energy can be stored in the smallest volume and thus can have a compact design.

4. Easy to ignite even under low pressure condition. But it must not ignite due to shock or pressure pulses.
5. Its constituents should be easy to handle.
6. Constituents of propellants must be locally available and cost-effective.
7. Processing of propellant should be simple and reproducible in nature.
8. Properties of propellant grain must not be physically and chemically unstable during storage and transport.
9. Propellant grain must be inexpensive to manufacture and easy to handle during operation.
10. Solid propellant grain must not react with atmospheric air/moisture.
11. Propellant grain must have high mechanical strength.
12. It must be smoke-free and nontoxic in nature.
13. It must be less prone to explosion hazard.

As discussed, solid propellants fall mainly into two categories: (1) homogeneous and (2) heterogeneous propellants, depending upon the physical entities of fuel and oxidizer in the propellant. Various types of homogeneous and heterogeneous solid propellants are discussed in the following.

6.4.1 Homogeneous Solid Propellants

In homogeneous propellants (also known as colloidal propellants), the oxidizing and reducing groups are present in the same molecule. They are thermoplastic in nature and are generally smokeless. The homogeneous propellants are mainly classified as *single base* and *double base* (see Figure 6.1). Single-base propellants, which contain only NC, are mainly used as gun propellants. NC $[(C_6H_7O_2–(ONO_2)_3]_n$ in gelatinized form constitutes the main ingredient in these propellants. NC is basically cellulose nitrate, which is produced by nitrating the cellulose materials $[(C_6H_7O_2(OH)_3)_n]$ using a nitrating agent (e.g., nitric acid) as per the following chemical reaction:

$$[C_6H_7O_2(OH)_3]_n + xHNO_3 \rightarrow (C_6H_7O_2)_n(OH)_{3n-x}(ONO_2)_x + xH_2O \tag{6.1}$$

The cellulose material mainly consists of a linear chain of carbon, hydrogen, and oxygen. During this process of nitration, certain portion of hydroxyl radical (OH) in the cellulose is replaced with nitrate radical ONO_2. Note that the energy content of NG is dependent on the extent of nitration and kind of cellulose materials. In earlier days, these were also known as flash paper, flash cotton and guncotton, and flash string, being flammable in nature. By nitrating cellulose, the extent of nitrogen content in NC can be enhanced to a maximum of 14.4% by mass. But in the case of propellant application, nitrogen content in NC varies generally from 11.0% to 13.3% by mass. Note that the heat of formation of NC decreases with increase in nitrogen content [3]. During autocatalytic reaction, NC decomposes into aldehydes (HCHO and CH_3CHO) and NO_2 gas due to breaking of weakest bond $O–NO_2$. This decomposing reaction can be considered to be first order between 363 and 448 K.

Double-base propellants are made by plasticizing NC with an energetic compound like NG. NG is formed from glycerin/polyol (alcohol containing multiple hydroxyl group; e.g., propane triol; $C_3H_5(OH)_3$) by replacing OH with nitrate radical ONO_2. The chemical structure of NG is shown in the following:

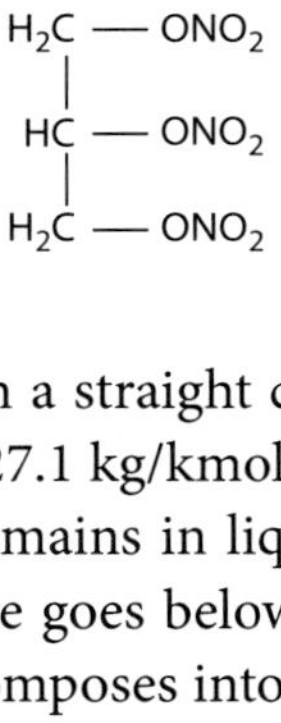

(6.2)

Note that it is an aliphatic with a straight chain structure that has a relatively low molecular mass of 227.1 kg/kmol as compared to NC. Note that it is slightly oxidizer-rich. It remains in liquid state at room temperature but solidifies when temperature goes below 286. NG at 418 K undergoes autocatalytic reaction and decomposes into aldehydes and NO_2 gas due to breaking of weakest bond $O–NO_2$. It has an activation energy of 109 kJ/kmol. It can undergo self-ignition at 491 K at certain critical concentration of NO_2. Note that both NG and NC contain both fuel and oxidizer together in their molecular structure and can be used as single-base propellants in the rocket engine. However, both are commonly used together in the rocket engine as double-base propellant. Generally, NC in the form of fine nodules or stripes are mixed in a sigma blade mixer along with the NG liquid and other ingredients to form a stiff dough at temperature around 60°C. Subsequently, this dough can be extruded through a suitable die of an extrusion press to form desired shaped propellant grain. Besides this, double-base propellant grain can be produced using casting method. Generally, extruded propellant grain has higher mechanical strength as compared to casted propellant grain. Depending upon the process of manufacture, DB propellants are classified as extruded double base (EDB) and cast double base (CDB). These propellants have burning rates in the range of 5–20 mm/s and have a heat of combustion around 3800–5000 kJ/kg. Specific impulse of the order of 230 s can be obtained from unmodified DB propellants. This double-base propellant is also known as colloidal/mixed propellant. Its trade name is cordite.

Let us consider a typical JPN ballistite double-base propellant with its constituents along with its respective functions as shown in Table 6.1. Note that ballistite is basically a smokeless double-base propellant made of two main explosives, namely, nitroglycerine (NG) and nitrocellulose (NC). It was first devised and patented by Alfred Nobel in 1888. The diethyl phthalate ($C_{12}H_{14}O_4$) is used as a nonexplosive plasticizer to improve mechanical properties of the propellant grain. The ethyl centralite ($C_{17}H_{20}N_2O$) acts as a stabilizer to counteract the autocatalytic decomposer of major constituents. The potassium sulfate (K_2SO_4) ensures smooth burning at low temperature to avoid combustion instabilities. The carbon is added to the transparent propellant to avert transmission of radiant energy, which may cause self-ignition around internal parts of propellant grain. The candelilla wax is used in this propellant as a lubricant for extrusion die that facilitates extraction process for maintaining accurate shape of grain. Besides this, metal powders in some cases are added to enhance higher performance of

TABLE 6.1 Typical DB Propellant JPN (Ballistic)

Material	%Weight	Purpose
Nitrocellulose (NC) (13.25%N)	51.22	Polymer (fuel)
Nitroglycerine (NG)	43	Explosive plasticizer (oxidizer)
Diethyl phthalate ($C_{12}H_{14}O_4$)	3.25	Nonexplosive plasticizer
Ethyl centralite ($C_{17}H_{20}N_2O$)	1.0	Stabilizer
Potassium sulfate (K_2SO_4)	1.25	Flash suppressing agent
Carbon black	0.2	Opacifier
Candelilla wax	0.08	Lubricant for extrusion process

Sources: Barrere, M. et al., *Rocket Propulsion*, Elsevier Publishing Company, New York, 1960; Kubota, N., *Propellants and Explosives, Thermochemical Aspects of Combustion*, Wiley-VCH, Weinheim, Germany, 2002.

propellant. In certain cases, NG in double-base propellants is substituted with other constituents such as di/tri-nitrodiethyleneglycol.

A third variety of homogeneous propellant is called *triple-base propellant*. This contains an organic compound known as nitroguanidine (NQ: $(NH_2)_2CNNO_2$) in addition to NC and NG. The advantage of triple base propellants is that they have significantly more energy than single-base propellants while their combustion temperatures still lie below 3000°C.

6.4.2 Heterogeneous Propellants

A heterogeneous propellant is one in which solid crystalline oxidizer (e.g., ammonium perchlorate NH_4ClO_4) and organic fuel, and metallic fuel powders are held together in a plastic (rubber) matrix. Generally, an organic plastic binder (fuel) surrounds the fine crystalline oxidizer particle. Note that this organic fuel acts as the main fuel component. Its main function is to produce heat while undergoing overall exothermic chemical reaction. Generally, this fuel is known as the binder because it binds the metal powders, solid crystalline and other ingredients in the propellant grain. As this kind of propellant is heterogeneous in nature, it is also termed as composite propellants (CP). Let us consider a typical composite propellant containing between 70% AP, 17% Al powder, and 12% elastomeric binder. Note that plasticizers, stabilizers, curing agent, bonding agents, burn rate catalysts, combustion instability suppressant, and antioxidants are used in small percentage (%) in CP. Typical ingredients of CPs are given in Table 6.1. Certain combinations of these ingredients can be provided in the propellant for optimized performance for either rocket engine or explosive applications. The selection of ingredients will be dependent on both optimized characteristics with desired mechanical properties of propellant grain. Note that CPs have a wide range of burning rates (7–20 mm/s) and densities (1700–1800 kg/m^3). The composite propellants can have higher specific impulse in the range of 250–300 s as compared to double-base propellants. The commonly used fuel (binder) and oxidizers in solid propellants are discussed briefly, as follows.

6.4.2.1 Solid Fuel (Binder)

Polymers, plastics, rubber, PVC are some of the binders commonly used for CP, as shown in Table 6.2. The mechanical properties of CP are dependent on the type of binder and its constituents. For selecting proper fuel, the following properties, namely, high heat of combustion, high combustion temperature, mechanical properties such as strength and elasticity, thermal properties, smooth burning characteristics, and good aging

TABLE 6.2 List of Typical Ingredients Used for Heterogeneous Propellants

Fuel (Binder)	Oxidizer	Plasticizer
PU: Polyurethane	*AP*: Ammonium perchlorate	*DOP*: Dioctyle phthalate
PVC: Polyvinyl chloride	*AN*: Ammonium nitrate	*DOA*: Dioctyl adipate
PBAN: Poly Butyl Acrylo Nitrate	*KP*: Potassium perchlorate	*IDP*: Isodecyl pelargonete
PS: Polysulfide	*NP*: Nitronium perchlorate	*Curing agent*
HTPB: Hydroxyl terminated polybutadiene	*ADN*: Ammonium dinitramide	*TDI*: Toluene-2,4-Di-Isocyanate
CTPB: Carboxyl terminated polybutadiene	*RDX*: Cyclotrimethylene trinitramine	*MAPO*: Tris(1-(2-methyl) Aziridinyl)phosphine oxide
Metal fuel: Aluminum, Magnesium, Beryllium, Boron	*HMX*: Cyclotetramethylene tetranitramine	*IPDI*: Iso-phorone di-isocyanate

characteristics are to be considered. It is preferable to have the binder in liquid state at ambient temperature and pressure, because it helps in the mixing of oxidizer and solid fuel. This mixing process is followed by a curing process, during which binder and fuel mixture turns into solid CP.

Generally, organic prepolymers or low-molecular-weight polymers containing fuel elements like carbon, hydrogen, and oxygen are used as binders in composite propellant. The binders should have workable viscosity, high heat of formation, low glass transition temperature, good stability, and compatibility with other ingredients of the formulation. Binders can be generally classified as thermoplastics and thermosetting. Some of the thermoplastics such as asphalt binders, polyvinyl chloride (PVC), and polystyrene are linear polymers. These were used in earlier propellant formulations. Unfortunately, the propellants made of these binders exhibit poor mechanical properties. In contrast, thermosetting binders are chemically cross-linked during the curing process. It is transformed into a tough and insoluble solid matrix that can withstand large thermal and mechanical stresses, even in large solid booster rocket engines. Some examples of thermosetting binders are polysulfide, polybutadienes with different functional groups such as carboxyl terminated polybutadiene (CTPB) and hydroxyl terminated polybutadiene (HTPB), polybutadiene–acrylic-acid (PBAA), and polybutadiene–acrylic acid–acrylonitrile (PBAN). Note that polybutadiene is basically a linear chain structure polymer. This polybutadiene consists of alkadienes with four carbon atoms and two double bonds which can be chemically represented as

$$-(CH_2{=}CH-CH{=}CH_2)_n- \tag{6.3}$$

where n is the number of butadiene groups. Note that when the polybutadiene chain attaches to poly–acrylo–nitrile group and acrylic acid groups, then polybutadiene–acrylic acid–acrylonitrile (PBAN) is formed. Its typical structure is given in the following:

$$-(CH_2=CH-CH=CH_2)_n-(\underset{\displaystyle COOH}{CH_2-\overset{|}{CH}})_x-(\underset{\displaystyle CN}{CH_2-\overset{|}{CH}})_y- \tag{6.4}$$

In order to enhance the mechanical properties of binder, randomness of the cross-linking is decreased by relocating the carboxyl group at the end of the polybutadiene chain ends shown as follows:

$$-(CH_2=CH-CH=CH_2)_{nl}-(\underset{\displaystyle COOH}{CH_2-\overset{|}{CH}})_{xl}-(CH_2=CH-CH=CH_2)_{n2}- \tag{6.5}$$

This binder is known as the carboxyl terminated polybutadiene (CTPB). When the carboxyl group at the end of the polybutadiene chain is replaced by hydroxyl (OH) radical, it is termed as hydroxyl terminated polybutadiene (HTPB). Its performance gets improved as compared to the PBAN and CTPB as OH is more reactive as compared to COOH. Hence among all polymers, HTPB is currently considered superior because of its outstanding processing characteristics and better mechanical properties. Some of the properties of HTPB made by ISRO are as follows: molecular weight: 2300–2900 kg/mol, hydroxyl value: 40–50 mg KOH/g, functionality: 1.8–2.5, viscosity: 40–65 cps at 30°C, specific gravity: 0.90–0.92, and T_g: −80°C. Recently, a high energetic binder like glycidyle azide polymer (GAP) is being developed. GAP provides higher performance as it contains more number of hydrogen atoms. Besides these polymers, metal powders, namely, aluminum, magnesium, boron, and beryllium, are used as fuel in CP to enhance energy content of propellant. These metal powders must be sufficiently fine (particle size must be 0.1–10 μm) so that complete combustion can be ensured. Aluminum powder being lighter is preferred over other metals. But metal oxides are formed during combustion, which have higher negative heat of formation. In recent times, metal hydrides are being considered over metal powders as more amount of hydrogen is released during combustion of metal hydrides. Note that the use of the metal powders decreases the mechanical properties of the propellant. This calls for higher proportion of binder to maintain same level of mechanical strength.

6.4.2.2 Solid Oxidizer

In CP, oxidizer plays a very important role in releasing heat during oxidization process and maintaining the grain's structural stability. In some cases, it accounts for more than 70% by weight of the total propellant weight. Ideally, the oxidizer should have high oxygen content and should be compatible with the binder and other ingredients in the propellant formulation. Some of the commonly used oxidizers are nitrates and perchlorates of ammonium, potassium, and so on, as shown in Table 6.3. For selecting a proper oxidizer, the following properties, namely, high heat of formation, high amount of available oxygen, high density, slightly hygroscopic, and smooth burning characteristics are to be considered. The most commonly used solid oxidizers are ammonium perchlorate (AP: NH_4ClO_4), ammonium nitrate (AN: NH_4NO_3), nitronium perchlorate (NP: NO_2ClO_4), potassium perchlorate (KP: $KClO_4$), potassium nitrate (KN: KNO_3), and so on. Among all these oxidizers, ammonium perchlorate (AP: NH_4ClO_4) is preferred as it has lower negative heat of formation and dissociates easily. Besides, this it is not very hygroscopic in nature and is compatible with most commonly used binders. Some of the common oxidizers and compounds that are currently pursued as promising oxidizers/ingredients in future high energy formulations are given in Table 6.4.

Although AP is a versatile and widely used oxidizer, the chlorinated combustion products formed during combustion of AP-based composition is a serious environmental concern. On the other hand, ammonium nitrate (AN) is an eco-friendly oxidizer. However, it is less energetic and is used mostly as a gas-generating propellant as its phase transition occurs around 32°C with a significant volume change. Hence, it is not used as a rocket propellant oxidizer in the rocket engine. Of course, it can be a promising substitute for AP only when used along with energetic binders like glycidyl azide polymer (GAP). High explosives made of ring structure nitramine compound, namely, research and development explosive (RDX ($C_3H_6N_3(NO_2)_3$): cyclotrimethylenetrinitramine). High melting explosive (HMX ($C_4H_8N_4(NO_2)_4$): cyclotetramethylenetetranitramine), as shown in Table 6.4, are used as high energetic smokeless propellants. In order to improve energy of the propellant, not more than 10%–20% of RDX/HMX is recommended as nitramines have less oxidizing power than AP. Besides, this addition of nitramines enhances the hazardousness of the propellant due to the presence of explosive materials.

TABLE 6.3 Some Oxidizers for Propellant Compositions

Oxidizer	Chemical Symbol	Molecular Weight	Density (kg/m³)	Oxygen Balance (%)	Heat of Formation (kJ/mol)
Ammonium perchlorate (AP)	NH_4ClO_4	117.5	1.95	34	−296
Ammonium nitrate (AN)	NH_4NO_3	80	1.72	20	−369
RDX	$C_3H_6N_3(NO_2)_3$	222	1.82	−21.6	70
HMX		296	1.91	−21.6	84
Ammonium dinitramide (ADN)	$C_4H_8N_4(NO_2)_4$ $NH_4N(NO_2)_2$	124	1.81	25.8	−150
HNIW (CL-20)	$C_6H_6 N_{12}O_{12}$	438	2.04	−10.9	372
Hydrazinium nitroformate (HNF)	$N_2H_5C(NO_2)_3^-$	183	1.83	13	−72

TABLE 6.4 Comparison between Solid and Liquid Propellants

No.	Solid Propellants	Liquid Propellants
1.	Low specific impulse	High specific impulse
2.	Easy to store, handle, and transport	Difficult to store, handle, and transport
3.	Simple to design and develop solid-propellant rocket engine	Complex to design and develop liquid-propellant rocket engine
4.	More economical	Less economical
5.	Difficult to test and calibrate solid-propellant rocket engine	Difficult to test and calibrate liquid-propellant rocket engine

6.4.2.3 Composite Modified Double-Base Propellant

In order to enhance energy and density levels of DB propellants, AP crystal can be added which also improves specific impulse of the rocket engine. Generally, oxidizers like AP and metallic fuels like aluminum are added to the fuel-rich DB propellants that reduce its fuel-richness and improve its performance. This kind of propellant is commonly known as composite modified double base (CMDB). The specific impulse I_{sp} of DB propellant can be enhanced by almost 60% with addition of CP, which is comparable to most of the common composite propellant formulations. These CMDB propellants are preferred in missile propulsion and upper stage of rocket engine as they are stronger as compared to the CP. The performance of DB propellant can be further enhanced by the addition of HMX/RDX in place of AP. In order to ease the casting of DB propellant, elastomeric binders are used, which are known as modified double-base propellants (EMCDB). Other ingredients such as inert plasticizers, stabilizers, darkening agents, burn rate modifiers, flash suppressors and platonizing agents are being added to these modified propellants.

6.4.2.4 Advanced Propellants

Several kinds of new propellants are being devised across the globe. Some of the advances and future directions in chemical propellants are discussed briefly as this subject is beyond the scope of this book.

Several new compounds that can act as burn rate catalysts, oxidizers, binders, and energetic ingredients or nanomaterials in rocket propellant formulations have been developed in recent times. Efforts have been made to develop minimum smoke, clean burning, high energy, and insensitive armaments for various propulsive devices. More energetic binders and oxidizers are being developed by using energetic groups, namely, azido (N_3–), nitramino (–$NHNO_2$–), nitro (NO_2–), and nitrato (–ONO_2–). The addition of these groups enhances the overall energy of the formulation, while increasing the overall oxygen balance.

Some of these energetic binders are glycidyl azide polymer (GAP), polyglycidyl nitrate (poly GLYN), nitrated polybutadiene (NHTPB), poly-nitratomethyl-1-3-methyl oxetane (poly NIMMO), poly 3,3-bis(azidomethyl)oxetane (BAMO), N–N bonded binders, and strained ring hydrocarbon binders. Among these energetic binders, GAP is considered to be the most prominent and effective in enhancing the performance of solid-propellant rocket engine, because it has a higher burn rate with positive heat of formation. Besides this, it has high density and insensitivity. As compared to conventional binders, GAP provides a better compromise between energetic performance and vulnerability of the energetic materials. But its main disadvantage is that it has higher glass transition temperature compared to HTPB.

Advanced oxidizers should have high density, high enthalpy of formation, high oxygen balance, and environmental compatibility. Ammonium dinitramide (ADN) is considered to be a promising oxidizer as it has higher heat of formation as compared to AP and chlorine-free combustion products. When it is mixed with energetic binders like GAP, it enhances specific impulse even at a lower solid loading of 80%. But it is not preferred as it has poor thermal stability and relatively high cost of production. Another advanced oxidizer is hexanitrohexaazaisowurzitane HNIW (CL20). It is one of the most powerful and dense single-component explosives. Although it is explosive by nature, it can be used in rocket propellant formulations in place of HMX. Hydrazinium nitroformate (HNF) is basically the salt of nitroform and hydrazine that is considered to be a promising oxidizer. This oxidizer with new energetic binders can have higher burn rates and specific impulse values as compared to conventional propellant. Many other energetic explosives with caged structure, namely, hydrazinium mono and diperchlorates, hydroxyl amine perchlorates, and difluramino compounds are being explored in order to enhance the performance of solid propellant. Interested readers can refer to advanced books on propellants and explosives [3].

In order to enhance burning rate of AP-based composite, ultrafine ammonium perchlorate and butacene catalysts are being used [6], which enhance burning rate of propellant even up to 100 mm/s, of course, in a very large range of pressure. Note that butacene is an HTPB prepolymer with attached ferrocene groups. In recent times, nanomaterials are being used to improve the performance of solid-propellant rockets as they alter the chemical kinetics rather than thermodynamics. The high surface area and short diffusion length of nanoscale particles are expected to enhance chemical kinetics.

In recent times, a novel solid propellant known as cryogenic solid propellant (CPS) has been conceived. It combines the simplicity of conventional solid propulsion with the high performance of liquid propulsion. In cryogenic solid rocket motors, cryogenic solids such as solid hydrogen (SH_2) and solid oxygen (SOX) can be used as propellants. It is expected to have specific impulse, density, and environmental protection provided technical impediments can be overcome in future [7].

6.5 LIQUID PROPELLANTS

Liquid-propellant rocket engines in spite of inherent complexities are preferred over the solid-propellant engines due to the added advantages (see Table 6.4) of liquid propellants. They have higher specific impulse and are capable of being throttled, shut down, and restarted easily. Liquid propellants consist of liquid fuel and liquid oxidizer and certain liquid additives. Several types of liquid propellants have been devised over the last six decades. Liquid hydrocarbons, liquid hydrogen, alcohols, and so on are examples of liquid propellants. Some of the examples of liquid oxidizers are liquid oxygen, nitric acid, and liquid fluorine. Liquid propellants can be classified based on the fuel–oxidizer arrangement, energy content, ignitability, and storability. Liquid propellants can be divided broadly into monopropellants and bipropellants. In liquid monopropellants, both fuel and oxidizer elements are located in the same molecular structure. Examples of monopropellants are hydrogen peroxide (H_2O_2) and hydrazine (N_2H_4). The monopropellant can be decomposed in the presence of a suitable catalyst into high-temperature and high-pressure gases. Monopropellants can be further divided into (1) simple and (2) composite. In simple monopropellant, fuel and oxidizers are contained in the same molecule. For example, methyl nitrate (CH_3NO_3) can be decomposed into CH_3O and NO_2. But the composite monopropellant consists of a mixture of oxidizer and fuel. For example, nitric acid and amyl acetate can undergo exothermic reactions to be used as composite monopropellant. In case of liquid bipropellant, fuel and oxidizer are mixed separately to have exothermic reactions. Liquid hydrogen and liquid oxygen are examples of liquid bipropellants. Based on the nature of ignitability, liquid propellants can be broadly divided into two categories: (1) hypergolic and (2) nonhypergolic. In case of hypergolic propellant, fuel and oxidizer when brought in contact will ignite spontaneously without any external ignition energy. Some hypergolic propellants are hydrogen–fluorine (H_2/F_2), hydrazine–nitric acid (N_2H_4/HNO_3), unsymmetrical dimethyl hydrazine–nitric acid (UDMH/HNO_3), and ammonia–fluorine (NH_3/F_2). Based on the energy contents, liquid propellants can be broadly classified into three categories: (1) low-energy, (2) medium-energy, and (3) high-energy propellants. Although the energy content of a propellant is dependent on the heat of combustion, in practice, this classification is based on the level of specific impulse. Let us now discuss the physical and chemical properties of certain liquid propellants.

6.5.1 Liquid Fuels

Several kinds of liquid fuels that can be used as propellant in rocket engines have been devised and developed. These fuel compounds contain mainly some atoms of carbon, hydrogen, nitrogen, boron, metal hydrides, organometallics, and so on. Some examples of liquid fuels are kerosene, furfurly/ethyl alcohols, hydrazine, aniline, amines, dimethyle hydrazine, xylidine, hydrogen, and ammonia whose properties are given Table 6.5. Some of these fuels are not being used in recent times. We will discuss in the following some of the commonly used liquid fuels in liquid-propellant rocket engine.

6.5.1.1 Hydrocarbon Fuels

Hydrocarbon fuels are a mixture of complex hydrocarbon chemicals that are basically refined from crude oil. Some of the fuels used for other engines, namely, kerosene, gasoline, jet fuels, can be used as propellants in rocket engines. For the rocket engine, a type of highly refined kerosene known as RP1 is devised, which is a mixture of saturated and unsaturated hydrocarbons with a narrow range of densities and vapor pressure. Generally, petroleum fuels are usually used in combination with liquid oxygen as the oxidizer. Note that RP1 and liquid oxygen are used as the propellant in the first-stage boosters of the Atlas, Titan, Delta II, and Saturn launch vehicles. Recently, Indian space research organization (ISRO) has developed a rocket petroleum fuel, which is known as ISROsene, with certain appropriate ratio of olefins and aromatics to avoid coking during its operation. Although petroleum fuels provide a specific impulse considerably less than cryogenic fuels, they are preferred due to their simplicity and cost-effectiveness. Besides, they are far superior to the hypergolic propellants.

In recent times, liquid methane (111 K) is considered to be a potential cryogenic hydrocarbon fuel due to its highly reproducible physical properties as compared to other petroleum fuels. It can be used along with

TABLE 6.5 Properties of Commonly Used Liquid Fuels

Fuel	Ammonia	Aniline	Ethyl Alcohol	Furfuryl Alcohol	Hydrazine	Dimethyl Hydrazine	Xylidine	Triethyl Amine	Kerosene	Hydrogen
Formula	NH_3	$C_6H_5NH_2$	C_2H_5OH	$C_5H_6O_2$	N_2H_4	$(CH_3)_2N_2H_2$	$C_8H_{11}N$	$(C_2H_5)_3N$	$C_{10}H_{20}$	H_2
Molecular weight (kg/kmol)	17.032	93.12	46.06	98.1	32.05	60.1	121.17	101.19	140	2.016
Density (kg/m)	680 (239.7 K)	1012 (298 K)	785 (298 K)	1129 (298 K)	1011 (288 K)	808 (288 K)	980 (288 K)	723 (288 K)	800 (at 298 K)	70.9 (at 20.5 K)
Melting point (K)	195.42	279.6	158.6	241.2	274.7	215	288.7	158.2	230	13.96
Boiling point (K)	239.8	457.6	351.7	444.2	386.7	354	490.2	362.5		2.039
Specific heat (kJ/kg·K)	4.4	2.01	2.59	2.43	3.14	2.69		1.89	2.0934	7.327 (at 14 K)
Heat of fusion (kJ/mol)	5.66		5.02		12.67	10.11				0.117230
Heat of vaporization (kJ/mol)	23.366530	44.34	38.6	40.11	42.71	35.03	45.81	31.44286		0.904348
Heat of formation (kJ/mol) (at 298 K)	−46.2	34.3	−278	−142.35	50.45	47.31	63.64	−154.912	−247.021	0
Viscosity (centipoises) to (kg/m·s)	255 (239.6 K)	6600 (283 K)	2370 (313 K)	1400 (293 K)	8100 (283 K)	1290 (274 K)	1450 (253 K)	347 (298 K)	1600 (at 288 K)	24 (at 14 K)
Thermal conductivity (kJ/ms·K) × 10^5	5.0	17.7	16.7		20.9	20.76	18.49	12.09	0.000155842	0.000066291

liquid oxygen in liquid rocket launch vehicles in future as it has higher performance than the state-of-the-art storable propellants, of course, with reduction in volume as compared to the LOX/LH_2 system. Its overall lower vehicle mass is low as compared to the commonly used hypergolic propellant. It is contemplated that this can be used for future Mars mission, because it can be manufactured partly from Martian in situ resources.

6.5.1.2 Hydrazine (N_2H_4)

Hydrazine (N_2H_4) is a colorless flammable but an excellent storable liquid with odor similar to ammonia gas. It can be used as a liquid fuel in the rocket engine, but it is quite toxic and unstable in nature to be used as a coolant. Hydrazine provides the best performance as a rocket fuel, but it has a high freezing point. It decomposes easily in the presence of a suitable catalyst and hence can be used as an excellent monopropellant. Some catalysts that can decompose hydrazine are iridium, iron, nickel, and cobalt. Iridium is found to be suitable for decomposition of hydrazine even at room temperature. But at higher temperatures beyond 450 K, several catalysts such as iron, nickel, and cobalt can be used to decompose hydrazine. Generally, hydrazine decomposes into gaseous ammonia and nitrogen, undergoing exothermic reactions leading to flame temperature of 1649 K. But ammonia gets decomposed further to form nitrogen and hydrogen with overall endothermic reaction, which results in a lower temperature. The global reaction model for the decomposition of hydrazine (N_2H_4) is given as follows:

$$N_2H_4 \rightarrow xNH_3 + (1 - x/2)\, N_2 + (2 - 3x/2)\, H_2 \tag{6.6}$$

where x is the degree of ammonia decomposition, which will be dependent on the catalyst type, size, geometry, chamber pressure, and residence time in the catalyst bed. When all ammonia is dissociated to hydrogen and nitrogen, then the flame temperature is around 867 K, which is not desirable. Hence, it is recommended to have least dissociation of ammonia for achieving higher specific impulse. In practice, 20%–40% ammonia dissociation is being preferred as it produces maximum specific impulse (189 s). Hydrazine can be used as a hypergolic bipropellant along with nitric acid (HNO_3) or nitrogen tetroxide (N_2O_4) as it has short ignition delay and can be ignited easily with these oxidizers.

Two more derivates of hydrazine fuels with similar physical and thermochemical properties, namely, monomethyl hydrazine (MMH) and unsymmetrical dimethyl hydrazine (UDMH), have been developed for liquid-propellant rocket engines. MMH ($CH_3N_2H_3$) is obtained when a methyl radical is substituted for a hydrogen atom in hydrazine. It is preferred over hydrazine as it is more stable and gives the best performance particularly when freezing point is an issue in spacecraft propulsion applications.

In UDMH [$(CH_3)_2N_2H_2$], two hydrogen atoms in hydrazine are substituted with two methyl radicals. Although it is the least efficient of the hydrazine derivatives as it has the lowest freezing point (215.9 K), higher boiling point (336.5 K), and has enough thermal stability to be used in large regenerative cooled engines. It is widely used with oxidizer N_2O_4 as it is more stable than hydrazine. It is used routinely in several countries like India, China, and Russia. However, UDHM is commonly used as blended fuel along with hydrazine. For example, Aerozine 50 is a mixture of 50% UDMH and 50% hydrazine, which has been used in lunar landing and takeoff engines. Aerozine 50 is almost as stable as UDMH and provides better performance.

6.5.1.3 *Liquid Hydrogen*

Generally, liquid hydrogen (LH_2) is used as the fuel, along with either liquid oxygen (LO_2) or liquid fluorine (LF) as the oxidizer. Note that all these liquids, namely, LH_2, LO_2, and LF, are known as cryogenic propellants as they are to be stored in cryogenic temperature to have liquid state. Liquid hydrogen (LH_2) and liquid oxygen (LO_2) are being used in several countries like India, Russia, Japan, United States, and Europe for space applications. Historically, it was used for the first time in the United States the upper stages of the Saturn V and Saturn 1B rockets, as well as the Centaur upper stage (1962). Recently, India has indigenously developed LH_2 and LO_2 engines for her ambitious space applications.

Note that hydrogen remains liquid at temperatures of −253°C (20 K). At this cryogenic temperature, liquid hydrogen remains in two isomeric forms: orthohydrogen and parahydrogen. In case of orthohydrogen, two proton spins are aligned in parallel, while in parahydrogen two proton spins are aligned in antiparallel manner. Furthermore, liquid hydrogen has a very low density (0.071 kg/L) and, hence, calls for a storage volume many times greater than other fuels. Besides, liquid hydrogen is difficult to store over long periods of time due to the low temperatures of cryogenic propellants. In spite of these disadvantages, LH_2 along with liquid oxygen (LO_2) is preferred in modern rocket engines as it delivers a specific impulse (I_{sp} = 430 s at sea level) about 30%–40% higher than most other rocket fuels. In order to overcome the problem of higher storage volume, hydrogen density can be enhanced by using a mixture of frozen (solid) hydrogen and liquid hydrogen. More research is required to overcome the problems of maintaining uniform mixture of (solid) hydrogen and liquid hydrogen.

6.5.1.4 Hydroxyl Ammonium Nitrate (NH_2OH*NO_3)

Hydroxyl ammonium nitrate (HAN) is a salt produced from hydroxylamine and nitric acid. It is also known as hydroxylamine nitrate (NH_3OHNO_3). In its pure form, it is a colorless and hygroscopic solid. It becomes a colorless and odorless liquid in aqueous solutions. The boiling point temperature of liquid HAN varies from 384 K to 418 K and its freezing point varies from 258 K to 229 K, depending on the extent of water in this solution. Note that the viscosity of aqueous HAN becomes higher with decrease in its water content. It is found to be more corrosive, toxic, and denser but less carcinogenic than hydrazine propellant. It has the potential of being used as monopropellant in liquid-propellant rocket engine as it decomposes easily, particularly in the presence of catalyst while undergoing overall exothermic chemical reactions. It contains both fuel and oxidizer similar to ammonium nitrate. The specific impulse of HAN monopropellant rocket engine varies between 220 and 265 s, depending on the water content and other blends of organic liquid fuels. Hydroxyl ammonium nitrate (HAN) fuel/oxidizer blend is being considered as green propellant in recent times by NASA [8]. As per NASA's claims, it offers nearly 50% higher performance for a given propellant tank volume compared to a conventional hydrazine system. This green propellant is less harmful to the environment, increases fuel efficiency, and diminishes operational hazards.

6.5.2 Liquid Oxidizers

Several kinds of liquid oxidizers that can be used as propellant in rocket engines have been devised and developed. These oxidizer compounds mainly contain some atoms of oxygen, fluorine, chlorine, hydrogen, nitrogen, boron, and so on. Some examples of liquid oxidizers (see Table 6.6) are hydrogen peroxide, nitric acid, liquid oxygen, ozone, nitrogen peroxide, liquid fluorine, chlorine trifluoride, chlorine pentafluoride, nitrogen trifluoride, and oxygen difluoride. Some of these oxidizers are not being used nowadays. Some common liquid oxidizers that are being used in liquid-propellant rocket engines are discussed in the following.

TABLE 6.6 Properties of Commonly Used Liquid Oxidizer

Oxidizer	Oxygen	Fluorine	Nitric Acid	Hydrogen Peroxide	Nitrogen Tetroxide
Formula	O_2	F_2	HNO_3	H_2O_2	N_2O_4
Molecular weight (kg/kmol)	32	38	63.02	34.016	92.016
Density (kg/m)	1141.5 (91.2 K)	1509 (85.2 K)	1520 (283 K)	1448 (293 K)	1450 (293 K)
Melting point (K)	54.39	55.2	231.5	273.5	261.9
Boiling point (K)	90.19	85.24	359	423.7	294.2
Specific heat (kJ/kg·K)	1.67	1.51	1.77	2.43	1.51
Heat of fusion (kJ/mol)	0.44	1.56	—	10.5339888	14.66
Heat of vaporization (kJ/mol)	6.82	6.32	30.36	54.470268	38.14
Heat of formation (kJ/mol)	0	0	−173.33	−187.74	−28.47 (liq)
Viscosity (centipoises) to (kg/m·s)	190 (53 K)	257	200 (269 K)	1300 (291 K)	4400 (288 K)
	870 (90 K)	—	450 (449 K)	—	—
Thermal conductivity (kJ/ms·K) × 10^5	2.1	2.5	27.45	62.8	—

6.5.2.1 Hydrogen Peroxide (H_2O_2)

This is the simplest peroxide that has been used as a viable liquid propellant in rocket engine application as it can decompose into water and oxygen by undergoing exothermic reaction. In its purest form, it is a colorless liquid with viscosity slightly higher than water. Hydrogen peroxide can be used either as a monopropellant or as an oxidizer of bipropellant. In rocket engine, highly concentrated (70%–98%) hydrogen peroxide is usually used as monopropellant, which is often referred to as the high test peroxide (HTP).

Note that the performance and density of HTP is quite close to that of nitric acid but less than that of liquid oxidizer. As compared to HNO_3, it is less toxic and corrosive but with a higher freezing point and lower stability. In the presence of catalyst, hydrogen peroxide is decomposed into water and oxygen, accompanied by the following chemical reactions:

$$H_2O_2 \rightarrow H_2O + 0.5O_2 \tag{6.7}$$

Some of the catalysts that can be used for this reaction to occur are platinum, silver, iron oxide, manganese oxide, and liquid permanganates. Hydrogen peroxide was first used as a major monopropellant in rocket engine. Hydrogen peroxide with 90% concentration can produce a maximum (theoretical) specific impulse of 154 s. As a hypergolic bipropellant, hydrogen peroxide (H_2O_2) can be used with hydrazine. But it is being used with kerosene and other fuels as nonhypergolic propellant. Note that hydrogen peroxide (H_2O_2) along with suitable fuel can produce specific impulses as high as 350 s. Although it produces a somewhat lower I_{sp} than liquid oxygen, it is preferred as it is storable, noncryogenic, and has more density. Besides, it can be used for regenerative cooling of rocket engines. In the past, it was employed successfully as an oxidizer in the Black Knight missile (British) and Me 163 rocket motors for aircraft booster (German) and F104 (USA). But nowadays, it is hardly used in practice.

6.5.2.2 Nitrogen Tetraoxide (N_2O_4)

It is a high-density yellow–brown liquid with specific gravity of 1.44 and is considered to be one of the most important rocket propellants. It is quite easy to store and is mildly corrosive. Nitrogen tetraoxide (N_2O_4) reacts spontaneously with most of the commonly used fuels. In the Titan missile, it was used as a liquid oxidizer along with a mixture of hydrazine and UDMH. It was used with MMH fuel in the Space Shuttle. Besides, it was used in the U.S. Gemini and Apollo spacecraft. It is also used in most of the geostationary satellites, and many deep-space probes. Russia's Proton booster rocket uses it as an oxidizer, while India's Vikas engines for PSLV and GSLV use it along with UDMH as fuel for space applications. These combinations are being used in most of the Chinese booster rocket engines. Since it is a liquid oxidizer, N_2O_4 along with UDMH/MMH as fuel is basically hypergolic in nature and storable over longer periods of times at reasonable temperatures and pressures.

The temperature range over which it exists as a liquid is quite narrow and hence it gets vaporized and frozen easily. Its freezing point temperature can be decreased by adding small amounts of either HNO_3 or NO. This mixture of NO and N_2O_4, commonly known as the mixed oxides of nitrogen (MON), has been developed as a viable liquid oxidizer with lower freezing point and lower corrosiveness. Note that the number included in the description of MON indicates the percentage of nitric oxide by weight. For example, MON25 indicates that MON25 contains 25% of NO by weight along with N_2O_4. Pure N_2O_4 has a freezing point of about −9°C, while that of MON25 is −55°C. MON is preferred in place of N_2O_4 in most spacecraft applications as the freezing point can be lowered easily. For example, MON3 containing 3% of NO by weight is used in the Space Shuttle reaction control system.

6.5.2.3 Nitric Acid (HNO_3)

Nitric acid (HNO_3) is considered as an oxidizer in rocket engine, particularly when freezing point is not an issue. Generally, HNO_3 is not used directly, rather nitric acid mixtures are being preferred in rocket engine. The nitric acid formulation most commonly used is the red fuming nitric acid (RFNA), which consists of HNO_3, 5%–15% N_2O_4, and 1.5%–2.5% H_2O. When the solution contains more than 86% HNO_3, it is referred to as **fuming nitric acid**. The red color of RFNA is due to the presence of the nitrogen dioxide that is formed during the breaking down of N_2O_4. RFNA, along with other substances like amine nitrates, can be used as monopropellant. But it is usually used as bipropellant in rocket engine. Note that HNO_3 is quite corrosive in nature. In order to reduce its corrosiveness, 0.4%–0.7% HF is added as a corrosion inhibitor and, hence, it is known as inhibited red-fuming nitric acid (IRFNA), which is being used in rocket engines. IRFNA along with aromatic amine fuels like aniline (C_6H_7N) and xylidine ($C_8H_{11}N$) had been used for missile applications. But IRFNA is quite toxic and volatile in nature as it contains large percentage of N_2O_4. In order to reduce its toxicity and volatility, less amount of N_2O_4 (less than 0.5%) is dissolved and it is known as white-fuming nitric acid (WFNA), which is considered as a safe and storable liquid oxidizer along with kerosene and hydrazine rocket fuel. Similarly, when 0.4%–0.7% HF is added to WFNA as a corrosion inhibitor, it is known as inhibited red-fuming nitric acid (IWFNA). Although WFNA has a somewhat less performance level as compared to the RFNA, it is used due to its nontoxic and low volatile nature. Nitric acid along with Aerozine has been used in the Titan family of launch vehicles and the second stage of the Delta II rocket engine. For tactical missiles, the US army had used IRFNA/UDMH. Note that Kosmos-3M is the most launched light orbital rocket with specific impulse of 291 s, in which IRFNA was used as an oxidizer along with UDMH fuel.

6.5.2.4 Liquid Oxygen

Liquid oxygen is the most widely used cryogenic liquid oxidizer propellant along with liquid hydrogen and kerosene for large rocket engine applications as it can provide higher specific impulse. It can be blended with alcohol, jet fuel, and gasoline. Its boiling temperature is 90 K at atmospheric pressure and needs cryogenic system. Hence, liquid oxygen tank and pipe line systems must be insulated properly. Each liquid oxygen tank must be provided with an external drainage system to avoid condensation of moisture. However it is preferred over other oxidizers due to its higher performance. Besides, it is noncorrosive and nontoxic in nature. It was used along with alcohol for the very first time in the V2 missile. Also, along with hydrogen it has found applications in Atlas, Thor, Titan, Saturn boosters, Space Shuttle, Centaur upper stage, GSLV(MkI & II), and so on.

Generally, liquid oxygen is not preferred in most modern ICBMs due to its cryogenic properties and poor deployability, because it calls for regular replenishment to replace boil-off, and hence cannot be launched quickly.

6.5.2.5 Liquid Fluorine

Liquid fluorine is considered as a super-oxidizer as it can react with almost all materials except nitrogen and substances that have already been fluorinated. Of course, its temperature has to be maintained below 85 K to be in liquid state. It is quite toxic and corrosive in nature. In spite of these disadvantages, fluorine produces very impressive engine performance. Although it was being considered in some of the NASA satellite and space missions, it has not been used in any operational rocket engine to date due to the problems of corrosion, high reactivity, and toxicity. It can also be mixed with liquid oxygen to improve the performance of LOX-burning engines; the resulting mixture is called FLOX. Some fluorine-containing compounds, such as chlorine pentafluoride, have also been considered for use as an “oxidizer” in deep-space applications.

6.5.3 Physical and Chemical Properties of Liquid Propellants

Calorific value: The calorific value of liquid fuels can be determined by using bomb calorimeter. The stainless steel container (bomb) is surrounded by a large temperature bath. The liquid fuel is burnt in the bomb in presence of oxygen. The change in temperature of water bath due to burning of a known quantity of fuel is measured using high-precision thermometer, which provides the calorific value of the fuel.

Specific gravity: The density of liquid fuel can be expressed in nondimensional form known as specific gravity (SG), which is defined as the ratio of mass density of fuel and mass density of water at the same temperature, given as follows:

$$SG = \frac{\rho_{fuel}}{\rho_{water}} \quad (\text{at the same temperature}) \tag{6.8}$$

In the rocket engine, liquid fuel and oxidizer with higher specific gravity are preferred to have large mass of propellant in a given tank volume, which in turn dictates the maximum flight velocity that a rocket engine can achieve. The density data of several liquid fuels are given in Table 6.5.

Autoignition temperature: This is the lowest temperature required to make combustion self-sustain without any external aid (spark or flame) in a standard container. The autoignition temperatures of typical liquid fuels are listed in Table 6.5.

Flash point: It is the minimum temperature at which liquid fuel will produce sufficient vapor to form a flammable mixture with air and can produce a momentary flame (flash) when an external heat source is brought in contact with the vapor. It indicates the maximum temperature at which liquid fuel can be stored without any fire hazard.

Fire point: In contrast to flash point, it is the minimum temperature at which liquid fuel produces sufficient vapors to form a flammable mixture with air that continuously supports combustion, establishing a flame instead of just flashing, even after ignition is withdrawn. Note that for the same fuel, fire point is always higher than flash point.

Freezing point: It is the temperature of propellant at which liquid is transformed from liquid phase to solid phase. Low freezing point of propellant is desirable as it allows the operation of the rocket engine at low-temperature environment. In some cases, chemical additives are used to decrease the freezing point temperature of propellants.

Viscosity: We know that *viscosity* of a liquid is the property that describes its resistance to flow. In other words, it is basically a measure of its resistance to gradual deformation by shear/tensile stress. The propellant must have low value of viscosity so that it can be injected and converted easily into spray with small droplet size distribution to have higher performance level. Besides this, viscosity of propellant should not vary too much with temperature.

Heat transfer properties: Mostly for liquid fuels that are used as coolant in rocket engines, it is important to have higher heat transfer properties, namely, specific heat and thermal conductivity. It is also important to have higher thermal diffusivity for higher heat transfer. Besides this, the propellant that is used as coolant for thrust chamber and nozzle must have higher boiling point. Typical heat transfer properties of certain propellants are provided in Table 6.5.

Thermal stability: The propellant should remain stable during storage for a longer period of time. Besides this, it should remain stable even with sudden change in temperature without any appreciable change in physical and chemical properties. The propellants like N_2O_4, H_2O_2, HNO_3 are preferred over the energetic propellants, namely, liquid oxidizer and fluorine, as they remain stable over wider temperature range.

6.5.4 Selection of Liquid Propellants

Several propellants have been developed over the years to meet the demands of various rocket engine applications. We have already discussed about various properties of propellants that can affect the performance of rocket engine. Note that proper selection of propellant for a certain application is quite cumbersome as propellants can have certain advantages and disadvantages. In other words, there is no ideal propellant that can meet all the requirements. Hence, one has to resort to compromise during selection of liquid propellants. However, three qualities, namely, energetic quality, kinetic quality, and utilization quality, can be considered to have certain basis of comparison for the selection of liquid propellants, enumerated as follows:

1. *Energetic quality*: The amount of energy released by the propellant and its use in nozzle can dictate the performance of rocket engines.

   a. High specific impulse can be achieved with higher release of energy per unit propellant.

   b. Low molecular weight of combustion products can provide higher specific impulse.

2. *Kinetic quality*: The kinetic quality depends on the amount of time required for complete release of energy in rocket engine. The time for injection, vaporization, ignition, and combustion will affect heat release during burning of liquid propellant. This quality can affect the ignition, combustion stability, and so on. Some of these qualities are

   a. Better ignition characteristics

   b. High combustion efficiency

   c. Stable combustion

   d. Possibility of recombination in the nozzle

3. *Utilization quality*: The utilization quality of propellant needs to be considered to assess the propellant from the utilization point of view. In other words, the utilization quality depends on the physico-chemical properties of the propellant. The problems in handling, for instance, maintenance, storage, and cost are to be considered while selecting propellant from the utilization point of view. Some of these utilization qualities are enumerated in the following:

   a. High density of propellant can reduce the size and weight of propellant tanks and feed systems, resulting in higher density specific impulse for rocket engine.

   b. Low viscosity liquid propellant can reduce the pressure drop across the feed and atomizer system. Besides, better spray is formed for propellant with low viscosity under same pressure drop across atomizer.

   c. Propellant with higher specific heat, thermal conductivity, and critical temperature for vaporization can act as an effective coolant for rocket engine.

d. Propellant with lower vapor pressure can reduce the weight of propellant tank since lower positive suction pump head is required.

e. Propellant with lower freezing point temperature can make the rocket engine operate even at low-temperature environment.

f. Propellant must be thermally and chemically stable over a long period of time.

g. Propellant must be compatible with tank material so far as corrosion is concerned.

h. It must have higher lubricant quality.

i. It must have lower risk of explosion and fire hazard.

j. It must be easy to handle and transport.

k. It must be cheap and abundantly available.

6.6 GEL PROPELLANTS

In order to overcome the problems of solid and liquid propellants, several kinds of gel propellants have been devised that can provide variable thrust strength and flexible operation over a wider range of operation. Gel propellants are considered as the modified liquid propellants, whose flow properties are significantly altered due to the addition of gellant. Generally, gel propellant can be defined as a substance that contains networks of solid matrix, enclosing a continuous liquid phase. Note that during synthesis of gel propellant, gellant forms a three-dimensional intermolecular network, enclosing the continuous liquid phase within it. As a result, gel propellant behaves like time-dependent non-Newtonian fluid and possesses thixotropic characteristics. Hence, gel propellants exhibit shear thinning behavior in which viscosity decreases with increasing applied shear stress and thixotropic behavior in which viscosity decreases in time at constant applied shear stress. This is advantageous for propulsion applications where negligibly small shear encountered in the storage tanks allows the gelled propellant to remain in a stable and highly viscous state. On the other hand, a large shear applied deliberately during injection reduces the propellant viscosity so that it can be conveniently supplied to the combustion chamber for atomization and combustion. These flow characteristics of gel propellants serve to overcome many problems associated with solid and liquid propellants; on-board or during flight. The potential benefits of gel propellants are listed in a following subsection. A variety of different gelling agents compatible with number of different liquid propellants are also discussed in the following section.

6.6.1 Common Gel Propellants and Gellants

Table 6.7 summarizes a number of gellant–propellant combinations, among which inorganic gellants like silica, gel well with both fuel and oxidizer, while other organic gellants like cellulose compounds (HEC, MEC, etc.), and agar-agar can gel only with fuels [9].

6.6.2 Advantages of Gel Propellants

Gel propellants have several advantages over solid and liquid propellants from various points of view, namely, safety, performance, and storage, which are discussed in the following [10,11].

TABLE 6.7 Common Gel Propellants and Gellants

Propellant (wt%)		**Gellant (wt%)**	**Additives**
Fuels	Hydrazine (N_2H_4)	Carbopol Colloidal silica	Al
	UDMH	Methyl cellulose (2%–4%) HEC (5.57%) Hydroxy propyl cellulose (7%) Agar-agar (0.64%–12%) SiO_2 (5%)	Al (5%–40%) Mg (5%40%)
	MMH	Sand (~5%)	Al (<60%)
		Hydroxy propyl cellulose (<5%) HEC (7%)	Al (0%–40%)
	N_2H_4/UDMH	Hydroxy propyl cellulose ethers	Al (10%–40%), Be, Boron
	H_2	BTMSE	Al (60%)
	RP-1	SiO_2 (3.5%–6.5%)	Al (0%–55%)
	Kerosene	Organophilic clay complex and propylene glycol (6%–7%)	Al (30%–40%)
Oxidizer	IRFNA	SiO_2 (3%–4%)	$LiNO_3$ (28%)
	RFNA	Sodium silicate (4.25%) Colloidal silica (4%)	AP (0%–30%)
	H_2O_2	SiO_2 (3.5%)	Boron carbide (19.3%)

Source: Varghese, T.L. et al., *Defense Sci. J.*, 45(1), 25, 1995.

6.6.2.1 Safety Aspects

Leaks and spill: The gel propellant has reduced leakage rate during storage or failure of feeding system as compared to liquids, because surface gets hardened while coming in contact with a gaseous environment. Even if spillage occurs due to accident, hazards due to explosion and toxicity are less likely to occur, as the volatility of gels is significantly lower than that of liquid propellant. However, in the case of hypergolic gel propellant, burning may occur at the fuel–oxidizer interface but chemical reaction ceases due to noncontact of fuel and oxidizer caused by the rheological nature of the gel propellant.

Sensitivity to impact, and electrostatic discharge: Gel propellants are less sensitive to deflagration, detonation, or explosion as compared to solid propellant when subjected to impact friction or electrostatic discharge. Combustion of gel propellant like liquid propellant is controlled and can be prevented easily.

Cracks: Gel propellants are fed into combustion chamber in liquid form and hence cracks in gel structure unlike in solid propellant would not lead to uncontrolled combustion and explosion.

6.6.2.2 Performance Aspects

Specific impulse: Its specific impulse is similar to that of the liquids and gets enhanced with addition of metal particles.

Density impulse: Its density impulse is enhanced significantly as compared to that of liquid propellant, depending on the metal additive and its loading.

Energy management: Its energy management is similar to that of liquid propellant.

6.6.2.3 Storage Aspects

Stability: Gel propellants have demonstrated long-term storage capability for even more than 10 years.

Packing: Gel propellants can be stored in tanks similar to liquids.

Particle sedimentation: Gel propellants have much lower particle sedimentation as compared to slurry propellant and phase separation is likely to occur only at very high acceleration levels.

6.6.3 Disadvantages of Gel Propellants

Gel propellants have certain disadvantages over solid and liquid propellants from various points of view, namely, feeding, atomization, burning, phase separation, physical stability, and cost, which are discussed as follows:

1. *Feeding process*: Gel propellants being non-Newtonian and thixotropic in nature need higher pressure than virgin liquid propellant for atomization of same mass propellant flow rate. As fuel and oxidizer have different rheological properties, feed line design has to be done separately. Besides this, the temperature dependence of rheological properties has to be considered for the design of feed system.
2. *Atomization process*: As gel propellant is non-Newtonian, it is quite challenging to atomize gel propellant. As a result, longer combustion chamber is to be provided for complete combustion of gel propellant.
3. *Burning process*: Gel propellant droplets evaporate at a slower rate, resulting in lower rate of burning. In addition to this and due to coarser spray, combustion efficiency gets reduced as compared to virgin liquid propellant. Hence, specific impulse is likely to be lower due to decrease in characteristic velocity. For example, the characteristic velocity of jet fuel–oxygen gel propellant is decreased by 4%–6%. Besides this, higher gellant concentration can lead to higher burning residue and can reduce combustion performance further.
4. *Phase separation and physical instability*: Solid phase in gel propellants can get separated from liquid phase at a very high in-flight acceleration rate. Besides this, loaded metal particles may be separated from liquid phase under adverse situations.
5. *Cost*: Gel propellants are more costly as compared to solid and liquid propellants.

6.7 HYBRID PROPELLANTS

In case of hybrid propellant rocket engines, fuel and oxidizer are used in different phases. Generally a typical hybrid propellant consists of a solid fuel and liquid oxidizer. Some of the common solid fuels used fuels in hybrid propellant rocket engines are polymers like HTPB, CTPB, etc., plastic fuels, namely, Polyvinyl chloride (PVC) or Polypropylene and hydrides of light metals, namely, beryllium, lithium, aluminum, etc., along with suitable binding polymer. Some of the common liquid oxidizer used in hybrid propellant rocket engines are LO_2, HNO_3, and N_2O_4. Besides these, high energetic oxidizers, namely, mixture of liquid oxygen and fluorine known as FLOX and chlorine–fluorine compounds, namely, ClF3, CLF5, etc., can be used in hybrid propellant rocket engines. Note that the performance level of hybrid propellant rocket engine gets enhanced with increases in percentage of LF in FLOX which is caused due to lower molecular mass of products. In first private manned spacecraft known as Space Ship One, HTPB and nitrous oxide (HNO_3) were used as hybrid propellants which had accidental explosion killing three people. Both Spacedev Streaker and Dream Chaser meant for suborbital and orbital human space applications use nitrous oxide and the HTPB as hybrid propellants. In certain rocket hybrid propellant engine, self-pressurizing nitrous oxide N_2O and HTPB as well as high test peroxide (HTP) and HTPB are used. US rockets has developed high test hydrogen peroxide H_2O_2, Hydroxyl-terminated polybutadiene (HTPB) and aluminum hybrids which can produce a sea level delivered specific impulse (I_{sp}) of 240.

REVIEW QUESTIONS

1. What do you mean by a propellant? What are the types of propellants you are aware of?
2. What are the properties of propellants?
3. Compare solid and liquid propellants, listing their respective advantages and disadvantages.
4. What do you mean by double-base propellant? How is it different from composite propellants?
5. What are the ingredients of a generic double base propellant? What are the functions of each ingredient?
6. What are the ingredients of a generic composite propellant? What are the functions of each ingredient?
7. What do you mean by a composite modified double-base propellant? Explain it by comparing it with both composite and double-base propellants.

8. What are the advanced solid propellants? Why are they used in recent times?
9. What do you mean by a monopropellant? How is it different from a bipropellant?
10. What do you mean by hypergolic propellant? How is it different from nonhypergolic propellants?
11. What are the criteria used for selecting liquid propellants?
12. What are the most important physical properties of liquid propellants used for selecting liquid propellants.
13. What are the physical and chemical properties of liquid propellants? Discuss each of these properties.
14. What do you mean by gel propellant? How is it different from liquid and solid propellants?
15. What do you mean by hybrid propellant? How is it different from solid and liquid propellants?

REFERENCES AND SUGGESTED READINGS

1. Barrere, M., Jaumotte, A., de Veubeke, B.F., and Vendenkerckhove, J., *Rocket Propulsion*, Elsevier Publishing Company, New York, 1960.
2. Sutton, G.P. and Biblarz, O., *Rocket Propulsion Elements*, 7th edn., John Wiley & Sons Inc., New York, 2001.
3. Kubota, N., *Propellants and Explosives, Thermochemical Aspects of Combustion*, Wiley VCH, Weinheim, Germany, 2002.
4. Wimpress, R.N., *Internal Ballistics of Solid Fuel Rockets*, McGraw Hill Book Co, New York, 1950.
5. Razdan, M.K. and Kuo, K.K., *Fundamentals of Solid Propellant Combustion*, AIAA, New York, 1983, Chap. 10, pp. 515–597.
6. Doriath, G., High burning rate solid rocket propellants, *International Journal of Energetic Materials and Chemical Propulsion*, 4(1–6), 646–660, 1997.
7. Lo, R.E., A novel kind of solid rocket propellant, *Aerospace Science and Technology*, 6, 359, 1998.
8. McLean, C.H., Hale, M.J., Deininger, W.D., Spores, R.A., Frate, D.T., Johnson, W.L., and Sheehy, J.A., Green propellant infusion mission program overview, *49th AIAA/ASME/SAE/ASEE Joint Propulsion Conference*, San Jose, CA, July 2013.
9. Varghese, T.L., Gaindhar, S.C., John, D., Josekutty, J., Rm., M., Rao, S.S., Ninan, K.N., and Krishnamurthy, V.N., Developmental studies on metallized UDMH and kerosene gels, *Defense Science Journal*, 45(1), 25–30, 1995.

10. Natan, B. and Rahimi, S., The status of gel propellants in year 2000, in *Combustion of Energetic Materials*, Kuo, K.K. and de Luca, L. (Eds.), Begel House, Boca Raton, FL, 2001.
11. Padwal, M.B. and Mishra, D.P., Synthesis of jet A1 gel fuel and its characterization for propulsion applications, *Fuel Processing Technology*, 106, 359–365, 2013.

CHAPTER 7

Solid-Propellant Rocket Engines

Mother Earth is a biological living trust. Let us not destroy her in the name of development armed with modern soulless science and technologies.

D.P. MISHRA

7.1 INTRODUCTION

Solid-propellant rocket engines (SPREs) have been used for various purposes and are historically believed to have been mastered by the Chinese and Indians after the invention of black powder. In the recent past, SPREs have undergone significant improvements. Hence, they have currently found a wide range of applications in the form of various propulsive devices, namely, spacecraft, missiles, aircraft, retro-rockets, and also gas-generating systems. They can easily provide thrust ranging from a few milli-Newtons to mega-Newtons.

SPREs are preferred over other chemical rocket engines, particularly whenever high reliability, quick operational readiness, and simple storage are required. Note that the SPRE does not contain any moving parts unlike the liquid-propellant rocket engine (LPRE; unless a thrust vector control system is used). On several occasions, it is also called a *solid-propellant rocket motor*. Hence, we will use these two terminologies, namely, motor and engine, interchangeably. In SPREs, the propellant is contained in the combustion chamber as a block of a certain shape, known as the grain.

Note that the propellant grain can be entirely stored in the combustion chamber for a longer period of time, even in the range of 5–15 years. These engines can be designed from very low (2 N) to high (10,000 kN) thrust levels. As a result, they find applications in satellite launchers, ballistic missiles, sounding rockets, assisted take-off, air-launched missiles, antitank weapons, and retrorockets.

Whenever very high thrusts are required for a short period of time (booster phase), SPREs are preferred over other two chemical rocket systems. With improvements in propellant chemistry and processing, it is possible to extend their operation and hence find applications during the sustainer phase as well. Even for a fairly long period of burning time, the SPRE system becomes lighter compared to an LPRE as the propellant stored within the combustion chamber gets consumed. Hence, it is employed for designing missiles. Note that its wall is protected from heating by the solid propellant itself. Due to its high reliability, even beyond 99%, and light weight, it has been used in antitank and antiaircraft missiles. The pressure in the combustion chamber of an SPRE is generally much higher compared to that in the LPREs and thus it has a higher thrust coefficient. Let us compare various features of an SPRE with those of an LPRE, as summarized in Table 7.1.

Some of the characteristic features of an SPRE are (1) high density, (2) specific impulse in the range of 210–290 s (vacuum), (3) simple system, and (4) no control over oxidizer to fuel ratio (O/F) once ignited. Note that the burning rate depends on the propellant composition and is sensitive to chamber conditions like pressure and temperature. These are to be seen in relation to liquid and hybrid rockets where higher performance is possible at the expense of simplicity.

TABLE 7.1 Comparison between a Solid-Propellant Rocket Engine and a Liquid-Propellant Rocket Engine

Characteristics	Solid-Propellant Engine	Liquid-Propellant Engine
Propellant	Solid	Liquid
Storage	Stable for 10–15 years	Stable for 1–2 years
Burning rate	Low	High
Chamber pressure	Higher (1–70 MPa)	Lower (0.3–10 MPa)
Chamber thickness	Low	High
Applications	Booster (generally)	Booster and sustainer
Thrust	High	Low
Reliability	Higher (99%)	Lower
Design	Simple	Complex

7.2 BASIC CONFIGURATION

Let us discuss briefly the main components of a typical solid rocket engine, shown in Figure 7.1. It consists of propellant grain, an igniter, a motor casing, a thermal insulator, a nozzle, and a thrust terminator. Generally, the motor casing that supports the propellant grain is made of high strength and temperature-resistant materials, and is coated with organic materials that not only act as a thermal insulator but also offer light bonding between the propellant grain and the chamber wall. The casing of the rocket engine is generally made of metal, namely, steel, titanium, and aluminum alloys that can withstand a high level of stress, particularly at high pressure and temperature. In recent times, designers have started using fiber-reinforced composite materials, which are much lighter than metal casing. The propellant grain of a particular shape is used to produce a certain type of thrust and time profile. The propellant grain in a typical solid rocket engine amounts to around 85% of the rocket engine's total mass. The propellant grain is supported generally by a wall, or retainer push (thermal insulators), or special grids or traps. The other main components of the motor are the insulated nozzle and the igniter. An igniter is generally mounted upstream of a combustion chamber, as shown in Figure 7.1. Its main function is to initiate the burning of the solid propellant. In order to initiate combustion, the igniter is switched on to produce hot gas with metal particles, which makes the virgin burning surface of the solid-propellant grains burn. Of course, even if combustion is initiated successfully, the igniter cannot be turned off till it is burnt out completely. Note that in a modern SPRE, a

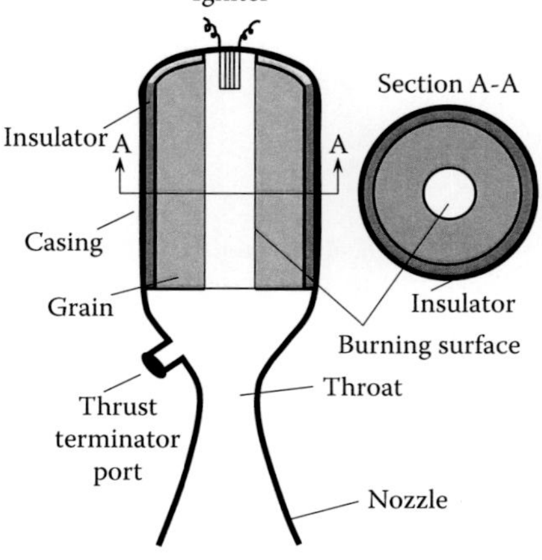

FIGURE 7.1 A typical solid-propellant rocket engine.

blowout diaphragm is employed to terminate its operation even before complete burning of the entire grain. As a result, it is possible now to shut down the operation of the SPRE. The high-pressure and high-temperature gas produced due to the burning of the propellant in the combustion chamber is expanded through the CD nozzle to produce thrust.

7.3 PHYSICAL PROCESSES OF SOLID-PROPELLANT BURNING

The burning of the solid propellant in the SPRE involves several complex processes including phase change, high-speed flow, chemical reactions, heat transfer, and mass transfer. Let us look briefly at the processes involved in the burning of the propellant as depicted in Figure 7.2. Generally, a flame is formed near the solid surface due to the gas phase exothermic reaction. Thus, heat from the flame (reaction) is transported to the propellant surface by three modes of heat transfer: conduction, convection, and radiation. The heat carried to the propellant surface is partially used for melting, evaporation, or sublimation of the propellant and some portion of the heat is conducted into the propellant to help raise its inner temperature.

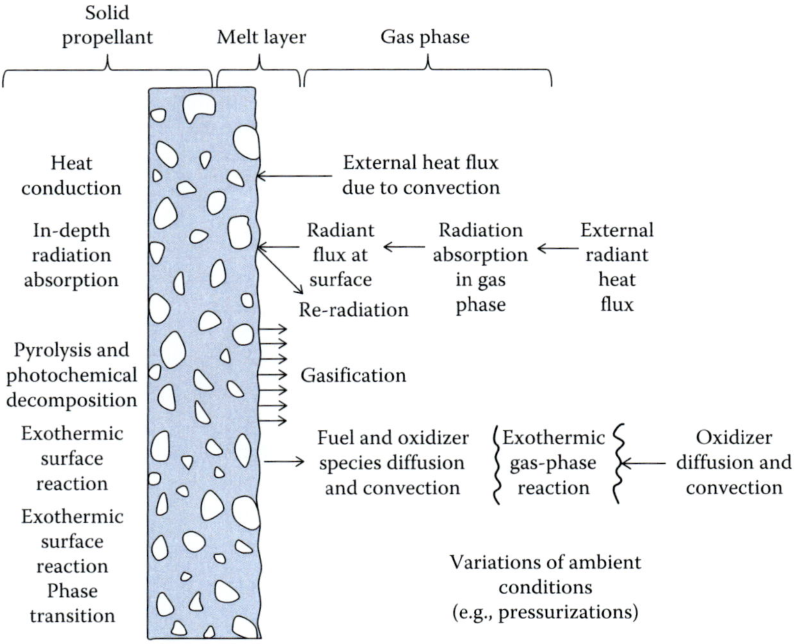

FIGURE 7.2 Burning processes during solid-propellant burning [1]. (From Timnat, Y.M., *Advanced Chemical Rocket Propulsion*, Academic Press, St Louis, MO, 1987.)

The heat energy helps in decomposing the solid into a gaseous state by a process known as pyrolysis (thermal decomposition). Besides this, there are also phase changes involving either exothermic or endothermic reactions. The gases coming out of the propellant surface mix and undergo chemical exothermic reactions within a very narrow zone, thus forming a flame near the propellant burning surface. From this discussion one can imagine the complex processes involved during solid-propellant burning. It may be noted that when the upper layer of the propellant is converted into gases, the virgin propellant is simultaneously getting prepared by heat conduction to decompose into gases, leading to regression of the propellant surface. Thus, we need to know the regression/burning rate of the propellant. This regression rate $\dot{r}$ (mm/s) is defined as the distance traveled per second by the regression front, perpendicular to the burning surface of the propellant grain, assuming that the ignition is homogeneous.

7.4 BURNING MECHANISM OF SOLID PROPELLANTS

7.4.1 Double-Base Propellants

The processes involved during the burning of a double-base (DB) propellant, which has a distinct mode of combustion, are depicted in Figure 7.3. As the DB propellant is homogeneous in nature, it is expected to have a one-dimensional homogeneous combustion flame structure along the

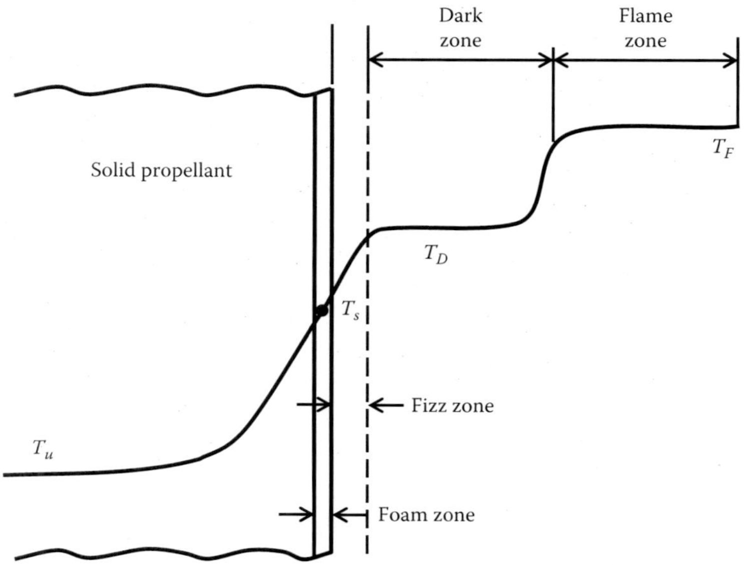

FIGURE 7.3 Various zones during the burning of a double-base propellant.

burning direction. However, four distinct zones, as shown in the figure, have been observed in experiments, particularly at low and moderate combustion chamber pressure. These are (1) the foam zone, (2) the fizz zone, (3) the dark zone, and (4) the flame zone.

When the propellant surface receives heat from a flame, the solid surface gets degraded thermally and undergoes exothermic reactions. This thermal degradation zone within the solid propellant or at the burning surface is known as the foam zone. Note that the interface between the solid surface and the burning surface is composed of a thin layer of solid/liquid/gas. In this foam zone, the decomposition of nitroglycerine (NG) and nitrocellulose (NC) takes place, during which solid phase reactions occur at very low temperature around 500 K, liberating gases like NO, NO_2, and aldehydes. Note that formaldehyde (H–C=O–H) and acetaldehyde (CH_3–C=O–H) can be formed in this zone. The aldehydes and NO_2 may react exothermally in the interface region. However, the overall reaction in this zone happens to be endothermic in nature. Note that the thickness of the foam zone is quite small in the range of 10–100 μm depending on the combustion chamber pressure level. Hence, it is expected to have this zone's temperature, which is the same as the burning surface temperature T_s.

The major portion of the gases, namely, NO, NO_2, and aldehydes, released in the foam zone gets mixed near the burning surface and reacts exothermically to produce nitric acid, carbon monoxide, carbon dioxide, water, hydrogen, and other organic materials. This zone is commonly known as the fizz zone, whose thickness varies from 1 to 10 μm. As these reactions take place exothermically, the temperature in this zone increases steeply from the surface temperature T_s to T_1, as shown in Figure 7.3. Hence, this fizz zone plays an important role in determining the burning rate of the propellant.

The dark zone is situated between the fizz zone and the luminous flame zone, and its size may vary from 0.1 to 2 cm. Although chemical reactions between products formed in the fizz zone do take place, the temperature in this zone remains almost constant, indicating that very little heat is being liberated into this zone. However, activated products are formed in this zone, which will be instrumental in forming the luminous zone. Hence, this zone is also known as the preparation zone. Note that this zone starts shrinking when the chamber pressure goes beyond 7 MPa.

When sufficient concentration of activated products (radicals) is formed in the dark zone, final reactions between NO, CO, and H_2 occur with a steep increase in temperature to about 3000 K. This zone is known as the flame zone and its size varies from 100 μm to 0.2 cm. This zone provides heat to the propellant surface if the thickness of the dark zone is not too large. The dark zone decreases with an increase in pressure and almost disappears at high pressures around 7 MPa. At this point, a significant amount of heat from the flame zone can be transferred to the burning surface, thus enhancing the burning rate. But the dark zone thickness increases very rapidly when P_{t2} is below 1 MPa. The reaction in the flame zone becomes more sluggish and can even cease completely while the propellant continues to get consumed at a slow rate. Then, it is more likely that the chamber pressure may drop further to a lower value (0.1 MPa). However, the propellant surface temperature T_s still remains high to have some finite burning rate. Hence, the gaseous products continue to form until a critical concentration of gas mixture is reached when spontaneous reignition of gaseous reactants occurs, reestablishing the flame. The chamber pressure P_{t2} may increase further or may decrease afterward with the extinction of the flame zone, leading to intermittent burning, also known as chuffing. Thus, a certain optimum pressure must be maintained for smooth operation of normal propellant burning. There is an upper limit of pressure above which combustion becomes erratic and unpredictable. For most solid propellants, the upper pressure limit is above 30 MPa. The pressure dependence of the burning rate is very complex and can therefore be determined only by conducting experiments.

7.4.2 Composite Propellant Combustion

We know that, unlike a DB propellant, the physical structure of a composite propellant (CP) is heterogeneous in nature. Hence, it is expected to have a heterogeneous combustion wave structure. Several models for combustion of CPs have been proposed by various researchers. However, we will discuss a model for CP combustion proposed by Beackstead, Dherr, and Price (BDP) [5], which advocates a triple-flame structure as shown in Figure 7.4. In this model, let us consider a single CP consisting of an oxidizer (ammonium perchlorate, AP) surrounded by polymeric fuel (binder) as shown in Figure 7.4. When the propellant surface is heated, the AP crystal embedded in the fuel gets decomposed into oxidizing gases while fuel is pyrolyzed to form hydrocarbon vapors. Note that the initial decomposition steps of both fuel and oxidizer at the propellant surface are endothermic. Sometimes a liquid layer is formed on the surface of the propellant. Note that either set of products formed during the condensed phase reactions of AP and fuel can result in less endothermic or exothermic reactions at the solid surface. Generally, the overall reaction on the burning

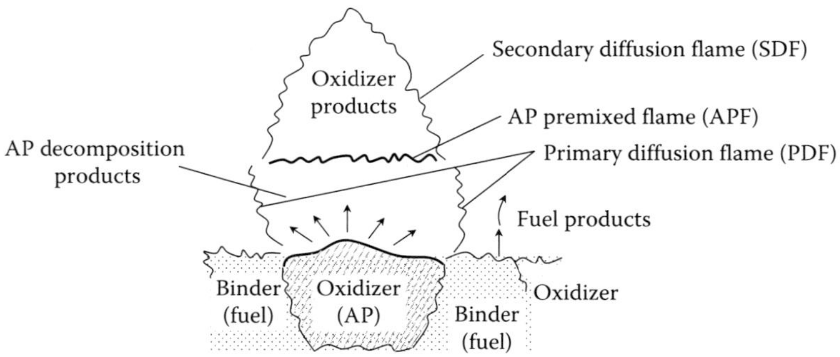

FIGURE 7.4 Various zones during the burning of a composite propellant.

surface happens to be exothermic in nature. The products formed during the overall endothermic reaction get mixed and undergo chemical reaction downstream of the regression surface.

As mentioned, the BDP model [4,5] for a CP advocates the occurrence of a triple-flame structure during the burning of the composite propellant as shown in Figure 7.4. Let us discuss the causes for these flames during the burning process of the CP. When the chamber pressure is above 1.9 MPa, AP can decompose and form oxidizer-rich products following chemical reactions at about 1400 K.

$$NH_4ClO_4(\text{AP flame}) \rightarrow HCLO_4 + NH_3 \rightarrow \text{Inert product} + \text{Oxidizing species} \quad (7.1)$$

At around 0.7 MPa of pressure, the AP decomposes and combines with fuel (binder) to establish a premixed flame above the oxidized crystal as shown in Figure 7.4, which is often known as an AP monopropellant flame (APPF). This APPF can be approximated as a one-dimensional premixed flame. Note that at lower pressure, the decomposed products of both fuel and oxidizer get sufficient time to mix before chemical reactions take place, forming this APFF. But at high pressure, chemical reaction time is small and mixing of fuel with oxidizer will limit the combustion process, resulting in the formation of a diffusion flame. Recall that at the same time, the fuel undergoes a pyrolysis process and produces CH_2, CH_4, C, and so on. These fuels products are mixed with $HClO_4$ products to produce combustion products as per the following representative chemical reaction:

$$\text{Fuel pyrolysis product } (CH_2, CH_4, C, \text{etc.}) + HClO_4 \rightarrow \text{Combustion products } (CO_2 + Cl_2 + H_2O + \cdots) \quad (7.2)$$

As the chamber pressure increases, the fuel products find it difficult to diffuse easily and mix with the oxidizer stream. As a result, two reaction paths, namely, (1) oxidizing products and ammonia from AP decomposition (Equation 7.1) and (2) fuel (binder) products and AP oxidizing products (Equation 7.2), become competitive. At higher pressure, chemical reactions are quite fast and thus oxidizer products react with ammonia to form a premixed flame before the binder products can mix with the oxidizer products. However, at the interface of oxidizer and fuel, a certain portion of the oxidizing products come into contact with the fuel pyrolysis products, and thus a primary diffusion flame (PDF) is formed at a temperature of 2800 K and all pressure, as shown in Figure 7.4. But at high pressure, the characteristic time for the gas phase reaction decreases and effects of PDF on regression rate decrease as APF and PDF almost merge into a single flame. Note that during APF formation, a large portion (around 30%) of oxidizer products remains unburnt as this flame is oxidizer-rich. These unburnt oxidizer products at 1400 K get mixed with fuel-rich pyrolysis products and undergo fast chemical reactions. As a result, a secondary diffusion flame (SDF) is formed, as shown in Figure 7.4, with a maximum temperature of 3200 K, which provides a large amount of heat to the regression surface, particularly at low pressure. But at high pressure, the effect of the SDF on the burning rate becomes less as its standoff distance increases compared to the APPF. Note that the diffusion of fuel and oxidizer becomes slower compared to the reaction, thus limiting the burning process. In summary, the entire burning process of the CP consists of one premixed flame (APPF) and two diffusion flames (PDF and SDF), which occur around 0.2–0.5 mm above the regression surface. The thickness of the combustion zone decreases with an increase in chamber pressure. The size and nature of the flames are affected by the size of AP particles. With a reduction in the size of AP particles, smaller flames are formed above the regression surface and thus high but uniform temperature distribution prevails on the regression surface. As a result, a higher burning rate can be achieved due to an increase in heat transfer from flame to regression surface.

7.5 MEASUREMENT OF PROPELLANT BURNING/REGRESSION RATE

We have learnt that the processes involved during propellant burning are quite complex. Hence, it is quite difficult to predict accurately the regression rate $\dot{r}$ of a solid propellant on a theoretical basis. Although several researchers have attempted to predict burning rate by modeling the physical processes involved in propellant burning, they have achieved very limited success in specific cases. Rather, an empirical relationship based on experimental data has been developed, which is being used to design and develop solid-propellant rocket motors. The linear burning rate depends on the pressure of the chamber, since gas phase reaction rates play a major role in providing heat to the burning surface [2,6,11].

Generally, the linear burning/regression rate $\dot{r}$ can be measured experimentally in two ways, by using (1) a strand burner, also known as a Crawford bomb, and (2) ballistic evaluation motor (BEM) over a certain range of pressure. Note that the BEM is basically a scaled-down model of a full-size rocket (1:10) and burn rates are inferred from the pressure–time plots. But the strand (Crawford) burner is the most widely used method for obtaining the linear burning rate $\dot{r}$. For both methods, the experimental burning rate $\dot{r}$ is empirically related to variables in order to commensurate with full-scale tests.

The schematic of a typical strand burner is depicted in Figure 7.5. It consists of a high-pressure chamber (bomb), long and thin strands with propellant holder to hold the propellant vertically, high-pressure N_2 gas cylinder and pipeline, igniter, surge tank, and so on. The propellant strand (5 mm diameter or 5 × 5 mm^2 size and 100–250 mm length) is mounted on a strand holder in a closed bomb. This propellant strand must be inhibited laterally such that it can regress one-dimensionally like a cigarette burning.

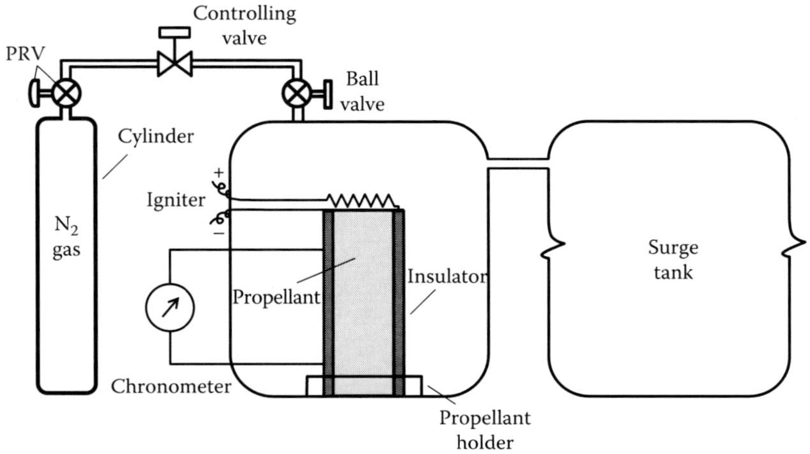

FIGURE 7.5 Schematic of strand burner setup. PRV, pressure regulating valve.

In order to pressurize the combustion bomb by N_2 gas cylinders, an inlet port with a pressure regulator and controlling valve is used as shown in Figure 7.5. The outlet port from the bomb is connected to a surge tank through a valve to maintain the desired pressure in it. During the burning of the propellant, the pressure inside the bomb may increase but has to be maintained at a constant with the help of a surge tank.

The initial ignition of the propellant is accomplished using a resistance coil placed on the top of the strand. As the propellant surface gets regressed downward, the decrease in the length of the propellant is measured either using the fuse-wire technique or by photography. In the fuse-wire method, embedded wire in the propellant strand provides a signal on the successive arrival of the burning front. By measuring the precise time taken by the front to reach each predetermined wire position in the strand, the average regression rate of the propellant can be easily determined. In the case of the photography method, the regression front movement is identified with time from the videographs that are used to determine the regression/burning rate. Recall that inhibitors are used around the propellant circumference to ensure that the burning of the propellant takes place along the height of the strand, thus providing the linear burning rate. Of course, a certain portion of the heat produced by its combustion is utilized for melting the thin layer of inhibitor, thus lowering the actual burning rate. Hence, it is important to compare the burning rate data from the strand burner with those of the BEM. For this purpose, a small-end burning rocket engine with a grain diameter of 5–10 cm can be used to measure the burning rate. In a BEM, the effects of pressure on burning rate can be obtained using different sizes of nozzles to vary the chamber pressure. Generally, the regression rate of a propellant from a strand burner matches well with that of the BEM. But for certain cases, the regression rate of the strand burner/BEM differs by even as much as 8%–12% from the regression rate prevailing in the actual rocket engine, which can be attributed to specific grain geometry and effects of turbulence and radiation heat transfer. In an actual rocket engine, the burning rate $\dot{r}$ of a propellant depends not only on its composition, but also on the following conditions prevailing during the operation of the SPRE, which are discussed in the following sections [3]:

1. Chamber pressure P_c as gas phase reaction plays an important role in providing heat to the propellant burning surface
2. Initial temperature of the propellant

3. Combustion gas temperature
4. Flow of hot gases on the propellant surface
5. Motor acceleration and spinning

7.5.1 Effect of Chamber Pressure on Burning Rate

The burning rate of the propellant is strongly dependent on the chamber pressure. The effect of the chamber pressure P_c on the regression/burning rate $\dot{r}$ of a particular propellant can be easily obtained using the strand burner method. By conducting several experiments at different pressure levels, an empirical relationship between the burning rate $\dot{r}$ and the chamber pressure P_c can be determined, which is known as the burning rate law. The most widely used and elegant empirical burning rate law known as Saint–Robert's law or Vieille's law is expressed as follows:

$$\dot{r} = aP_c^n \tag{7.3}$$

where

- a is the empirical constant influenced by the initial temperature of the propellant grain
- P_c is the chamber pressure
- n is known as the pressure exponent or combustion index, dependent on the constituent of the propellant

This relationship can be represented as a straight line in a logarithm plot as shown in Figure 7.5. This was mainly developed for internal ballistic rocket motors. The more general form of this empirical burning rate law is expressed as follows:

$$\dot{r} = aP_c^n + b \tag{7.4}$$

where a and b are the empirical constants influenced by the initial temperature of the propellant grain for a particular propellant. For very high chamber pressure, particularly such as that prevailing in artillery, a linear burning law known as Muraour's law has been developed:

$$\dot{r} = aP_c + b \tag{7.5}$$

Note that these relationships can be applied for both composite and DB propellants. However, these empirical burning rate laws are valid only within their respective ranges of chamber pressure and initial temperature of grain. All the other correlations for burning rate are meant for steady-state conditions. However, erratic burning may occur below certain chamber pressures that might not have been captured by the empirical relation for regression rate. Besides this, specific units for which these relationships are obtained must be employed while referring to them.

Let us consider the experimental data of burning law for both DB and composite modified double-base (AP-CMDB) propellants as shown in Figure 7.6. It may be noted in this figure that the burning rate increases with chamber pressure P_c and with increasing concentration of AP. The regression rate increases with decreasing particle size. The DB propellant has a higher burning rate compared to the CP for the same pressure. Hence, the DB propellant has a higher combustion index n in the range of 0.6–0.8 compared to that of 0.3–0.5 of the CP. Depending on the value of this combustion index n, a solid propellant can be categorized as normal burning ($n > 0$), plateau burning ($n = 0$), or mesa burning ($n > 0$) as shown in Figure 7.7. It may be observed from this figure that for a normal-burning propellant, the burning rate increases while for a mesa-burning propellant, the burning rate decreases with chamber pressure P_c for a given initial temperature of grain. But in the case of a plateau-burning propellant,

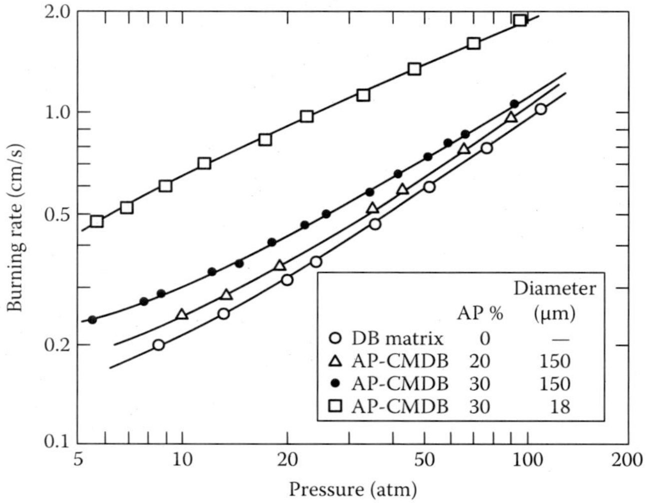

FIGURE 7.6 Variation of regression/burning rate $\dot{r}$ with chamber pressure. (From Kubota, N., *Propellants and Explosives, Thermochemical Aspects of Combustion*, Wiley-VCH, Weinheim, Germany, 2002.)

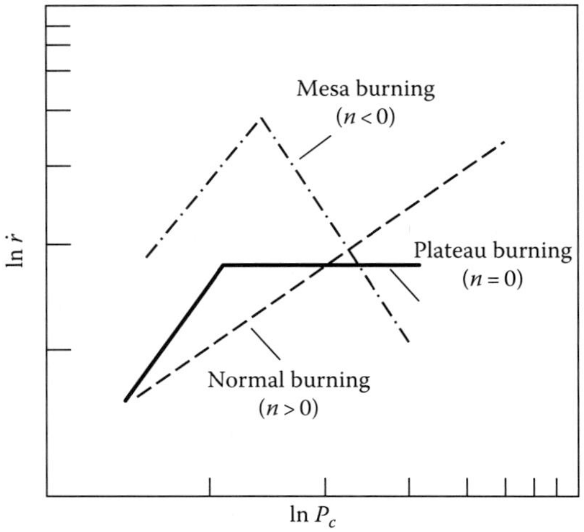

FIGURE 7.7 Variation of regression/burning rate, $\dot{r}$ with chamber pressure: normal burning ($n > 0$), plateau burning ($n = 0$), and mesa burning ($n > 0$).

the burning rate remains almost constant with an increase in chamber pressure P_c. A much more complicated correlation can be interpreted in terms of these three burning modes.

Example 7.1

During testing of a new propellant in a strand burner, the regression rate at a chamber pressure of 7 and 17 MPa are found to be 25 and 45 mm/s, respectively. If the regression rate happens to follow Saint–Robert's law, determine the chamber pressure when it regresses at 35 mm/s.

Solution

As per Saint–Robert's law, the regression rate of the propellant is related to chamber pressure by Equation 7.3:

$$\dot{r} = aP_c^n$$

$$\text{At } P_c = 7 \text{ MPa}, \quad 0.025 = a(7\times10^6)^n \tag{A}$$

$$\text{At } P_c = 7 \text{ MPa}, \quad 0.045 = a(7\times10^6)^n \tag{B}$$

Using Equation B by (A) we have

$$\frac{0.045}{0.025}=\frac{\left(17\times10^{6}\right)^{n}}{\left(7\times10^{6}\right)^{n}}; \quad \rightarrow n=\frac{\ln(1.8)}{\ln(2.43)}=0.662$$

The constant a in Saint–Robert's law (Equation 7.3) can be determined as

$$a=\frac{\dot{r}}{P_c^{n}}=\frac{0.025}{\left(7\times10^{6}\right)^{0.662}}=7.35\times10^{-7}$$

For $\dot{r}=35\,\text{mm/s}$, the chamber pressure can be evaluated as

$$P_c=\left(\frac{\dot{r}}{a}\right)^{\frac{1}{n}}=\left(\frac{0.035}{7.35\times10^{-7}}\right)^{\frac{1}{0.662}}=11.57\times10^{6}$$
$$=11.57\text{ MPa}$$

7.5.2 Effects of Grain Temperature on Burning Rate

The temperature of the solid propellant increases from the unburnt temperature at the left end T_u to the surface temperature T_s as shown in Figure 7.3. The propellant gets heated up during combustion as heat is conducted from the propellant burning surface to the unburnt propellant end. However, the initial unburnt temperature of the propellant T_u has a strong influence on the burning rate $\dot{r}$ as depicted in Figure 7.8; this is caused by surface kinetics due to condensed phase heat transfer. We will appreciate this point better when we discuss the modeling of propellant combustion. This dependence of $\dot{r}$ on the unburnt temperature can be expressed by the following empirical relationship:

$$a=\frac{A}{T_I-T_u} \tag{7.6}$$

where

A is the empirical constant
T_I is the auto/self-ignition temperature of the propellant
T_u is the unburnt temperature of the propellant grain

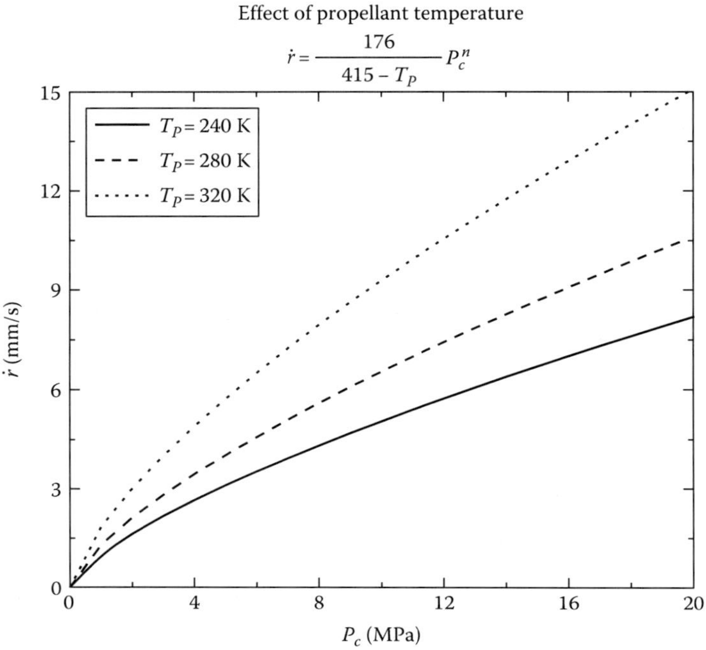

FIGURE 7.8 Variation in regression/burning rate $\dot{r}$ with chamber pressure for three unburnt temperature cases.

The sensitivity of the burning rate to variations in propellant temperature T_u at constant chamber pressure P_c is defined as follows:

$$\sigma_P = \left(\frac{\partial \ln \dot{r}}{\partial T_u}\right)_{P_c} = \frac{1}{\dot{r}}\left(\frac{\partial \dot{r}}{\partial T_u}\right)_{P_c} = \frac{1}{T_I - T_u} \tag{7.7}$$

where σ_P is the temperature sensitivity coefficient of the burning rate expressed as a percentage change of burning rate per degree change in propellant temperature at a particular value of chamber pressure. The value of σ_P varies from 0.1% to 1.0%/K for most propellants. Note that the value of σ_P depends on composition, combustion mechanism, and burning of propellant. Of course, the value of σ_P for a DB propellant will be on the higher side compared to that of a CP. Higher sensitivity is desirable for missile-type applications compared to launch vehicle applications. It is interesting to note from Equation 7.7 that if the unburnt temperature (T_u) of the propellant grain happens to be closer to the auto/self-ignition temperature of the propellant T_I, the burning rate will increase at a much faster rate, leading to the occurrence of an explosion, which must be avoided in the case of rocket engines.

Example 7.2

The burning rate of a particular propellant is given by $\dot{r} = aP_c^n$ in which the regression rate $\dot{r}$ is measured in mm/s, P_c is the chamber pressure in MPa, a is constant, and n = 0.68. When the propellant unburnt temperature (T_0) is 10°C and chamber pressure is 4.5 MPa, the regression rate happens to be 25 mm/s. If the temperature sensitivity of the burning rate happens to be 0.006 per °C, determine the new burning rate at the unburnt temperature (T) of 50°C.

Solution

By using Equation 7.7, the temperature sensitivity coefficient of the burning rate is expressed as follows:

$$\sigma_P = \frac{1}{\dot{r}}\left(\frac{\partial \dot{r}}{\partial T_u}\right)_{P_c} = 0.006$$

The burning rate can be determined at the unburnt temperature of 50°C by integrating this equation as follows:

$$\dot{r} = \dot{r}_{T_0} \exp\left[\sigma_P\left(T - T_0\right)\right] = 25\exp\left[0.006\left(50 - 10\right)\right] = 31.78 \text{ mm/s}$$

7.5.3 Effect of Gas Flow Rate

We know that in a rocket engine, gaseous products along with metal/metal oxides, formed during the burning of a solid propellant, are moved along the lateral surface of the grain. When this gas velocity exceeds threshold velocity, it enhances the burning rate of the propellant under quiescent atmospheric condition. This dependence of the burning rate on the lateral velocity of the combustion products is known as erosive burning. The experimental data for a DB propellant are shown in Figure 7.9. This indicates that the burning rate of the propellant remains almost constant below a certain velocity. For example, for a low-energy propellant, the burning

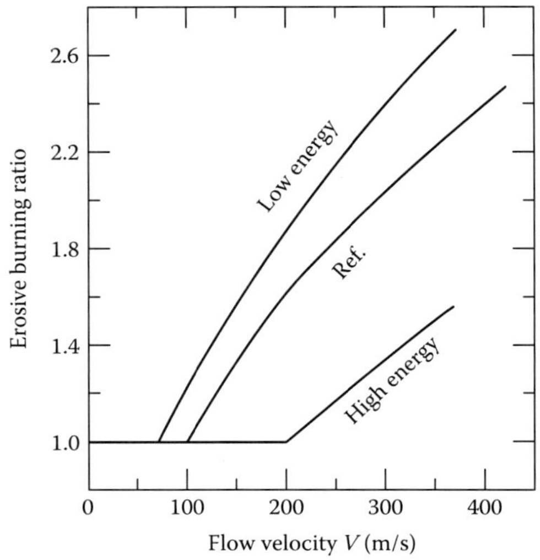

FIGURE 7.9 Erosive burning ratio and threshold velocity for different double-base propellants. (From Kubota, N., *Propellants and Explosives, Thermochemical Aspects of Combustion*, Wiley-VCH, Weinheim, Germany, 2002.)

rate remains constant till it attains a velocity of 70 m/s. Beyond this threshold velocity, the burning rate increases with gas velocity V over the propellant. This threshold velocity will depend on the type of propellant and its energetic quality. It is interesting to note that a high-energy propellant is less sensitive to convective flow velocity compared to a low-energy propellant. It has been observed that CPs exhibit less erosion than DB propellants. When the gas velocity on the propellant burning surface exceeds a certain threshold value in the range of 50–200 m/s, the propellant grain surface attains a higher temperature due to higher heat flux by the thinner thermal boundary layer, leading to higher burning rate. This existence of a threshold velocity V_{Th} for erosive burning to occur can be expressed by a simplistic linear relationship as follows:

$$\frac{\dot{r}_E}{\dot{r}} = 1 + K_E\left(V - V_{Th}\right)H\left[\left(V - V_{Th}\right)\right] \tag{7.8}$$

where

$\dot{r}_E$ and $\dot{r}$ are the burning rates with and without erosive burning, respectively
K_E is the erosion coefficient
V_{Th} is the threshold velocity
$H[(V - V_{Th})]$ is the Heaviside step function

Note that the convective term for erosive burning becomes zero for negative values of $(V - V_{Th})$. The values of the erosion coefficients K_E, V_{Th}, and V are dependent on the ratio of the port area to the throat area (A_p/A_t), as it influences the extent of erosive effect. When this area ratio (A_p/A_t) lies between 2 and 5, the velocities in the port can attain high subsonic velocity, which can augment the burning rate to an extent of 25%–100%. It is expected that the gas velocity over the burning surface increases from fore end to aft end as the mass flow rate of combustion gases is larger at the aft end (nozzle end) than at the fore end. This results in a change in burning rate along the length of the grain and pressure differences at the two ends. Thus, it can cause tapering of propellant grain at the nozzle end, leading to the early burnout of the web and mechanical failure of the grain. Of course, erosive burning is often likely to occur soon after ignition due to the smaller port area compared to the propellant surface area. Some of the conclusions from experiments on erosive burning are (1) higher burning rate propellants have lower erosion coefficients, (2) erosive burning occurs beyond a certain threshold mass flux, and (3) the extent of erosive burning is dependent on propellant compositions.

Extensive experimental and modeling studies on erosive burning for both DB and CP have been compiled by Mukunda and Paul [7] with an objective of proposing a universal relationship between burning rate with and without erosive burning. They proposed a single parameter G, known as the critical nondimensional mass flux ratio that encompasses all the effects of erosive burning, defined as follows:

$$G = \frac{\rho_g V}{\rho_P \dot{r}} \left[\frac{Re_p}{1000} \right]^{-0.125} \tag{7.9}$$

where

Re_p is the Reynolds number of the gas in the port
ρ_P is the density of the propellant
$\dot{r}$ is the nonerosive burning rate
V is the free stream gas viscosity

Similar to Equation 7.8, a new relationship between burning rate of erosive and nonerosive burning was formulated by Mukunda and Paul [7]:

$$\frac{\dot{r}_E}{\dot{r}} = 1 + 0.023\left(G^{0.8} - G_{Th}^{0.8}\right) H\left[\left(G - G_{Th}\right)\right]; \quad G_{Th} = 35 \tag{7.10}$$

Note that this expression correlates most of the reported experimental data within the experimental accuracies. Of course, the concept of threshold mass flux is preserved in this correlation. Besides this, experimental observation of enhancement in erosive burning rate for lower burning rate propellants has been embedded in the definition of mass flux ratio *G*. In other words, a decrease in burning rate leads to an increase in *G* and thus results in an increase in erosive burning rate. This expression is quite handy in designing SPREs.

7.5.4 Effects of Transients on Burning Rate

During the operations of a rocket engine, chamber pressure does not stay constant; rather it changes with time. Hence, the burning rate of the propellant will change with time as the chamber pressure varies. Let us consider a very simple model for the analysis of the effect of the transient chamber pressure on the burning of the propellant assuming that the instantaneous burning rate $\dot{r}(t)$ is related to the steady-state burning rate $\dot{r}$:

$$\dot{r}(t) = \dot{r} + \left(\frac{d\dot{r}}{dt}\right)dt + \cdots \tag{7.11}$$

Note that the expression can be expanded further as per Maclaurin series. But we will consider only the first term in this series for simplicity. The expression in Equation 7.11 can be recast in terms of chamber pressure:

$$\dot{r}(t) = \dot{r} + \left(\frac{d\dot{r}}{dP_c}\right)\frac{dP_c}{dt}dt + \cdots \tag{7.12}$$

Substituting the expression for the steady-state burning rate as per Saint–Robert's law (Equation 7.3) and the characteristics of time *dt* in terms of thermal diffusivity α_t and considering one-dimensional unsteady heat transfer on the propellant surface, we get

$$\dot{r}(t) = \dot{r} + a\ nP_c^{n-1}\frac{dP_c}{dt}\frac{\alpha_t}{\dot{r}^2} + \cdots = \dot{r}\left[1 + \frac{n\alpha_t}{\dot{r}^2}\frac{1}{P_c}\frac{dP_c}{dt} + \cdots\right] \tag{7.13}$$

This simple relationship is used with an empirical constant, *C* as follows:

$$\frac{\dot{r}(t)}{\dot{r}} = \left[1 + C\frac{n\alpha_t}{\dot{r}^2}\frac{1}{P_c}\frac{dP_c}{dt}\right] \tag{7.14}$$

This simple and elegant expression is used profusely, particularly for designing rocket engine, as other sophisticated models have not demonstrated any significant improvements in its prediction. Note that the constant C happens to be very close to unity as verified by experimental data. Hence, for all practical purposes, this constant of unity value can be used without incurring exorbitant errors in its prediction. It is interesting to note that this elegant expression indicates that during the sharp pressure rise, the burning rate must be much higher than the steady-state burning rate depending on the value of instantaneous chamber pressure. Of course, the burning rate will also decrease sharply with a decrease in chamber pressure. If this fall in chamber pressure is quite high, the propellant may be extinguished, which is generally known as depressurization extinction.

7.5.5 Effects of Acceleration on Burning Rate

The burning rate of the propellant in a rocket engine gets affected by the acceleration of the vehicle during its operation. These can be longitudinal or lateral acceleration or due to the spinning of the rocket engine around its longitudinal axis. For a particular propellant composition and grain, the change in burning rate depends on the magnitude and orientation of the acceleration vector with respect to the burning surface and the pressure level. Generally, when the acceleration vector forms an angle of 60°–90° with respect to the burning surface, the burning rate gets enhanced considerably. The burning rate of the internal burning grain gets altered more compared to the end burning grain for the same level of spinning during its operation. The effects of acceleration on the burning rate for CP (AP [57.3%], polyurethane [25%]; aluminum [17.7% by mass]) are depicted in Figure 7.11 [9] for three different pressure ranges. It may be observed that acceleration must have a threshold value to alter the burning rate of a propellant. Generally, it must be greater than $10g$ to have a tangible effect on the burning rate. It may also be observed that the effect of acceleration is dependent on chamber pressure and composition of the propellant. A similar observation was made by Northam [9], who demonstrated that the composition of the propellant is one of the controlling variables in the augmentation of the burning rate during its operation. The slow burning rate gets more affected by acceleration compared to the high burning rate propellant. Besides, it has been observed that even the burning rate exponent n apart from the burning rate for a particular propellant composition changes when subjected to acceleration. Generally, the burning rate exponent n gets enhanced with acceleration.

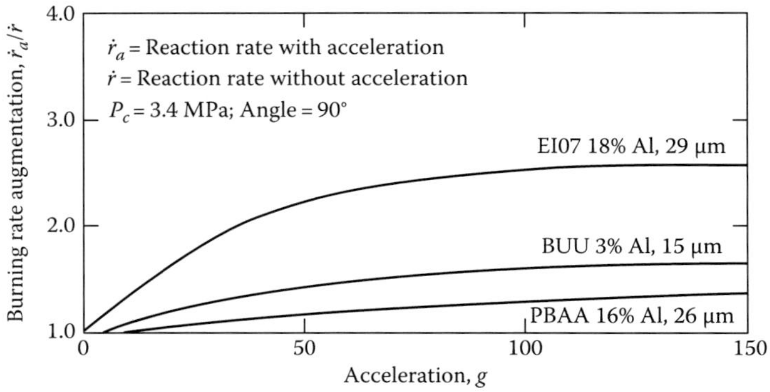

FIGURE 7.10 Effect of acceleration on the burning rate for three different propellants [8]. (From Anderson, J.B., Investigation of the effect of acceleration on the burning rate of composite propellants, PhD thesis, United States Naval Postgraduate School, New York, 1996.)

The propellant surface may crack due to induced stress caused by its rapid acceleration or due to a sharp chamber pressure gradient. As a result, there will be an augmentation in the burning rate due to an increase in surface area. It is quite difficult to ascertain how much augmentation in the burning rate is ascribed to the strain developed on the propellant and to what extent the prevailing flame structure on the propellant surface is due to the acceleration effect, as both occur simultaneously. There is no satisfactory theory till date that can predict the effects of acceleration on the burning rate; hence, engineers depend on experimental data for design purposes (Figure 7.10).

7.5.6 Other Methods of Augmenting Burning Rate

It is a common notion that the burning rate in a rocket engine is equal to the measured value of the burning rate in a strand burner at a particular pressure and grain temperature. But during the operation of a rocket engine, high-velocity/high-mass flow occurs over the propellant surface, which augments the burning rate measured in a strand burner at a particular pressure. This kind of burning is known as erosive burning, which is discussed in detail in Section 7.5.5. Besides this, the burning rate is dependent on the rate of change of chamber pressure, which is likely to occur during the actual operation of a rocket engine. Recall that the burning rate is highly dependent on the chamber pressure P_c. Moreover, the burning rate is affected by acceleration during the motion of the rocket engine. In order to evaluate this aspect, the ballistic evaluation method (BEM) is generally used over a wide range of operating conditions. All these aspects of burning rate augmentation are discussed in the three following sections.

7.5.6.1 Particle Size Effects

The effect of particle size and initial temperature on the burning rate for a typical AP (40%)–HTPB (20%)–aluminum (40% by mass) propellant at two different initial propellant temperatures is shown in Figure 7.11 for bimodal fine and coarse AP particles (fn: 3,20 and cn: 200,400 μm). It may be noted that the linear burning rate gets enhanced with decreases in AP particle size. As a result, there will be a decrease in the mixing distance between the oxidizer and fuel as discussed earlier, leading to the lowering of the flame standoff distance. When the flame stands near the propellant surface, a higher rate of heat transfer can enhance the burning rate. In addition, the burning rate increases with initial propellant temperature. However, any decrease in particle size also causes an adverse increase in the pressure index. It may be noted that the burning rate increases while the temperature sensitivity decreases with a decrease in particle size.

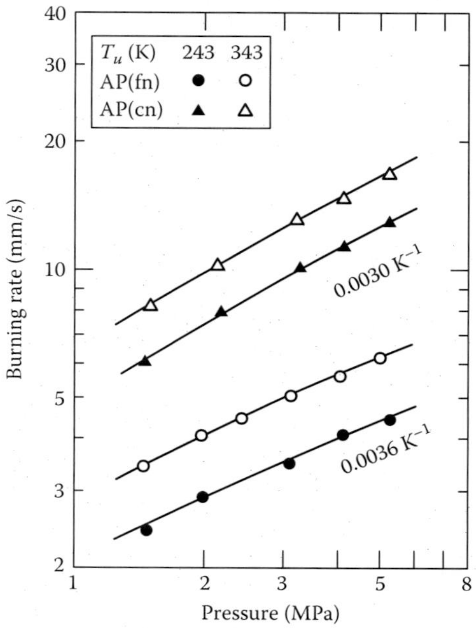

FIGURE 7.11 Burning rate and temperature sensitivity of AP–HTPB composite propellants composed of bimodal fine (fn: 3,20 μm) or coarse (fc: 200,400 μm) AP particles. (From Kubota, N., *Propellants and Explosives, Thermochemical Aspects of Combustion*, Wiley-VCH, Weinheim, Germany, 2002.)

7.5.6.2 Burning Rate Modifiers

Significant changes in burning rate can be brought about by adding a small percentage (around 0.5%) of additives to the propellant formulation. Some of the reported burn rate modifiers for AP-based propellants are iron oxide, copper oxide, manganese oxide, ferrocene, and copper chromate. Compounds like lithium oxides and fluorides, on the other hand, can reduce the burning rate of these propellants. DB propellants are also very sensitive to burn rate modifiers. Addition of lead stearate or lead salicylate in small percentages can enhance the burning rate of DB propellants significantly. Sometimes while modifying the burn rate, the addition of catalysts also alters the pressure index.

7.6 THERMAL MODEL FOR SOLID-PROPELLANT BURNING

We have already discussed the burning mechanism of both DB and CP in Section 7.4. Let us understand how the combustion index affects the burning law by considering a thermal model for propellant burning as depicted in Figure 7.12. The propellant gets converted into gases from the regressed surface and diffuses along the vertical direction (see Figure 7.12). It is assumed that these premixed gases react and form a one-dimensional flame at a certain standoff distance x^* from the regression surface of the propellant. The gas temperature increases from the surface temperature T_s at the propellant surface to the flame temperature T_F at the standoff distance x^*. We can assume that the heat is transferred from the flame to the propellant surface by heat conduction only. Assuming that the entire

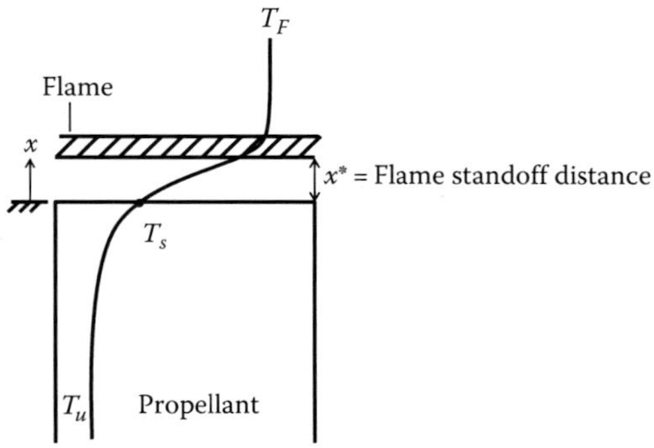

FIGURE 7.12 A simplistic model for solid-propellant burning.

process takes place under steady-state condition, we conduct an energy balance at the surface of the propellant as follows:

$$k_g \left[\frac{dT}{dx} \right]_s = \rho_P \dot{r} \left[C\left(T_s - T_u\right) + \Delta H_v \right] \tag{7.15}$$

where

$\rho_P \dot{r}$ is the product of propellant density and regression/burning rate
ΔH_v is the heat of phase change from solid to gas
k_g is the thermal conductivity of gas
C is the specific heat of the propellant
T_s is the surface temperature of propellant
T_u is the unburnt temperature of the propellant

In this expression, heat flux from the flame surface due to heat conduction is utilized to regress the propellant. Note that assuming the temperature profile to be linear, we can replace dT by the temperature difference $(T_F - T_s)$ and substitute dx by the flame standoff distance x^*. Note that the flame standoff distance x^* can be considered to be the same as the flame thickness δ_F defined in the one-dimensional premixed flame [10] and can be expressed as follows:

$$x^* = \delta_F = \sqrt{\frac{k_g}{C_{Pg} \dot{m}'''}} \tag{7.16}$$

where

C_{Pg} is the specific heat of the gaseous product
$\dot{m}'''$ is the average reaction rate (kg/m^3·s) of the propellant, which is expressed as

$$\dot{m}''' = A_F Y_F^{n1} Y_{Ox}^{n2} P^n e^{-E/RT} \tag{7.17}$$

where

A_F is the pre-exponential factor
Y_F and Y_{Ox} are the mass fraction of fuel and oxidizer, respectively
E is the activation energy
R is the universal gas constant
T is the temperature and three exponents, namely, $n1$, $n2$, and n, are for fuel, oxidizer, and pressure P, respectively

Combining Equations 7.15 through 7.17, we obtain a relation for the regression rate $\dot{r}$:

$$\dot{r} = \frac{k_g \left[\frac{dT}{dx}\right]_s}{\rho_P\left[C(T_s - T_u) + \Delta H_v\right]} = \frac{\sqrt{k_g C_{Pg} A_F Y_F^{n1} Y_{Ox}^{n2} P^n e^{-E/RT}}}{\rho_P\left[C(T_s - T_u) + \Delta H_v\right]} \tag{7.18}$$

where n is the overall order of reaction, which varies from 1.6 to 1.8. From this expression, it may be noted that the regression rate $\dot{r}$ is proportional to pressure $P^{n/2}$. From this simplistic model, it can be shown that the combustion index m in Saint–Robert's law can vary from 0.8 to 0.9, particularly for a DB propellant as its combustion takes place in the premixed flame mode.

7.7 SOLID-PROPELLANT ROCKET ENGINE OPERATION

We know that the SPRE filled with gas from burning propellant grain with the help of an ignition system can be conceived as a pressure vessel, which produces thrust when its high-temperature, high-pressure gas is expanded through the exhaust nozzle. Note that combustion is restricted to the selected areas of the propellant grain. When a propellant grain is ignited, heat liberated during ignition is transferred by the combination of conduction, convection, and radiation to the unburnt propellant (see Figure 7.2). Let us understand the processes that occur during the ignition of a solid propellant in the next section.

7.7.1 Ignition of a Solid Propellant

We know that ignition in an SPRE is a transient process by which burning of solid-propellant grain is initiated, leading to self-sustained combustion. This is a very critical phenomenon as the success of a rocket engine is dependent on the successful initiation of combustion, particularly during flight. Ignition must take only a fraction of a second during which local ignition of the propellant occurs, leading to the spread of the flame along the propellant grain surface, building up the chamber pressure. In a successful ignition, the chamber pressure increases with time to an equilibrium value as shown in Figure 7.13. The processes involved during the ignition of a solid propellant include a series of complex rapid events. Generally, three types of energy stimuli, namely, thermal (conduction, convection, and radiation heat transfer), chemical (e.g., hypergolic liquid), and mechanical (e.g., high-velocity impact), are

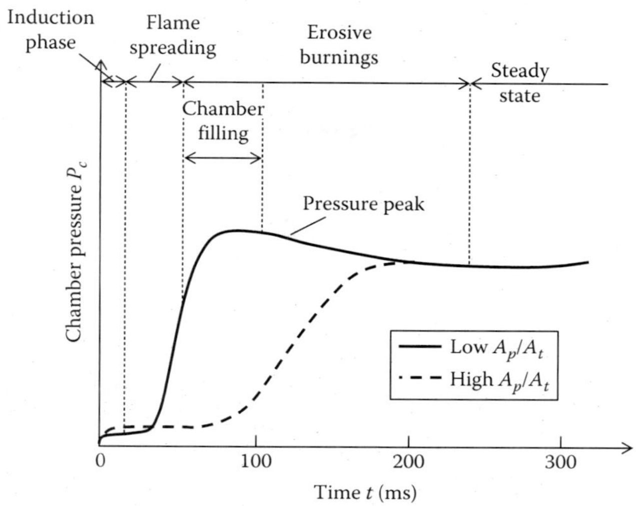

FIGURE 7.13 Description of ignition process in a solid-propellant rocket engine.

used for the ignition of a solid propellant. As mentioned earlier, ignition is basically a complex physiochemical process as shown in Figure 7.2. Once the solid propellant receives energy stimuli, it can transfer heat to an adjacent surface, which is known as *inert heating*. After this period of inert heating, a certain portion of the solid phase starts disintegrating around the local ignition point due to pyrolysis. In some cases, a molten layer is formed on the solid surface. The gasification of this layer or direct sublimation of the solid propellant can also take place due to a number of processes, such as heat conduction, absorption of radiation below the surface, chemical reactions below the surface, and surface pyrolysis. Note that chemical reactions can take place either in the gas phase (homogeneous reaction) or solid/liquid phase (heterogeneous reaction). Some of these reactions can be exothermic and some endothermic in nature. But the sum total of heat release during these chemical reactions must be higher than the total heat loss for achieving a steady ignition, leading to successful initiation of self-sustained combustion of a solid propellant. Of course, there will be an increase in temperature of the virgin solid propellant and heat losses due to convection and radiation. In the end, a runaway condition is achieved when the local flame spreads over the entire propellant grain surface by filling the entire chamber with high-temperature product gas and thus the combustion chamber attains its equilibrium pressure. This is generally accompanied by high chemical reaction rates, heat release, and steady burning of the solid propellant. All these processes in actual rocket engines take place within a few microseconds to milliseconds. The entire ignition process can be broadly divided into three phases—(1) induction phase, (2) flame spreading, (3) chamber filling—as depicted in Figure 7.13 by considering typical pressure–time history curves for two cases: (1) low A_p/A_t and (2) high A_p/A_t ratio. Just after the ignition phase of an internally burning grain is complete, gas velocity along the burning surface becomes very high so that heat transfer to the propellant surface gets enhanced considerably, leading to the occurrence of erosive burning as discussed earlier. Note that erosive burning is likely to take place for low A_p/A_t ratio as shown in Figure 7.13. Note that phase I in this figure indicates ignition time lag, which is caused by inert heating of the virgin propellant. The ignition time lag is defined as the time period between initiation of ignition stimuli and first burning of the grain surface. During this inert heating, pressure remains almost constant at a very low value. Phase II is the flame-spreading interval that spans from the first local ignition of the grain surface till entire surface is ignited along with flame spreading. Phase III, the chamber-filling interval, is the time required to reach equilibrium pressure. The successful attainment of equilibrium chamber pressure is dependent on (1) igniter composition, and its gas temperature; (2) flow created due to igniter; (3) heat transfer due to convection, and radiation between igniter gas and grain surface; (4) propellant composition; (5) flame-spreading rate over grain surface; and (6) the dynamics of filling the chamber volume with hot gas. In short, the heat transfer gas phase mixing (consisting of both diffusion and convection) and chemical kinetics are the key processes that influence the final ignition. In other words, the ignition delay, time elapsed between applications of external stimuli and steady burning, is dependent on three characteristic times: (1) inert heating time, (2) mixing time, and (3) reaction time. In actual rocket burning, it is quite difficult to differentiate among these three phases as they may overlap. Note that ignition delay is an important parameter in propellant combustion. Experimental results have shown that ignition delay (t_{ig}), defined as the time lapse from the firing signal to the chamber pressure attains 10%–20% of the motor peak pressure, and varies inversely with the incident heat flux (Q'') in the form $t_{ig} \sim 1/(Q'')^n$ with n ranging from 1.5 to 3. Besides this, the rate of heat flux and time of ignition are also dependent on the gaseous composition over the burning surface, the gas pressure above, and gas velocity across

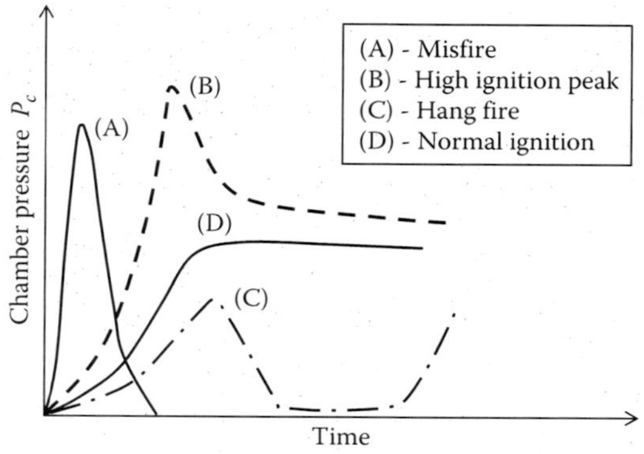

FIGURE 7.14 Description of combustion phenomena in a solid-propellant rocket engine in terms of chamber pressure and time.

the propellant surface. In general, ignition delay becomes shorter with an increase in both heat flux and chamber pressure.

During this ignition process, several situations may arise as depicted in Figure 7.14. Misfire (see curve A in Figure 7.14) is likely to occur when heat withdrawal from the propellant surface by conduction exceeds the net heat generated by the exothermic reactions. The ignition can be termed as normal only when it leads to a smooth and rapid pressure transition to steady-state burning as depicted in Figure 7.14 (see curve D). However, if the ignition energy is too strong, it can result in high initial pressure peaks (see curve B in Figure 7.14). This situation must be avoided as it may result in fissure of grains due to the formation of shocks and impact forces. In contrast, low ignition energy may lead to hang fire, and chuffing (see curve C in Figure 7.14) due to lower chamber pressure, which must be avoided.

7.7.2 Action Time and Burn Time

A typical variation of chamber pressure with time is shown in Figure 7.15. It may be noted that chamber pressure changes very slowly over a small period of time, which is known as ignition delay time t_d. Beyond this time period, chamber pressure rises rapidly due to the generation of burnt gases as a result of combustion of propellants in both igniter and main propellant. Generally, igniter propellant mass contributes to 30% of the pressure rise in the initial phase. On the burning of the propellant, the chamber pressure builds up and attains a stable operating condition as shown in Figure 7.15. In some cases, particularly for upper-stage rocket

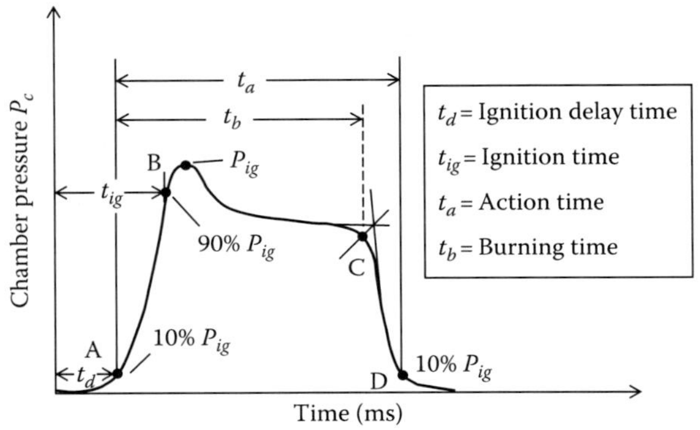

FIGURE 7.15 Description of combustion phenomena in a solid-propellant rocket engine in terms of chamber pressure and time.

engines, for successful ignition of the solid propellant, chamber pressure has to be built up rapidly by sealing the nozzle with a rupture disk, which is designed to fail at a predetermined pressure, typically 70%–90% of the operating pressure. If higher ignition energy is used, the propellant grain generates an excess amount of combustible gas that leads to abnormal ignition as shown in Figure 7.15. The time duration corresponding to 90% of ignition pressure (point B in Figure 7.15) is known as the ignition time t_{ig}. Subsequently, the chamber pressure drops to the steady-state pressure value, which remains almost constant as shown in Figure 7.13 for the major portion of its operation. Note that in some cases chamber pressure keeps increasing or decreasing or remains constant depending on the variation in the burning area of grain with time. Toward the end of propellant burning, the chamber pressure slowly drops to ambient pressure as shown in Figure 7.13 as all of the propellant is almost burnt out except the residual portion known as sliver, which cannot be utilized for developing any thrust. This portion of the curve that indicates the end of the burning phase is known as the tail-off phase. Note that the tail-off phase becomes slower, particularly for highly loaded propellant grain, in which case erosive burning is likely to take place. The intersecting point between the tangent to grain burning and the tangent to the tail-off phase is labeled as point C in Figure 7.15, which indicates the end of the burning phase. Hence, the total burning time is defined as the time between point A and point C (see Figure 7.15). The rocket can provide thrust beyond the end of propellant burning till the chamber pressure becomes equal to around

10% P_{ig} (point A in Figure 7.15). Hence, the action time t_a is defined as the time duration between the ignition delay time (point A) and the time corresponding to 10% P_{ig} on the tail-off curve (point D in Figure 7.15). Note that the action time is always more than the burning time t_b. The difference between action time and burning time is more predominant (4%–6%) in the case of highly loaded erosive burning grain compared to small erosive burning grain (1%–2%).

7.8 INTERNAL BALLISTICS OF SPRE

We can model the operation of a typical rocket engine by making the following assumptions:

1. Uniform pressure across the entire combustion chamber.
2. Combustion products can be treated as ideal gas.
3. Burning rate remains constant over entire burning surface and is governed by $\dot{r} = aP_c^n$.
4. The time for the mass to flow through the nozzle is negligible compared to the rate of pressure change within the combustion chamber even during transient state.

We know that the operation of a rocket engine depends on the gaseous mass generation during the burning of the propellant, the mass expelled through the nozzle, and the accumulation of gas in the combustion chamber. Hence, we need to apply the continuity equation to the control volume (CV) for a rocket engine as shown in Figure 7.16.

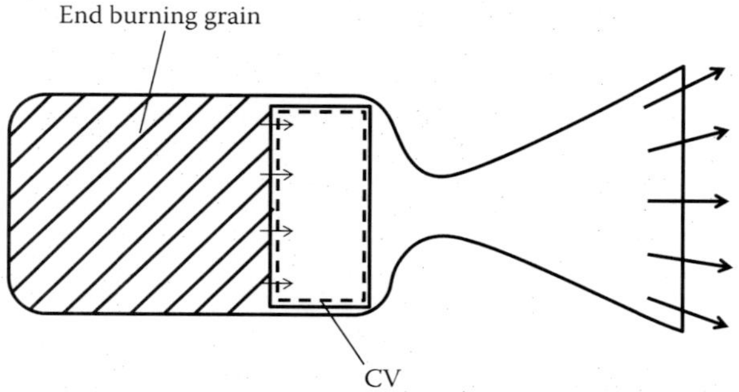

FIGURE 7.16 Schematic diagram of CV in a rocket engine.

$$\frac{dm}{dt} = \dot{m}_g - \dot{m}_n \tag{7.19}$$

where

the first term dm/dt in Equation 7.19 represents the rate of mass accumulation in a rocket engine

$\dot{m}_g$ is the mass generation due to propellant burning

$\dot{m}_n$ is the mass flow rate leaving the nozzle

Besides this, the gas generation rate $\dot{m}_g$ is also dependent on the chamber pressure P_c since the linear burning rate $\dot{r}$ is dependent on P_c. Hence, the mass flow rate of gas generated due to the burning of the propellant is given by

$$\dot{m}_p = \dot{m}_g = \rho_p A_b \dot{r} = \rho_p A_b a P_c^n \tag{7.20}$$

where

$\dot{m}_g$ is the mass generation due to propellant burning, which is equal to the mass of propellant $\dot{m}_p$ per unit time that is consumed during its burning

ρ_P is the density of the propellant

A_b is the burning surface area

P_c is the chamber pressure

Assuming the nozzle flow is choked, the mass flow rate through the nozzle can be expressed as follows:

$$\dot{m}_n = \frac{A_t P_c \Gamma}{\sqrt{RT_c}}; \quad \Gamma = \sqrt{\gamma}\left(\frac{2}{\gamma+1}\right)^{\frac{(\gamma+1)}{2(\gamma-1)}} \tag{7.21}$$

where

T_c is the chamber temperature

A_t is the throat area

γ is the specific heat ratio

By combining Equations 7.19 through 7.21, we get

$$\frac{dm}{dt} = \frac{d\left(P_c V_c / RT_c\right)}{dt} = \rho_P A_b a P_c^n - \frac{A_t P_c \Gamma}{\sqrt{RT_c}} \tag{7.22}$$

where
- V_c is the chamber volume
- R is the specific gas constant

By assuming that T_c does not vary with time, we get

$$\frac{d\left(P_c V_c / RT_c\right)}{dt} = \frac{V_c}{RT_c}\frac{dP_c}{dt} + \frac{P_c}{RT_c}\frac{dV_c}{dt} = \rho_P A_b a P_c^n - \frac{A_t P_c \Gamma}{\sqrt{RT_c}} \tag{7.23}$$

Assuming an ideal gas, and recognizing that chamber volume change with time is caused due to the burning rate of the propellant, we obtain

$$\frac{P_c}{RT_c}\frac{dV_c}{dt} = \rho_g A_b a P_c^n \tag{7.24}$$

By combining Equations 7.23 and 7.24, we have

$$\frac{V_c}{RT_c}\frac{dP_c}{dt} = A_b a P_c^n \left(\rho_P - \rho_g\right) - \frac{A_t P_c \Gamma}{\sqrt{RT_c}} \tag{7.25}$$

This expression indicates that the chamber pressure will vary with time during its operation whenever there is an imbalance between the gas generation rate and the mass flow rate leaving the nozzle. Under steady-state condition, the term (dP_c/dt) on the left-hand side of Equation 7.25 will be zero. Then, for steady-state operation, Equation 7.25 becomes

$$P_c = \left[\frac{A_b}{A_t}\frac{a\left(\rho_P - \rho_g\right)}{\Gamma / \sqrt{RT_c}}\right]^{1/(1-n)} \tag{7.26}$$

By substituting $C^* = \sqrt{RT_c}/\Gamma$ in Equation 7.26 and recognizing that ρ_g is quite small compared to propellant density ρ_P, we derive an expression for the equilibrium chamber pressure P_c under quasi-steady-state operation of an SPRE:

$$P_c = \left(a \rho_P C^* AR \right)^{1/(1-n)} \tag{7.27}$$

where AR is the ratio of the burning surface area and the throat area A_b/A_t. Generally, for most solid rockets, AR ranges from 200 to 1000. We observe that the chamber pressure P_c is proportional to AR when the combustion exponent n is zero. If the combustion index n happens to be greater than 1, the chamber pressure becomes infinity, leading to explosion. Hence, such a propellant with n greater than 1 is not designed for rocket engine applications. When n lies between 0 and 1, any small change in burning surface area and, in turn, area ratio AR, will have a significant effect on the chamber area, which can be appreciated by logarithmic differentiation of Equation 7.27:

$$\frac{dP_c}{P_c} = \frac{1}{1-n} \frac{d(AR)}{AR} \tag{7.28}$$

For achieving a constant chamber pressure, the burning surface area A_b must remain constant. We know that if P_c remains constant, we can get a constant thrust. Recall that the thrust F is proportional to A_t, as $F = P_c A_t C_F$, where C_F is the thrust coefficient and P_c is the chamber pressure, which depends on A_b. The variation of A_b with time depends on the burning rate $\dot{r}$ and the initial geometry of the propellant grain configuration. Thus, the geometric variation of grain can provide increasing (progressive), constant (neutral), and decreasing (regressive) thrust with time. We will discuss in detail the various propellant grain configurations in the subsequent section.

Recall that we had defined a term known as temperature sensitivity coefficient σ_{P_c} for regression rate at constant chamber pressure (see Equation 7.7). Now let us define a temperature sensitivity coefficient for chamber pressure that expresses the chamber pressure change with respect to the initial temperature:

$$\left(\pi_P \right)_{AR} = \frac{1}{P_c} \left(\frac{\partial P_c}{\partial T_u} \right)_{AR} \tag{7.29}$$

Generally, $(\pi_P)_{AR}$ is used as a propellant variable during the design phase. Its value varies from 5×10^{-3} to 5×10^{-4} K^{-1} for CP and from 10^{-2} to 10^{-3} K^{-1}

for DB propellants. There is another temperature sensitivity coefficient that indicates relative variation in burning rate with initial temperature at a fixed value of AR, which is defined in Equation 7.7:

$$\sigma_P = \left(\frac{\partial \ln \dot{r}}{\partial T_u}\right)_{P_c} = \frac{1}{\dot{r}}\left(\frac{\partial \dot{r}}{\partial T_u}\right)_{P_c} \tag{7.30}$$

We know that $\dot{r} = aP_c^n = \dfrac{A}{T_I - T_u} P_c^n$. By substituting this expression for $\dot{r}$ in Equation 7.30, we obtain

$$\left(\pi_{\dot{r}}\right)_{AR} = \frac{1}{\dot{r}}\left(\frac{\partial \dot{r}}{\partial T_u}\right)_{P_c} = \frac{1}{a}\left(\frac{da}{dT_u}\right) = \frac{1}{T_I - T_u} \tag{7.31}$$

Note that when the unburnt temperature of propellant T_u tends toward the self-ignition temperature of the propellant T_I, the burning rate $\dot{r}$ becomes highly sensitive to temperature and may lead to an explosion. Hence, the propellant should have a high value of self-ignition temperature T_I with respect to its expected operational temperature. Let us now find the relationship among three temperature sensitivity coefficients:

$$\left(\sigma_P\right)_{AR} = \left(\sigma_{\dot{r}}\right)_{AR} = \sigma_{AR} = \frac{1}{1-n}\left(\sigma_{\dot{r}}\right)_{P_c} \tag{7.32}$$

From this relationship, it may be observed that the sensitivity of chamber pressure to initial temperature of the grain for a fixed value of AR is higher than the sensitivity of burning rate $(\pi_{\dot{r}})_{AR}$ by a factor $1/(1 - n)$. Hence, it is important to have low values of n apart from a high self-ignition temperature for safe use of propellants in rocket engines. Recall that for stable operation of the rocket engine, the pressure coefficient n must be less than unity.

Example 7.3

In a rocket engine, a solid propellant with $n = 0.68$, as in Example 7.2, is used, which has a thrust coefficient C_F of 0.95, and a characteristic velocity C^* of 1850 m/s. When the unburnt propellant temperature is 10°C and the chamber pressure is 4.5 MPa, the regression rate happens to be 25 mm/s. If the temperature sensitivity of the burning rate happens to be 0.006/°C and the burning surface area to throat area AR is 65, determine the chamber pressure and thrust at the unburnt temperature of 50°C assuming that C_F and C^* remain almost constant? Consider the throat area as 450×10^{-6} m^2 and propellant density as 1350 kg/m^3.

Solution

By using Equation 7.27, the chamber pressure P_c under quasi-steady-state operation of an SPRE can be determined as

$$P_c = \left(a\rho_P C^* AR\right)^{1/(1-n)}$$

We need to evaluate the constant a by using Equation 7.3 as follows:

$$a = \frac{\dot{r}}{P_c^n} = \frac{31.78 \times 10^{-3}}{\left(4.5 \times 10^6\right)^{0.68}} = 0.94 \times 10^{-6}$$

By substituting values in Equation 7.27, we can estimate P_c as

$$P_c = \left(0.94 \times 10^{-6} \times 1350 \times 1850 \times 65\right)^{1/(1-0.68)} = 6.7 \times 10^6 \text{ Pa} = 6.7 \text{ MPa}$$

The thrust can be determined as

$$F = C_F P_c A_t = 0.95 \times 6.7 \times 10^6 \times 450 \times 10^{-6} = 2864.25 \text{ N}$$

7.8.1 Stability of SPRE Operation

We have learnt just now that for the combustion process in the rocket motor to be stable, the burning rate exponent "n" must be less than unity. Let us understand the process of unstable combustion physically. For this purpose, we assume that the combustion temperature T_c remains constant. Then the mass flow rate through the fixed throat area nozzle is proportional to the chamber pressure P_c, which is depicted in Figure 7.17 as a straight line passing through its origin. We know that the mass flow rate $\dot{m}_g$ generated due to the burning of the propellant is proportional to P_c^n. For cases

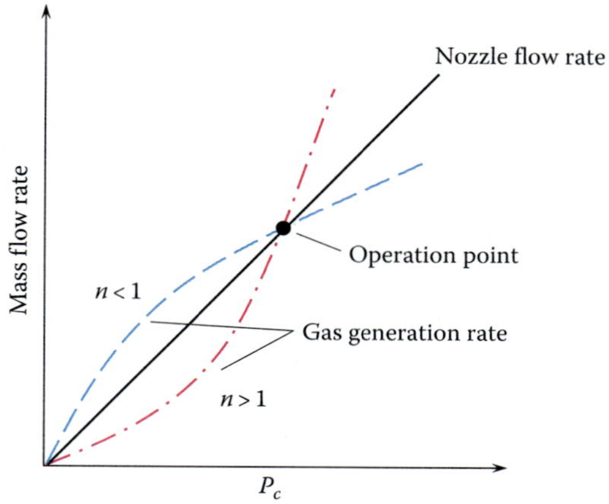

FIGURE 7.17 Stable operation of an SPRE.

where $n > 1$ and $n < 1$, the mass flow rate generated $\dot{m}_g$ with chamber pressure is depicted in Figure 7.17 by two different curves. The crossing point S between the mass flow rate generated $\dot{m}_g$ and the mass flow rate through the nozzle $\dot{m}_n$, known as the stable point, corresponds to the stable operation of the rocket engine because there would not be any accumulation of mass in the chamber. In other words, the gas mass flow rate generated $\dot{m}_g$ is equal to the mass flow rate passing through the nozzle, leading to a constant chamber pressure. Let us consider that there is a small change in chamber pressure by P_c' over its steady value. Then the mass flow through the nozzle increases as shown in Figure 7.17. For $n < 1$, the mass generation rate $\dot{m}_g$ is less than the mass passing through the nozzle $\dot{m}_n$, which will result in a decrease in chamber pressure. Hence, the chamber pressure will return to its original stable point. If the chamber pressure drops down by P_c' over its steady value, then mass will be accumulated in the chamber due to an imbalance, which will bring the system to its original stable point of operation. In contrast, for $n > 1$, it can be shown by a similar argument that any small perturbation in chamber pressure or any other variable can drive the system to an unstable condition from its stable operation point.

7.9 PROPELLANT GRAIN CONFIGURATION

The propellant grain is one of the important parts of the combustion chamber in an SPRE. We know that the propellant grain is basically the body of the propellant in a particular shape and size that burns on its exposed surface to generate hot gas in a rocket engine. The shape and size of the

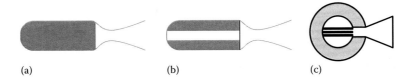

(a) (b) (c)

FIGURE 7.18 Classification of solid-propellant grain configurations: (a) end burning grain, (b) side burning grain, and (c) noncylindrical grain.

propellant grain, apart from its composition and burn rate, dictate the performance of the rocket engine. The propellant grain looks like hard plastic or rubber of a particular shape and size. Generally, a single grain is used in most rocket engines. But in some rocket engines, multiple grains can be used with different burning rates achieved from different propellant compositions to realize a certain thrust profile. Of course the propellant grain is designed to meet specific mission requirements. As shown in Figure 7.18, propellant types can be broadly divided into three categories: (1) end burning, (2) side (internal) burning, and (3) noncylindrical grains.

1. *End burning grain*: This is the simplest cylindrical grain in which combustion takes place across the transverse cross section of the chamber as shown in Figure 7.18a. Hence, it is sometimes also called cigarette burning grain. It is generally used for low-thrust, low-performance rocket engines with long burning time as it is limited by a particular cross-sectional burning surface area. Of course it is quite easy and natural to get a constant thrust during the entire duration, even beyond 10 min. But for a longer duration of operation, thick chamber walls must be used to protect the casing from heating excessively due to direct exposure of high-pressure and high-temperature combustion gases. As a result, the chamber diameter and weight become quite exorbitant even for medium-thrust rocket engines. Moreover, the center of gravity of the propellant grain shifts while it is burning, making its motion unstable. Hence, its use is restricted to small-thrust rocket engine applications. In recent times, high-thrust and high-performance side burning grain has been devised with the advent of fast-burning propellants.

2. *Side burning grain*: In side burning grain, combustion takes place on the lateral surface of the grain and the end surfaces are protected from burning in contrast to the end burning grain. It may be observed in

Figure 7.18 that burning takes place on the internal surface of the straight cylindrical grain and covers a higher burning surface area compared to end burning grain for the same length and diameter. During the burning, high-pressure hot gases are not exposed to the wall of the casing till the end of the burning and thus the weight of the casing is quite low compared to that for end burning grain for the same size of rocket engine. Hence, side (internal) burning propellant grains are used profusely in most modern rocket engines. Recall that the mass generation rate due to the burning of the propellant surface in the combustion chamber can be expressed as follows:

$$\dot{m}_P = \dot{m}_g = \rho_P \dot{r} A_b \tag{7.33}$$

where

ρ_P is the density of the propellant
A_b is the burning surface area

In the case of a cylindrical side burning grain, as the burning surface regresses outward from the inner side of the grain till it reaches the casing wall, the burning surface area A_b increases with time, resulting in an increase in chamber pressure. Hence, this kind of propellant grain is known as progressive burning grain. If the burning surface area A_b remains constant during the entire period of the burning, as in the case of a typical end burning grain, then the chamber pressure remains almost constant with time and is termed neutral burning grain. But if the burning surface area A_b decreases with time, the chamber pressure decreases with time and thus it is termed regressive burning grain (Figure 7.19). In other words, according to chamber pressure/thrust versus time history, propellant grains can be classified further into three types: (1) progressive, (2) neutral, and (3) regressive, as shown in Figure 7.19.

7.10 EVOLUTION OF BURNING SURFACE

The burning surface evolution can be simply viewed as the evolution of the burning perimeter, which will evidently vary with the geometry of the grain. Generally, the burning of any isotropic material (propellant) proceeds in a direction along the local normal to the instantaneous burning surface. In other words, even a curved surface will propagate as a parallel curved surface as shown in Figure 7.20. An initially convex edge (cusp) (Figure 7.20a) toward a gas phase will become an arc of a circle or sphere

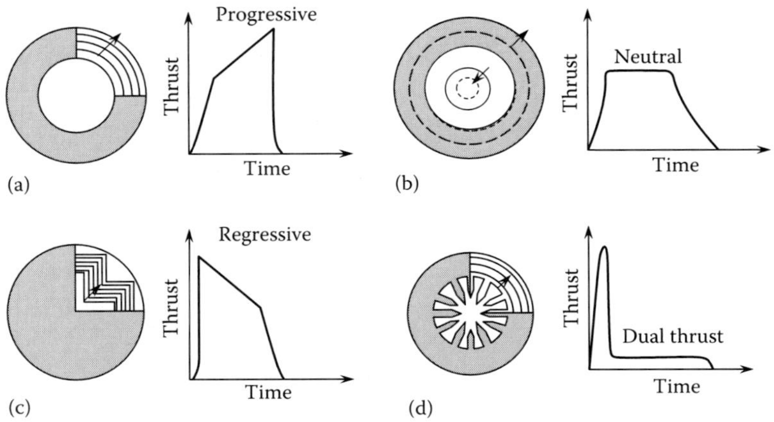

FIGURE 7.19 Classification of propellant grains according to thrust–time history: (a) tubular—progressive burning, (b) tube with rod—neutral burning, (c) star—regressive burning, and (d) multifin—dual thrust burning.

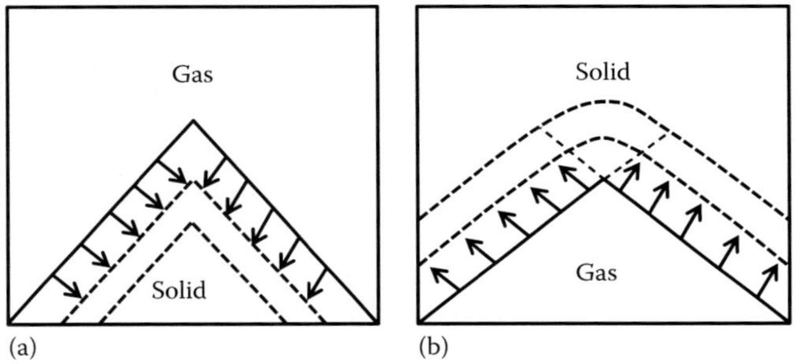

FIGURE 7.20 Effect of burning on (a) convex and (b) concave cusps.

with its center at the original cusp. This phenomenological law is often known as Piobert's law. It simply states that there is no preferential direction for burning to propagate into its surface. Rather it defines a regression rate vector and thus the evolution of the burning surface can be determined easily. Based on this principle, one can determine the evolution of the burning surface area A_b with time as a function of the distance burnt from the knowledge of the burning rate for a particular propellant. Let us consider an example of a tubular grain as shown in Figure 7.21 for which the burning surface area A_b can be easily determined from the grain length L and the burning perimeter S_b as follows:

$$A_b = LS_b \tag{7.34}$$

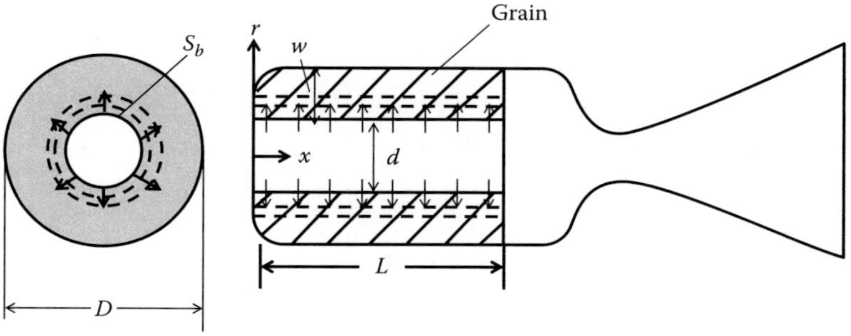

FIGURE 7.21 Evolution of burning surface in a tubular grain.

In this tubular grain, the ends are restricted as shown in Figure 7.21. Hence, it burns progressively as the burning perimeter S_b increases linearly with the distance burnt r. Of course, if both ends are not restricted, propellant burning becomes three-dimensional in nature. However, a tubular grain with restricted ends can also exhibit neutral burning if burning takes place radially both at its inner and outer surfaces, as shown in Figure 7.19a. But in this case, the chamber wall is exposed directly to high-pressure and high-temperature combustion gas. In order to overcome this problem, a tube and rod configuration (Figure 7.19b) has been devised that can provide neutral burning. But both configurations offer support to the propellant grain such that structural integrity can be maintained during its entire burning period.

Before delving into the analysis of the burning surface area of any specific grain geometries, we will have to get familiarized with various terms used for defining a typical propellant grain. The thickness of the grain that determines the burning time is known as the web thickness w, which can be defined as follows:

$$w = \int_0^{t_b} \dot{r} dt \tag{7.35}$$

Physically, it can be conceived as the minimum thickness of the grain from its initial burning surface to the insulated case wall or to the intersection of another burning surface. In Figure 7.21 for a tubular grain, the web thickness w is equal to the thickness of the grain, but in the case of the end burning grain, w is equal to the length of the grain. Generally, it is convenient to use a dimensional quantity for web thickness. For this purpose, web thickness is normalized by half of the outer diameter D of the grain, which is known as the web fraction w_f that is given by

$$w_f = \frac{2w}{D} \tag{7.36}$$

The dimensionless distance burnt $\bar{r}$ is defined as

$$\bar{r} = \frac{2\int_0^{t_b} \dot{r}dt}{D} \tag{7.37}$$

Similarly, the dimensionless burning surface area $\bar{A}_b$ can be defined as

$$\bar{A}_b = \frac{4A_b}{\pi D^2} \tag{7.38}$$

The volumetric loading fraction VF is defined as follows:

$$VF = \frac{V_p}{V_c} \tag{7.39}$$

where

V_p is the volume of the propellant in the grain

V_c is the volume of the combustion chamber, of course excluding the nozzle available for the propellant, insulation, and restrictors

It may be noted from this definition that the volume loading fraction has to be maximized for utilizing the chamber volume effectively. The initial throat to port area ratio J_i is defined as follows:

$$J_i = \frac{A_t}{A_{pi}} \tag{7.40}$$

where A_{pi} is the initial port area. Generally, J_i is preferably less than unity. However, a higher initial throat to port area ratio J_i can be more vulnerable to erosive burning, which must be avoided.

7.10.1 Star Grain

The star grain configuration is the most preferred radially burning cylindrical grain in SPREs as it has both a progressive burning tube and regressive burning star points. Hence, by varying these two aspects, the designer has more options of designing a suitable thrust law to meet the requirements of the mission. Even a neutral thrust law can be easily obtained with the help of both progressive and regressive surfaces in the grain. A typical pointed star grain, shown in Figure 7.22, can be specified by four independent variables: number of star points n, web thickness w, opening star point angle θ, maximum inner diameter d, and outer diameter D. The evolution of the propellant surface in this star grain is shown in Figure 7.22. Note that the propellant surface AB in the figure regresses and evolves into A_1CB_1. This surface is repeated $2n$ times in this grain. Hence, it will be prudent to consider π/n portion of the grain configuration as shown in Figure 7.22 to estimate the evolution of the burning surface. It may be observed from Figure 7.22 that the segment AB becomes A_1CB_1 as it regresses by the dimensionless burnt distance $2y/d$. At the same time, it can also be appreciated that as the initial star point A regresses, it evolves to A_1C as the start point takes on a convex shape as per Piobert's law of regression, while the segment AB decreases to CB_1. The change in total burning surface area of this grain can be obtained by a net change in perimeter multiplied by depth (L) and

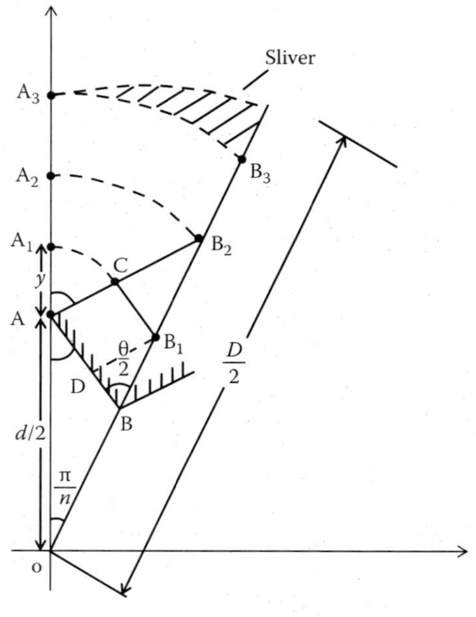

FIGURE 7.22 Evolution of burning surface in a star grain.

number of segments ($2n$). Now we need to evaluate change in the initial burning segment AB, for which we will have to determine various angles shown in Figure 7.22:

$$\angle \mathrm{AOB} = \frac{\pi}{n}; \quad \angle \mathrm{OAB} = -\frac{\pi}{n} + \frac{\theta}{2}; \quad \angle \mathrm{A_1AC} = \frac{\pi}{2} + \frac{\pi}{n} - \frac{\theta}{2} \tag{7.41}$$

Then the net change (dS) in the initial segment AB is determined easily as

$$dS = \left(\mathrm{A_1C} + \mathrm{CB_1}\right) - \mathrm{AB} = \left[\left(\frac{\pi}{2} + \frac{\pi}{n} - \frac{\theta}{2}\right) - \cot\frac{\theta}{2}\right] y; \tag{7.42}$$

The net change in the burning surface area A_b with regression of the propellant surface can be expressed as

$$dA_b = \left[\left(\frac{\pi}{2} + \frac{\pi}{n} - \frac{\theta}{2}\right) - \cot\frac{\theta}{2}\right] 2nyL \tag{7.43}$$

where L is the length of the grain. If there is no net change in the burning surface area A_b, it can produce a neutral grain. We can determine the condition for a neutral burning star grain easily by setting Equation 7.43 to zero as follows:

$$\left[\left(\frac{\pi}{2} + \frac{\pi}{n} - \frac{\theta}{2}\right) - \cot\frac{\theta}{2}\right] = 0 \tag{7.44}$$

For a particular star grain, the neutral star grain angle $\theta_{neutral}$ can be obtained by solving this transcendental equation. Note that the star grain angle θ is a function of the number of the star n. For example, the neutral star grain angle $\theta_{neutral}$ is equal to 67° and 85° for $n = 6$ and 12, respectively. When the star grain angle θ is greater than $\theta_{neutral}$, it results in regressive burning. In contrast, propellant burning becomes progressive when θ is less than $\theta_{neutral}$. By choosing proper θ, all three modes of burning can be easily obtained.

Let us now derive the expression for the burning surface area of the entire star grain in terms of independent parameters by using Equation 7.45 as follows:

$$\bar{A}_b = \frac{4A_b}{d^2} = \left[S + \left(\frac{\pi}{2} + \frac{\pi}{n} - \frac{\theta}{2}\right)y - y\cot\frac{\theta}{2}\right]\frac{8nL}{d^2};$$

$$\text{whereas } S = \frac{d}{2}\left(\frac{\sin\frac{\pi}{n}}{\sin\frac{\theta}{2}}\right) \tag{7.45}$$

The variation in estimated dimensionless burning surface area $\bar{A}_b$ is shown in Figure 7.23 with the dimensionless distance burnt y/d for a star grain with $n = 6$ and three representative values of star angle, $\theta = 60°, 67°, 85°$. It may be observed that when the star angle θ is equal to 67°, the burning surface area remains almost constant from the beginning of its burning, indicating the occurrence of neutral burning. However, progressive burning occurs for $\theta = 60°$ due to an increase in burning surface area. In contrast, for $\theta = 85°$, the burning surface area decreases, indicating a regressive burning surface. It indicates that, for this case, all three modes of burning, namely, regressive, neutral, and progressive, can occur in the first phase. Subsequently, when the star points disappear, the second phase begins during which the port area

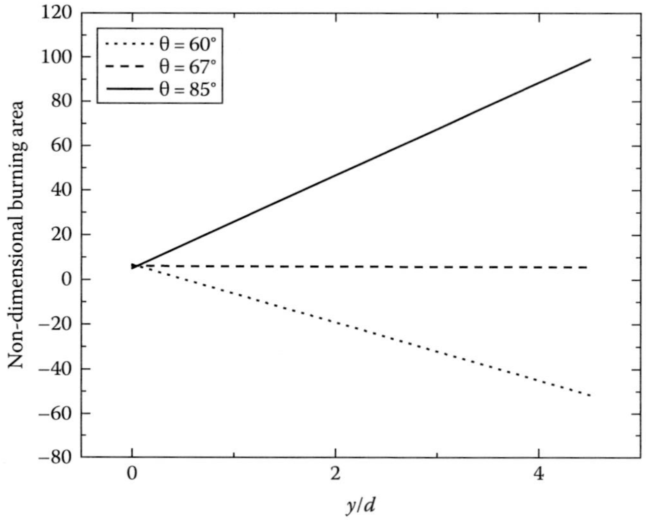

FIGURE 7.23 Normalized perimeter with burnt distance in a star grain.

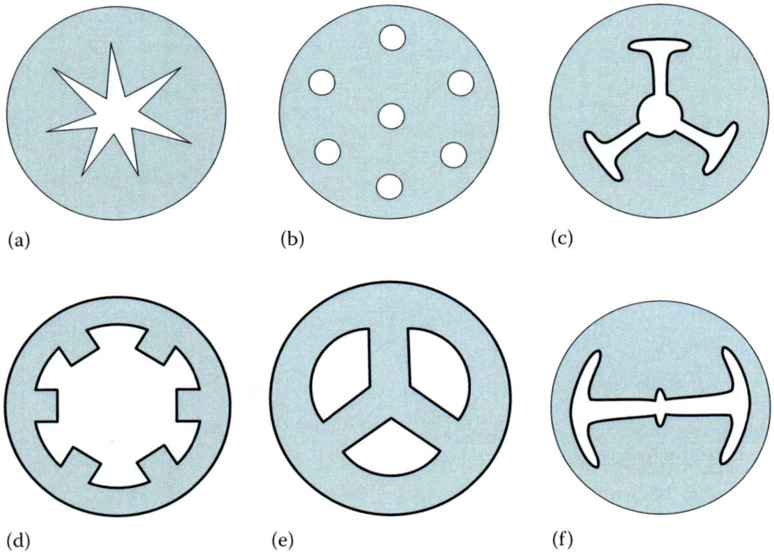

FIGURE 7.24 Various types of grain configurations: (a) star (neutral), (b) multi perforated, (c) dog bone, (d) wagon wheel, (e) three conical port, and (f) double anchor.

becomes strongly progressive irrespective of the nature of initial burning (regressive/neutral/progressive). Interested readers may refer an advanced book [11] for a detailed analysis of the second phase of burning for star grain.

Some of the other well-known side burning grain geometries are shown in Figure 7.24. In internal burning grain, neutrality can be obtained in configurations like star, forked, wagon wheel, and so on, where convex cusps provide control on the type of A_b evolution. In most of the high volumetric loading grains like wagon wheel or dendrite, one can also expect an erosive burning effect in the aft end regions of the port. A possible way of reducing these erosive effects is to taper the port of the grain.

7.10.1.1 Three-Dimensional Grains

In the last section, several kinds of two-dimensional grains were discussed in detail, which mainly consist of simple geometrical shapes, namely, rods, tubes, wedges, and so on. We have learnt that most of these two-dimensional propellant grains combine two or more of these basic surfaces to obtain the desired shapes for a particular thrust law. However, there are several three-dimensional grains that can also be devised from basic surfaces to obtain a certain thrust law in actual practices of missile and upper-stage rocket applications working at relatively high loading density and length to diameter ratio. Some examples of these three-dimensional grains are tubular grains with unrestricted ends, slotted tube, finocyl (fin-in-a-cylinder), conocyl (cone-in-a-cylinder), and spherical with internal slots as shown in Figure 7.25. The slotted tube grain is basically a tubular grain with slots over a certain portion of the entire length as shown in Figure 7.25a. It may be noted that the slotted portion of the grain exhibits a regressive burning surface while the cylindrical portion provides a progressive burning surface. As a result, the propellant surface regresses in radial, tangential, and longitudinal directions. The burning surface evolution is dependent on the number of slots n, slot length l, width of slots w_s, and length of grain L. In this kind of grain, erosive burning due to the occurrence of initial high velocity of gas can be avoided as the port area of the slotted portion is increased with respect to the tubular section. The finocyl, basically the fin-in-cylinder, as shown in Figure 7.25b is another type of three-dimensional grain that is profusely used in modern intercontinental ballistic missile (ICBM). It provides longer burning time for low L/D ratios. The axial slots in finocyl provide a regressive burning surface. It is preferred over conocyl as it is easier to fabricate. Another three-dimensional grain is the conocyl (cone-in-cylinder), which is used in missile applications. The conical propellant surface burns regressively while the central perforation burns

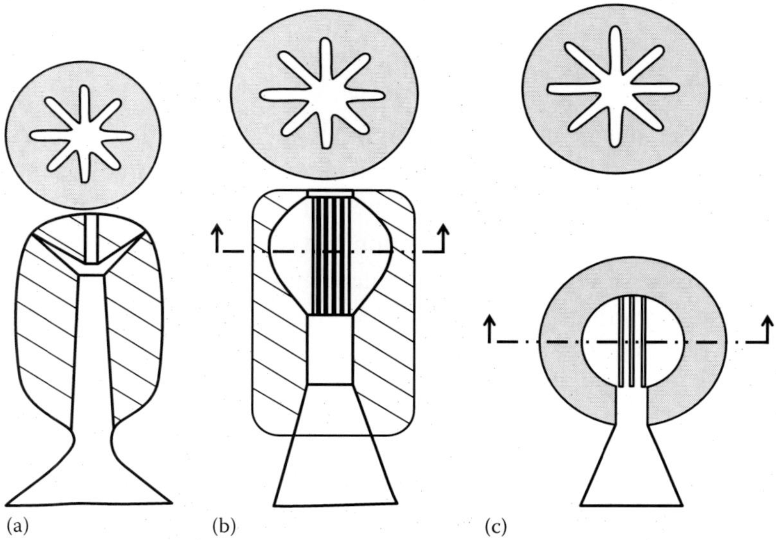

FIGURE 7.25 Various types of three-dimensional grain configuration: (a) conocyl design, (b) finocyl design, and (c) spherical design.

progressively. Interestingly, this grain can provide neutral burning for a range of L/D ratios up to the value of 4. In this case, care must be taken to insulate a certain portion of the combustion chamber that gets exposed to hot gas prematurely before the end of its operation. It can also provide an added flow channel in the forward end and hence is preferred over the finocyl grain, particularly for system requirements, namely, high thrust reversal and thrust termination. The spherical grain shown in Figure 7.25c is an important class of three-dimensional grain that can provide volumetric loading fraction. However, this type of grain encounters higher aerodynamic drag. Hence, it can be used for certain special applications, for example, in apogee-kick motors, final-stage motors, and rocket engines for interplanetary spacecraft where aerodynamic drag does not pose any problem. Apart from these three-dimensional grains, several other configurations can be conceived and developed. Although it may be easy to conceive three-dimensional grains, their manufacturing poses several challenges for designers. The burning surface evolution in these three-dimensional grains needs complex computer models.

In practice, a particular grain geometry is selected to have high volumetric loading fraction and desired thrust time profile with minimum erosive burning. Quite often, high loading fraction is also associated with high sliver fractions. One method of producing minimum or no sliver is to adopt dual composition or dual burning rate grains.

Example 7.4

A designer has proposed to use propellant grain with rectangular cross section (0.5×0.3) with an outer diameter of 1.1 m and length of 1.5 m. If the density of the propellant is 1450 kg/m^3, determine the sliver mass of the propellant grain and the percentage of the sliver mass compared to the initial propellant mass at the end of its operation. By restricting the fraction of sliver mass to 5%, can it be used for practical application?

Solution

In this rectangular cross section of propellant grain, the corner point A in Figure 7.25 evolves as a quadrant of a circle. But the flat surface A_c regresses parallel to the initial surface. The web thickness δ can be determined from the geometry shown in the figure as follows:

$$\delta = \text{OB} - \text{OA} = \frac{1}{2}\left(D - \sqrt{a^2 + b^2}\right)$$
$$= \frac{1}{2}\left(1.1 - \sqrt{0.5^2 + 0.3^2}\right) = 0.26 \text{ m}$$

The sliver volume V_S can be determined as

$$V_S = \left[\frac{\pi}{4}D^2 - \left(a \times b + 2(a+b)\delta + \frac{\pi}{2}\delta^2\right)\right]L$$
$$= \left[\frac{3.14}{4}1.1^2 - \left(0.5 \times 0.3 + 2(0.5+0.3) \times 0.26 + 3.14 \times 0.26^2\right)\right]1.5$$
$$= 0.26 \text{ m}^3$$

The sliver mass can be determined as

$$m_S = \rho_P V_S = 1450 \times 0.257 = 373.2 \text{ kg}$$

The total mass of the propellant is estimated as

$$m_P = \eta_P\left[\frac{\pi}{4}D^2 - (a \times b)\right]L = 1450\left[\frac{3.14}{4}(1.1)^2 - (0.5 \times 0.3)\right]$$
$$\times 1.5 = 1739.7 \text{ kg/m}^3$$

The fractional sliver mass can be determined as

$$\%m_S = \frac{m_P - m_S}{m_P} \times 100 = \frac{1739.7 - 373.2}{1739.7} \times 100 = 7.86$$

As the fractional sliver mass is greater than 5%, it cannot be used in the present design.

7.11 IGNITION SYSTEM

We know that the function of an igniter is to initiate a smooth combustion in a specific time interval without any undesirable pressure spikes. A typical ignition system has two major components: (1) initiator and (2) energy release system. Basically, the initiator transfers thermal energy from electrical/mechanical /optical (high power laser) to the main charge for ignition to take place. In this initiator, a safety system is also introduced depending on mission requirement. Several types of initiators have been designed and developed for reliable and safe ignition of SPREs. Some of the initial stimuli for ignition can be provided by several kinds of bridges, namely, electrical resistance, semiconductor, laser, mechanical shock wave, and so on. The most widely used initiation system is the electrical initiator (squib), which can either be a hot wire or an exploding bridge wire. The details of a typical electrical initiator used in a pyrotechnic igniter are shown in Figure 7.26a. It may be noted that this squib is surrounded immediately by a transfer/relay propellant (charge). The conventional squib uses a nichrome wire of small diameter (0.01–0.015 mm) attached to two electrical terminals. On the passage of electrical current through the squibs, it gets heated up sufficiently to cause the deflagration of the heat-sensitive charge in contact with the wire. In contrast, the bridge wire explodes whenever it becomes red hot with passage of electrical current, because bridge wires are generally coated with explosive primer (mercuric fulminate or lead azide). Of course, the response of the ignition current can be controlled by current passing through the wire, selecting suitable material and size of wire, and composition of relay charge. In practice, the electrical current passing through the squib is stipulated by fixing "no fire" and "all fire" current levels. Note that "no fire" current is the maximum current that can be applied to the circuit without firing this initiator system. In other words, "no fire" is the safe current level that can be passed through the bridge wire to check continuity and circuit connection. This provides a large safety margin against any premature electromagnetic field induced by power lines or radio frequency sources. But the "all fire" is the current that is to be supplied through

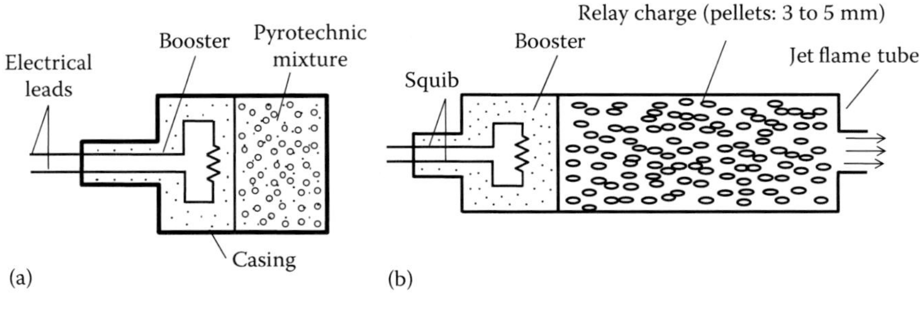

FIGURE 7.26 Schematic of pyrotechnic igniters: (a) basket igniter; (b) jet flame pyrotechnic igniter.

the squib for successful initiation of ignition. Although, the squib must fire at the "all fire" current, the firing will take a longer time on the order of 50 ms. Hence, higher current known as recommended fire current (RFC) is always specified. For a typical electrical initiator, 0.1, 0.5, and 0.5 amp are used for "no fire," "all fire," and RFC, respectively. Normally, squib charge is formulated such that its autoignition temperature is above the wire temperature at "no fire" current and is below the wire temperature at "all fire" current. Besides this, the temperature should be such that the bridge wire can be melted out once the squib is ignited to protect the battery for further drainage of its energy. The energy released due to the burning of the squib charge is transported to the relay charge. Generally, the relay/transfer charge consists of a complex mixture of pyrophoric metal powders or granules like boron, magnesium, aluminum, and oxidizers like potassium perchlorate and potassium nitrate, whose size ranges from 0.5 to 2 mm. As this charge is in powder form, it transports energy at a faster rate and results in combustion that spans from 1 to 10 ms before the combustion front moves to booster charge. Generally, the propellant used for relay charge is also used for booster charge, of course with larger pellet sizes ranging from 5 to 10 mm. The mass of the booster charge is dependent on the free volume and pressure of the combustion chamber to be built up during the ignition phase, which normally ranges from 30% to 40% of equilibrium pressure.

The energy release system is the portion of the igniter that initiates combustion over the propellant grain of the rocket engine. This ensures supply of the requisite amount of heat energy for a solid propellant to ignite and induces chamber pressure to reach stable equilibrium pressure smoothly for successful ignition. Based on the methods by which heat energy can be liberated in an energy release system, igniters can be classified as pyrotechnic and pyrogen types. If, on initiation of ignition, the composition of pyrotechnic charge is released directly to the propellant grain of the rocket engine, it is called a pyrotechnic igniter. A pyrogen igniter, on the other hand, is a small motor that is used to ignite a larger rocket motor. In this case, exhaust gases due to rapid combustion of fast burning grain from a small motor that is located inside the main rocket engine initiate combustion on the surface of the main rocket engine; this is called a pyrogen igniter. A pyrotechnic igniter is used for small rocket engines while a pyrogen igniter is preferred for large rocket motors. Generally, pyrogen igniters are used widely in major launch vehicles all over the world due to their higher level of reproducibility. A comparison between pyrotechnic and pyrogen igniters is made in Table 7.2.

TABLE 7.2 Comparison between Pyrotechnic and Pyrogen Igniters

	Pyrotechnic	Pyrogen
1. Main charge	Pyrotechnic propellant	Fast-burning propellant
2. Configuration	Powders/pellets	Cast as grain
3. Action time	Shorter burn time (100–200 ms)	Longer burn time (250–1000 ms)
4. Applications	In small solid-propellant rocket engines	In large solid-propellant rocket engines
5. Relative merits/demerits	1. Performance moderately controllable 2. Thrust law depends on test chamber volume 3. Simple construction and processing 4. Lower cost	1. Performance closely controllable 2. Thrust law independent of test chamber volume 3. Complex construction and processing 4. Higher cost

7.11.1 Pyrotechnic Igniter

Generally, two types of pyrotechnic igniters—(1) basket igniter and (2) jet flame igniter—as shown in Figure 7.26 are used for both small and medium-sized rocket engines. A basket-type igniter, being the simplest among all designs, is employed in small rocket engines. Normally, it constitutes a squib, transfer charge, booster charge, and main charge as discussed earlier. All these ingredients are enclosed in a malleable metal or hermetically sealed plastic (cellulose/polymer) container. A metal container poses the problem of grain fissure or nozzle plugging while a plastic container presents the problem of storage. Moreover, this container must maintain its structural rigidity till the ignition occurs over the entire charge. A certain number of holes are provided in the container for allowing hot products to disperse easily in all directions to ignite the entire main propellant grain. In any circumstance, the container must not burst due to the buildup of inside pressure, which can create fissures in the main propellant grain. Generally, for medium-sized rocket engines, a jet flame igniter is preferred over the basket-type pyrotechnic igniter. It has components similar to that of the basket-type igniter. The only difference is that it is contained inside a nozzle in contrast to a compact basket. The hot gas produced by ignition is ejected from the nozzle end as a jet that can spread over the entire propellant grain easily. Several multiple jets are also used for ensuring the spread of hot gas from the igniter along several directions to reach the entire portion of propellant grain. As a result, the ignition delay for medium-sized rocket engines can be reduced.

The pyrotechnic mixture consists of binder and oxidizers (e.g., nitrates: KN, AN, or perchlorates: KP, AP) and metals such as Al, Ti, Boron (B), or

Zr in the form of a powder or pellet. The composition of this pyrotechnic mixture can be chosen such that it can generate the requisite flow rate of high-temperature gas containing hot particulates, which will impinge upon the solid-propellant grain to cause a number of local ignition spots. In general, some of the requirements of an igniter propellant are fast and high heat release, high gas evolution per unit mass, smooth operation over a wide range of temperature and pressure, and low ignition delays. In order to keep the mass and volume minimal and achieve the desired mass flow rate, propellants with high burn rates and regressive burning grain configuration are used in igniter motors. Proper selection of charge composition and arrangement in the ignition system must ensure long-term safe storage and should be insensitive to random electrical and other disturbances. The most commonly used pyrotechnic mixture is black powder, which is composed of 74% potassium nitrate (KNO_3), 15% charcoal (C), and 11% sulfur (S) with an average density of 900–1150 kg/m^3 depending on the level of granulation. The chemical reactions during the burning of black powder can be represented as follows:

$$\begin{aligned}&74KNO_3+96C+30S+16H_2O\rightarrow 35N_2+56CO_2+16CO+3CH_4+2H_2S\\&\quad+4H_2+19K_2CO_3+7K_2SO_4+8K_2S_2O_3+2K_2S+2KSCN\\&\quad+\left(NH_4\right)2CO_3+C+S\end{aligned}\tag{7.46}$$

It may be noted from this chemical reaction that the first six terms are gases while the rest are in solid phase. Hence, 1 kg of black powder provides 0.4 kg of gases of average molecular weight of 34.75 and 0.6 kg of solid materials. This combustion of black powder produces an adiabatic flame temperature of 2590 K. Of course, several other compositions can be devised easily based on different combinations of binder, oxidizer, and metal powder as they are routinely formulated for main grain. Of course, binders are used to shape ignition grains in the form of pellets, granules, and cast blocks. Keep in mind that propellants for the igniter must have a higher burning rate, typically in the range of 15–30 mm/s. In some cases, certain burn rate accelerators, namely, ferric oxides and copper chromites, are used to enhance the burning rate. Although a higher proportion of metallic fuel can provide higher flame temperature and calorific value, it is not favored as it reduces gaseous contents. The metal oxides formed during combustion can inhibit the propellant surface and thus hinder the ignition process. Let us appreciate the differences between rocket engines and igniter propellant formulations as given in Table 7.3.

TABLE 7.3 Comparison of Rocket Engine and Igniter Propellants

	Igniter Propellant	Rocket Engine Propellant
Burning rate	15–30 mm/s	3–10 mm/s
Oxidizer	Coarse in particle size Low percentage	Fine in particle size High percentage
Metallic fuel	Low percentage	High percentage
Calorific value	Low	High
Combustion products	80%–90% gases	40%–50% gases
Burn rate accelerator	High percentage	Low percentage

It is important to provide a certain minimum amount of charge for successful ignition. This is dependent on the burning surface area to be ignited, the entire chamber volume to be filled with hot gases, and the proportion of solid fraction *SF*. Based on this, the mass of the charge can be determined as

$$m_{ig} \geq \frac{1}{1-SF}\frac{MW_{ig}V_cP_{cig}}{R_uT_{ig}} \tag{7.47}$$

where

- SF is the fraction of solid materials in the combustion products
- V is the initial free chamber volume, which includes the convergent portion of the exhaust nozzle
- P_{cig} is the chamber pressure, which is around 30%–40% of equilibrium P_c
- MW is the molecular weight of the gaseous products
- T_{ig} is the flame temperature at constant volume

Note that this equation is basically derived from the ideal gas law for gaseous products of combustion at the end of ignition. This method of determining the mass of the ignition charge can be used satisfactorily for small to medium-sized rocket engines. However, for large rocket engines, where the free volume would be large in the range of 20–50 m^3, advanced analysis based on the principles of aero-thermo-chemical dynamics can be used.

7.11.2 Pyrogen Igniter

We know that a pyrogen igniter is mainly a small rocket engine that produces hot combustion gases from the burning of the ignition propellant to initiate combustion on the propellant grain of the large rocket engine.

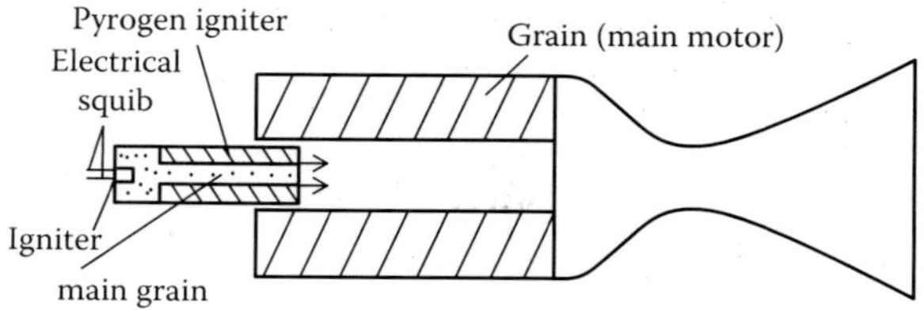

FIGURE 7.27 Schematic of pyrogen igniters.

A typical pyrogen igniter is shown in Figure 7.27 which consists of initiator and booster charge, similar to that of a pyrotechnic igniter, with internal and external ignition grain, casing, nozzle, head end, and nozzle end. The head end is generally made of metal, which acts as an interface for the initiator, pressure tapping adaptor, and so on. Insulating material is provided to protect the metallic head end and certain parts of the igniter casing and nozzle end, which are likely to get exposed during the firing. Normally, igniter nozzles are convergent/orifice type unlike the convergent–divergent (CD) nozzle of a main rocket engine. The exit of the pyrogen igniter can have multiple holes for hot gases to be ejected and spread over the large grain of the main rocket engine. The plume ignition flame and jet of hot gases ejected from the igniter must not impinge too much upon the propellant surface as it can cause fissures on the grain. Note that the ignition of a large rocket engine is mainly controlled by the extent of heat transfer due to the convection of hot gases ejected from the pyrogen igniter compared to the radiative heat transfer emitted by a pyrotechnic igniter.

7.12 MODELING OF FLOW IN A SIDE BURNING GRAIN OF ROCKET ENGINE

We know that most modern rocket engines use either cylindrical or nearly cylindrical propellant grain because of its various advantages, namely, small cross section, light weight, wide range of thrust laws, and so on, over end burning grain. Of course, the cross section of the propellant grain can have any shape to provide the desired thrust law. In most cases, a cross-sectional area of the flow channel, known as the port area, remains almost constant along the grain length. Hence, the gas velocity increases and is accompanied by pressure drop toward the nozzle end of the grain. In some other cases, the port area may vary along the grain length, for which three-dimensional flow will prevail in the combustion chamber. However, for simplicity, we will

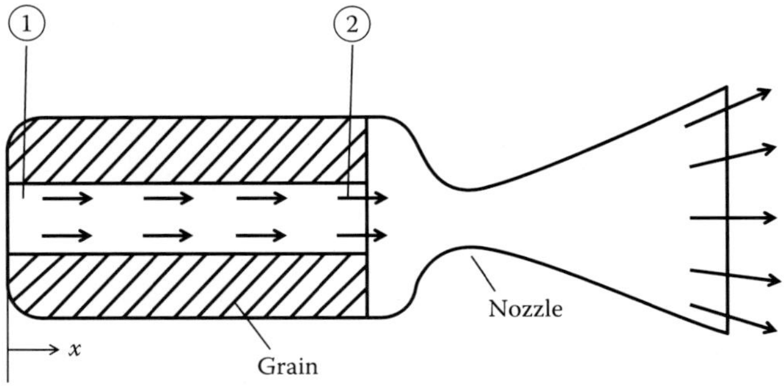

FIGURE 7.28 Schematic of flow in a side burning grain.

conduct the following analysis of flow by considering a cylindrical port combustion chamber as shown in Figure 7.28 with the following assumptions:

1. The burning rate $\dot{r}$ remains constant along the grain length.
2. The perimeter S_b remains constant along the grain length. The pressure drop between stations (1) and (2) will be compensated by the erosive burning effect. Hence, the average burning rate can be determined at station (1) as $\bar{\dot{r}} = aP_1^n \approx aP_2^n\left[1+K_E(V-V_{Th})\right]$
3. Change in A_b, S_b, and V_c per unit time due to regression of the propellant is quite small so that a quasi-steady-state process can be assumed.
4. Total enthalpy remains constant along the grain length.
5. Thermo-chemical properties of gas, namely, C_p, γ, and so on, remain almost constant.
6. The gas temperature at the burning surface across the entire domain is assumed to be equal to the flame temperature.

By mass conservation, we obtain

$$\rho_2 V_2 A_p = \bar{\dot{r}}\rho_P A_b = \dot{m}_p \tag{7.48}$$

By neglecting the friction, we can invoke conservation of momentum principle to obtain

$$d(\rho V^2) = -dP \tag{7.49}$$

By integrating Equation 7.49 between stations (1) and (2), we obtain

$$\rho_2 V_2^2 = P_1 - P_2 \tag{7.50}$$

By energy conservation, we obtain

$$C_P T_c = C_P T_2 + \frac{V_2^2}{2} = \left(\frac{\gamma}{\gamma - 1}\right)\frac{P_2}{\rho_2} + \frac{V_2^2}{2} \tag{7.51}$$

By multiplying Equation 7.51 by V_2^2, we obtain

$$\left(C_P T_c\right)V_2^2 = V_2^2\left(\frac{\gamma}{\gamma - 1}\right)\frac{P_2}{\rho_2} + \frac{\left(V_2^2\right)^4}{2} \tag{7.52}$$

By using Equations 7.50 and 7.52, we obtain

$$\left(C_P T_c\right)V_2^2 = \left(\frac{\gamma}{\gamma - 1}\right)\frac{P_2}{\rho_2}\left(\frac{P_1 - P_2}{\rho_2}\right) + \frac{1}{2}\left(\frac{P_1 - P_2}{\rho_2}\right)^2 \tag{7.53}$$

By using Equations 7.48 and 7.53, we obtain

$$\left(C_P T_c\right)\left[\frac{\bar{\dot{r}}\rho_p A_b}{P_2 A_p}\right]^2 = \left(\frac{\gamma}{\gamma - 1}\right)\frac{P_2}{\rho_2}\left(\frac{P_1 - P_2}{\rho_2}\right) + \frac{1}{2}\left(\frac{P_1 - P_2}{\rho_2}\right)^2 \tag{7.54}$$

By multiplying Equation 7.54 ρ_2^2/P_2^2 and rearranging it, we obtain

$$\frac{1}{2}\left[\frac{P_1}{P_2} - 1\right]^2 + \frac{\gamma}{\gamma - 1}\left[\frac{P_1}{P_2} - 1\right] = C_P T_c\left[\frac{\bar{\dot{r}}\rho_p A_b}{P_2 A_p}\right]^2 \tag{7.55}$$

where

A_p is the port area

A_b is the burning area

This quadratic equation can be solved easily as follows:

$$\frac{P_1}{P_2} - 1 = \frac{\gamma}{\gamma - 1}\left\{-1 + \sqrt{1 + 2\frac{\gamma - 1}{\gamma}\left[\frac{\sqrt{RT_c}\rho_P A_b \bar{\dot{r}}}{P_2 A_p}\right]^2}\right\} \tag{7.56}$$

Note that we have considered the positive value as $P_1 > P_2$. By expanding Equation 7.56 and considering only lower-order terms, we obtain

$$\frac{P_1}{P_2} = 1 + \left[\frac{\sqrt{RT_c}\rho_P A_b \bar{\dot{r}}}{P_2 A_p} \right]^2 \quad (7.57)$$

For choked flow condition in the nozzle and steady burning process, we know that

$$\rho_P \bar{\dot{r}} A_b = \Gamma \frac{P_c A_t}{\sqrt{RT_c}} = \rho_2 V_2 A_p \quad (7.58)$$

where P_c is the effective stagnation pressure in the combustion chamber. By using Equations 7.57 and 7.58 and expanding the series, we obtain

$$\frac{P_1}{P_2} \approx 1 + (\Gamma J)^2; \quad \Gamma = \sqrt{\gamma \left(\frac{2}{\gamma + 1} \right)^{\frac{\gamma+1}{\gamma-1}}} \quad (7.59)$$

The density at station (2) can be related to P_c and T_c as

$$\rho_2 = \frac{\rho_2}{\rho_c} \rho_c = \left(\frac{P_2}{P_c} \right)^{\frac{1}{\gamma}} \frac{P_c}{RT_c} \quad (7.60)$$

We can relate stagnation pressure to P_2 by considering isentropic recompression from state (2) as follows:

$$V_2^2 = 2C_p T_c \left(1 - \frac{T_2}{T_c} \right) = \frac{2\gamma RT_c}{\gamma - 1} \left(1 - \left[\frac{P_2}{P_c} \right]^{\frac{\gamma-1}{\gamma}} \right) \quad (7.61)$$

By eliminating ρ_2 and V_2 in Equation 7.58 and using $J = A_t/A_p$ along with Equations 7.60 and 7.61, we obtain

$$\Gamma J = \left(\frac{P_2}{P_c} \right)^{\frac{1}{\gamma}} \sqrt{\frac{2\gamma}{\gamma - 1} \left(1 - \left[\frac{P_2}{P_c} \right]^{\frac{\gamma-1}{\gamma}} \right)} \quad (7.62)$$

Note that the pressure ratio P_2/P_c in Equation 7.62 is quite close to unity. Hence, Equation 7.62 can be simplified by expanding and neglecting higher-order terms:

$$\frac{P_2}{P_c} \approx 1 - \frac{(\Gamma J)^2}{2} \tag{7.63}$$

Let us find the relationship between P_1 and P_c as follows:

$$P_1 = \frac{P_2}{P_1}\frac{P_2}{P_c}P_c = \left(1+(\Gamma J)^2\right)\left(1-\frac{(\Gamma J)^2}{2}\right)P_c \approx \left(1+\frac{(\Gamma J)^2}{2}\right)P_c \tag{7.64}$$

Note that Equation 7.64 is derived by neglecting the fourth-order term. By using Equations 7.63 and 7.64 we can show that the combustion chamber pressure is approximately equal to

$$P_c = \frac{P_1 + P_2}{2} \tag{7.65}$$

By combining Equations 7.57 and 7.63, we can derive the relationship for local pressure, P at any location x from the front end of side burning grain as follows:

$$\frac{P_1}{P} = 1 + \left[\frac{\sqrt{RT_c}\,\rho_P x S_b \bar{\dot{r}}}{\left(1-0.5(\Gamma J)^2\right)P_c A_p}\right]^2 \tag{7.66}$$

where S_b is the port perimeter and $x = A_b/S_b$ = length of grain.

The variation of local pressure with respect to front-end pressure (P/P_1) along the axial distance for a cylindrical port of 0.2 m in diameter is shown in Figure 7.29. Note that the pressure decreases with axial distance when the gas velocity increases due to acceleration of flow along with mass addition during the burning of a propellant.

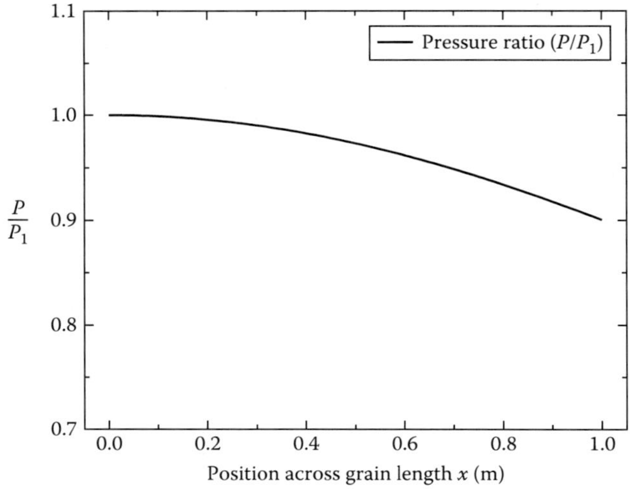

FIGURE 7.29 Variation of pressure for a side burning cylindrical grain along axial direction.

REVIEW QUESTIONS

1. What are the various components of a typical solid-propellant rocket engine (SPRE)?
2. What are the advantages of SPRE over LPRE?
3. Describe the physical processes during solid-propellant burning with a neat sketch.
4. What are the zones during DB solid-propellant burning? Show them with a neat sketch.
5. What are the zones during campsite solid-propellant burning? Show them with a neat sketch.
6. What is a strand burner? Show its components and their functions.
7. What is meant by regression rate?
8. What is meant by grain?
9. Why is radial burning grain preferred over end burning grain?
10. What is Saint–Robert's law? How is it different from Muraour's law?
11. How is mesa burning different from plateau burning?

12. Define the temperature sensitivity coefficient of burning rate. Derive an expression for it using Saint–Robert's law.

13. What is erosive burning? What are the ways by which erosive burning can be avoided?

14. Derive an expression for unsteady regression rate using Saint–Robert's law.

15. What are the effects of acceleration on the linear regression rate?

16. What are the ways of augmenting the regression rate?

17. Assuming you have a thermal model, derive an expression for the regression rate.

18. Describe the ignition processes in an SPRE.

19. What do you mean by ignition delay time? How is it different from ignition time?

20. What do you mean by action time? How is it different from burn time?

21. Define the temperature sensitivity coefficient for chamber pressure. Derive an expression for it using Saint–Robert's law.

22. What are the types of grain you are aware of based on their geometry?

23. What are the differences between end and side burning grains? Why is side burning grain preferred for missile application?

24. What is regressive burning grain? How is it different from progressive burning grain?

25. Why is neutral burning grain preferred over other types of grain?

26. A designer has chosen a star burning grain for progressive burning in his rocket engine. Unfortunately, he observed neutral burning during its testing. What are the reasons for this behavior?

27. What are the three-dimensional grains you are aware? Describe them with relative merits and demerits.

28. How is a pyrotechnic igniter different from a pyrogen igniter? Enumerate their relative merits and demerits.

PROBLEMS

7.1 The regression rates of a solid propellant are found to be 22 and 39 mm/s at chamber pressure of 5 and 15 MPa, respectively. Determine the chamber pressure when the regression rate is 27 mm/s as per Saint–Robert's law.

7.2 The regression rate happens to be 29 mm/s at an unburnt temperature of 10°C, and chamber pressure of 6.5 MPa, which is governed by Saint–Robert's law. If the temperature sensitivity of the burning rate happens to be 0.006 per °C and pressure index is 0.65, determine the new burning rate at unburnt temperature of 45°C?

7.3 A side burning hollow solid-propellant grain of 5 m length with inner and outer diameter 0.37 and 0.82 m, respectively is to be designed. The burning rate is 1.2 cm/s. The combustion chamber pressure and gas temperature downstream of the grain are 2.17 MPa and 2580 K, respectively. Estimate (1) the static pressure ratio p_2/p_1 and (2) the Mach number M_2 downstream of the grain. Take ρ_P = propellant density = 1875 kg/m³. $\gamma = 1.2$, $MW = 20$.

7.4 A booster solid-propellant rocket engine with characteristic velocity of 1675 m/s, thrust coefficient of 1.2, and chamber pressure of 8.9 MPa is used to carry a payload of 220 kg. The initial total mass of the rocket engine happens to be 10,500 kg. The propellant with a density of 1685 kg/m³ follows the regression rate law $\dot{r} = aP_c^n$ where $a = 1.75$, P_c is the chamber pressure in atmosphere, $n = 0.45$, and $\dot{r}$ is in mm/s. If it is subjected to an initial acceleration of 1.6*g*, determine the throat diameter of the booster rocket and the initial burning area.

7.5 In a rocket engine, a solid propellant with pressure index of $n = 0.56$ is used. The thrust coefficient C_F and characteristic velocity C^* happen to be 1.05 and 2025 m/s, respectively. When the propellant unburnt temperature is 15°C, and chamber pressure is 5.5 MPa, the regression rate happens to be 34 mm/s. If the temperature sensitivity of the burning rate happens to be 0.008 per °C and the burning surface area to throat area AR is 45, determine the chamber pressure and thrust at the unburnt temperature of 48°C assuming C_F and C^* remain almost constant. Consider the throat area as 375×10^{-6} m² and propellant density as 1295 kg/m³.

7.6 A designer has proposed to use propellant grain with rectangular cross section 0.5 × 0.5 with an outer diameter of 1.5 m and length of 2.5 m. If the density of the propellant is 1295 kg/m³, determine the sliver mass of the propellant grain and the percentage of sliver mass compared to the initial propellant mass at the end of its operation.

7.7 An end burning rocket employs DB propellant grain ($\rho_P = 1900$ kg/m³) with a diameter of 125 mm and generates 205 N over 215 s with the characteristic velocity of 1350 m/s and thrust coefficient of 1.2. The pressure/combustion index and burning constant of this propellant are 0.7 and 3.2 mm/s, respectively, for the following regression rate expression: $\dot{r} = a\left(P_c/P_R\right)^n; P_R = 70$ MPa

Determine the length of grain, chamber pressure, and throat diameter of nozzle.

7.8 An end burning rocket employs DB propellant grain with a density of 1438 kg/m³ and length of 245 mm, which produces 1.5 kN over 175 s with the characteristic velocity of 1250 m/s and thrust coefficient of 0.98. The pressure index n and burning constant of this propellant are 0.7 and 3.2 mm/s, respectively, for the following regression rate expression: $\dot{r} = a\left(P_c/P_R\right)^n; P_R = 70$ MPa. Determine the diameter of the propellant grain, the throat diameter, and the chamber pressure.

7.9 The burning rate of a particular propellant is given by $\dot{r} = \dfrac{A}{\left(T_I - T_u\right)P_c^n}$

in which $\dot{r}$ is in mm/s, P_c is in MPa, T is in K, and A = 176, T_I = 415 K, n = 0.716. When the propellant initial temperature is 20°C, the chamber pressure is 3 MPa (steady) for 4 min. If the same sample propellant grain in the same rocket is heated to 45°C, what would be the new steady-state pressure and burning period?

7.10 A radial burning diverging cone cylindrical grain with outer diameter of 430 mm and length of 0.5 m has a port diameter of 120 mm at the aft end and 150 mm at the rear end of the grain. If the throat diameter happens to be 45 mm, determine the maximum web thickness, initial chamber pressure, and mass of sliver. Note that this solid-propellant rocket engine has the characteristic velocity of 1595 m/s with a propellant density of 1785 kg/m³. Its regression rate law follows $\dot{r} = aP_c^n$ where a = 0.14, P_c is the chamber pressure in atmosphere, n = 0.45, and $\dot{r}$ is in mm/s.

7.11 During the burning of a solid propellant with a specific heat of 925 J/kg·K, a diffusion flame with a temperature of 3200 K occurs at 150 μm from its surface with a temperature of 750 K. The thermal conductivity of gas generated during the burning of the propellant is 0.65 W/m·K and the heat of pyrolysis is 56 kJ/kg. Determine the regression of the solid propellant considering a thermal model for its combustion.

7.12 A solid-propellant rocket engine is designed to operate in a chamber pressure and temperature of 5.6 MPa and 3250 K, respectively. This solid propellant has a density of 1635 kg/m^3 that produces gas with a molecular mass of 25 kg/kmol and specific heat ratio γ of 1.25. Determine the burning to the throat area if its regression rate law follows $\dot{r} = aP_c^n$ where a = 1.25, P_c is the chamber pressure in atmosphere, n = 0.75, and $\dot{r}$ is in mm/s. When the burning to throat area increases at a rate of 0.9% per second, determine the time taken for the chamber pressure to be tripled.

7.13 A radial burning cylindrical grain with a port diameter of 120 mm, outer diameter of 425 mm, and length of 8.5 m is used to produce thrust by expanding hot gases at 3200 K and molecular mass of 25 kg/kmol in a nozzle with a throat diameter of 55 mm. Note that this solid-propellant rocket engine has the characteristic velocity of 1675 m/s with a propellant density of 1635 kg/m^3. Its regression rate law follows $\dot{r} = aP_c^n$ where $a = 1.25 \times 10^{-5}$, P_c is the chamber pressure in atmosphere, n = 0.35, and $\dot{r}$ is in m/s. if the chamber pressure happens to be 50 atm, determine the static pressure at (1) the front end and (2) the rear end of this grain, and the static pressure ratio along with the length of the grain. Take γ = 1.25.

7.14 A solid-propellant rocket engine with an initial propellant temperature of 25°C can be operated for 5 min with a steady chamber pressure of 5 MPa. Its regression rate law follows $\dot{r} = \dfrac{A}{T_I - T_u} P_c^n$ where A = 172, T_I = 403 K, P_c is the chamber pressure in MPa, n = 0.75, and $\dot{r}$ is in mm/s. Determine the steady chamber pressure when the initial propellant temperature becomes 50°C.

7.15 A solid-propellant rocket engine with an initial propellant temperature of 15°C can be operated with a steady chamber pressure of

3.5 MPa. Its regression rate law follows $\dot{r} = \frac{A}{T_I - T_u} P_c^n$, where $A = 165$, $T_I = 413$ K, P_c is the chamber pressure in MPa, $n = 0.65$, and $\dot{r}$ is in mm/s. Determine the steady chamber pressure when the burning surface area is increased by 15% due to sudden crack in the propellant grain.

REFERENCES

1. Timnat, Y.M., *Advanced Chemical Rocket Propulsion*, Academic Press, St Louis, MO, 1987.
2. Sutton, G.P. and Ross, D.M., *Rocket Propulsion Elements*, 7th edn., John Wiley & Sons Inc, New York, 2001.
3. Kubota, N., *Propellants and Explosives, Thermochemical Aspects of Combustion*, Wiley VCH, Weinheim, Germany, 2002.
4. Beckstead, M.W., Derr, R.I., and Price, C.F., A model of composite solid-propellant combustion based on multiple flames, *AIAA Journal*, 8, 2200, 1970.
5. Price, E.W., Sambamurthi, J.K., Sigman, R.K., and Panyam, R.R., Combustion of ammonium perchlorate polymer sandwiches, *Combustion and Flame, V*, 63, 381–413, 1986.
6. Razdan, M.K. and Kuo, K.K., *Fundamentals of Solid Propellant Combustion*, AIAA, New York, 1983, Chap. 10, pp. 515–597.
7. Mukunda, H.S. and Paul, P.J., Universal behavior of the erosive burning behavior in solid propellants, *Combustion and Flame*, 109, 224–236, 1997.
8. Anderson, J.B., Investigation of the effect of acceleration on the burning rate of composite propellants, PhD thesis, United States Naval Postgraduate School, New York, 1996.
9. Northam, G.B., Effects of propellant composition variables on the acceleration induced burning rate augmentation, NASA TN D-6923, Washington, DC, 1972.
10. Mishra, D.P., *Experimental Combustion*, CRC Press, New York, 2014.
11. Barrere, M., Jaumotte, A., de Veubeke, B.F., and Vandenkerckhove, J., *Rocket Propulsion*, Elsevier Publishing Company, New York, 1960.

CHAPTER 8

Liquid-Propellant Rocket Engines

> Towering genius disdains a beaten path. It seeks regions hitherto unexplored.
>
> ABRAHAM LINCOLN

8.1 INTRODUCTION

Liquid propellants have been used for a long time in rocket engines due to their higher specific impulse. They were first conceived way back in 1915 by Robert Goddard, who demonstrated the launching of a liquid rocket engine using liquid oxygen and gasoline as propellants on March 26, 1926, whose flight lasted for only 2.5 s covering a distance of 56 m. Although dubbed an unsuccessful event by some critics, it importantly demonstrated that it is possible to design and develop a liquid-propellant rocket engine (LPRE). Subsequently, LPREs underwent significant improvements that outweigh the performance of typical solid-propellant rockets. Hence, they currently find wide application in various propulsive devices such as spacecraft, missiles, retro-rockets, and gas-generating systems. It can easily provide thrust levels ranging from a few Newton to several hundred Newton. It is preferred in interplanetary mission and other important areas due to its efficiency, reliability, and cost-effectiveness over other chemical propulsive systems.

A liquid rocket engine employs liquid propellants that are fed, either through pressurized tanks or by using a pump, into a combustion chamber.

The propellants usually consist of a liquid oxidizer and a liquid fuel. In the combustion chamber, the propellants chemically react (burn) to form hot gases, which are then accelerated and ejected at high velocity through a nozzle, thereby imparting momentum to the engine. Momentum is the product of mass and velocity. The thrust force of a rocket motor is the reaction experienced by the motor structure due to the ejection of the high velocity matter. This is the same phenomenon that pushes a garden hose backward as water squirts from the nozzle or makes a gun recoil when fired.

Some of the characteristic features of LPRE as compared to solid-propellant rocket engine (SPRE) are (1) compactness, (2) higher specific impulse, (3) complex system, (4) throttle capability, and (5) longer burning time. These are to be seen in relation to solid and hybrid rockets where higher performance is possible at the expense of simplicity of the solid-propellant rocket engine. It has several disadvantages over other chemical rocket engines. It is difficult to control the liquid-propellant rocket engine as its center mass gets too close to the center of drag. As the propellant is a very large proportion of the mass of the vehicle, the center of mass shifts significantly rearward when a liquid propellant is used. Besides this, liquid propellants are subject to *slosh*, which may lead to loss of control of the vehicle. The liquid-propellant engines are more prone to fire/explosion hazard as compared to solid rocket engines, because liquid propellants may leak, possibly leading to the formation of an explosive mixture. Besides, turbo-pump system is quite complex to design and maintain. The liquid-propellant engine needs considerable preparation before its launch. Hence, for military applications, LPSE is not preferred over SPRE.

8.2 BASIC CONFIGURATION

The main components of a typical liquid-propellant rocket engine, shown in Figure 8.1, consist of thrust chamber, injector, igniter, combustion chamber, nozzle, propellant tank, propellant feed system, and cooling system [1]. The combustion chamber along with nozzle is commonly known as the thrust chamber. It houses injectors and igniter, which atomizes the liquid propellants, mixes and ignites, leading to the combustion of liquid propellants. The main function of the combustion chamber is to produce high-temperature and high-pressure gas due to burning of liquid propellants. Hence, it must have arrays of propellant injectors that can produce fine spray such that both fuel and oxidizer can be vaporized and mixed

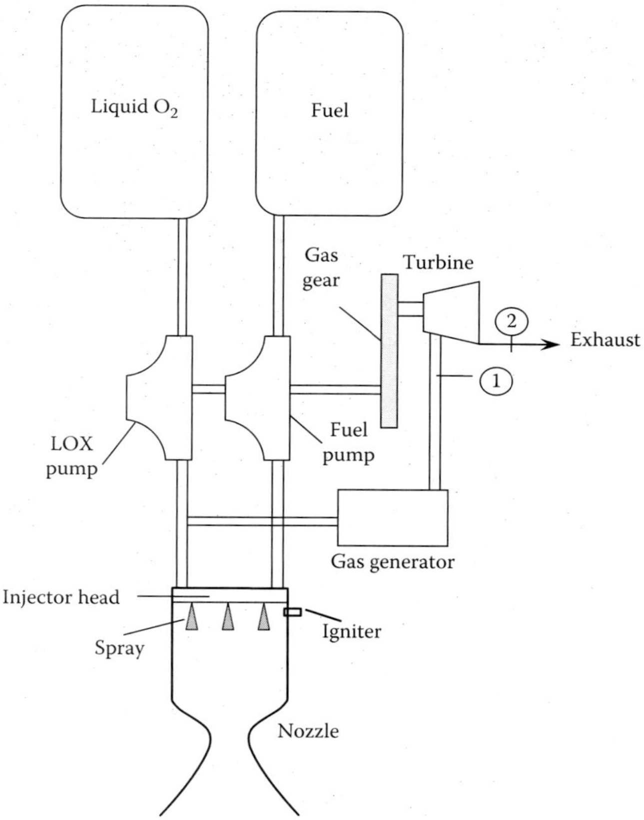

FIGURE 8.1 A typical liquid-propellant rocket engine.

well for combustion to take place within a shorter zone of the combustion chamber. In order to have better quality spray requisite average droplet size and distribution, propellant must be fed into the injectors at high pressure. For this purpose, the propellant feed system needs to be designed and developed properly. Several propellant feed systems have been developed, which will be discussed in subsequent sections. Of course, the propellant must be stored in high-pressure tank, particularly in small rocket engine. However, turbo-pumps are used to feed high-pressure propellants in larger LPRE. The function of ignition system is to provide certain initial ignition energy for initiation of combustion. Of course, in certain liquid propellants known as hypergolic propellants, there is no need to have an ignition system. As high temperature is required to produce high thrust, cooling system is being used routinely to cool both combustion chamber and nozzle. The high-pressure and high-temperature gas is expanded in convergent–divergent nozzle to produce thrust.

8.3 TYPES OF LIQUID-PROPELLANT ROCKET ENGINES

Several types of liquid-propellant rocket engine have evolved from the time of its inception way back in 1926. Based on single liquid or two liquids, it can be broadly classified into two types: (1) monopropellant rocket engine and (2) bipropellant rocket engine [2–5].

8.3.1 Monopropellant Rocket Engines

In case of monopropellant rocket engine, a single liquid propellant is used, in which monopropellant gets decomposed with the help of a suitable catalyst into hot gases that are expanded in the nozzle to produce requisite thrust. The great advantage of this system is the elimination of the oxidizer system altogether, making it a very simple system. However, their applications are restricted to low-thrust and low-duration flight conditions during each firing. A schematic of a typical monopropellant rocket engine is shown in Figure 8.2, in which liquid propellant is injected into a catalyst bed and decomposes into high-pressure and high-temperature gas. These hot gases are expanded in the convergent–divergent nozzle to produce requisite thrust. Generally, monopropellant is a slightly unstable chemical that decomposes easily exothermally to produce hot gas. Some of the monopropellants used in rocket engines are hydrazine (N_2H_4), hydrogen peroxide (H_2O_2), hydroxylammonium nitrate (HAN), and propylene glycol denitrate (PGDN). Among all the monopropellants, hydrazine (N_2H_4) is considered to have desirable properties as it has higher specific impulse and lower density. Although it was quite cumbersome to ignite it, with the advent of a better catalyst such as iridium pallet, it is possible to ignite

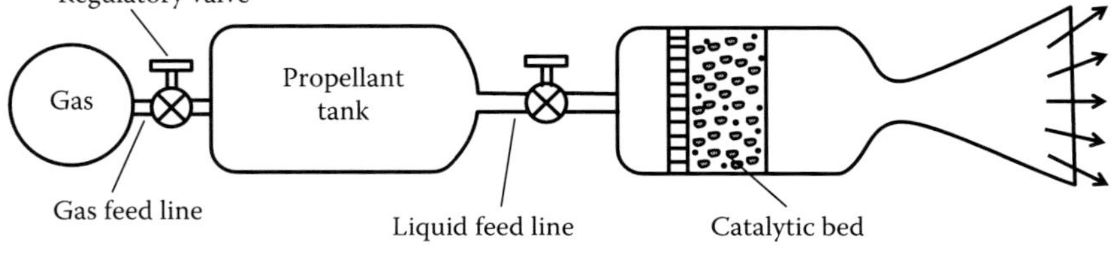

FIGURE 8.2 Schematic of monopropellant LPR engine.

it easily. In the presence of iridium catalyst, hydrazine can be decomposed first into ammonia and nitrogen, as given in the following:

$$3N_2H_4 \rightarrow 4NH_3 + N_2 - 336.28 \text{ kJ} \quad (8.1)$$

Note that this reaction is exothermic in nature. As a result, the ammonia gets dissociated further into nitrogen and hydrogen, as given in the following:

$$NH_3 \rightarrow 2N_2 + 6H_2 + 184.4 \text{ kJ} \quad (8.2)$$

As this reaction is endothermic in nature, the temperature of gases decreases with increase in degree of ammonia dissociation. Interestingly, the molecular mass of product gases decreases with increase in ammonia dissociation. Note that the extent of ammonia dissociation depends on the residence time of hydrazine remaining contact with catalyst, size and configuration of catalyst bed. The used catalyst has ensured almost spontaneous restart capability of the hydrazine monopropellant along with relative stability, clean exhaust, and low flame temperature, which has made it the most preferred among all other monopropellants.

8.3.2 Bipropellant Rocket Engines

The bipropellant liquid rocket engines use one liquid propellant as fuel and another as an oxidizer, as shown in Figure 8.3. This kind of rocket engine offers higher performance as compared to monopropellant engines

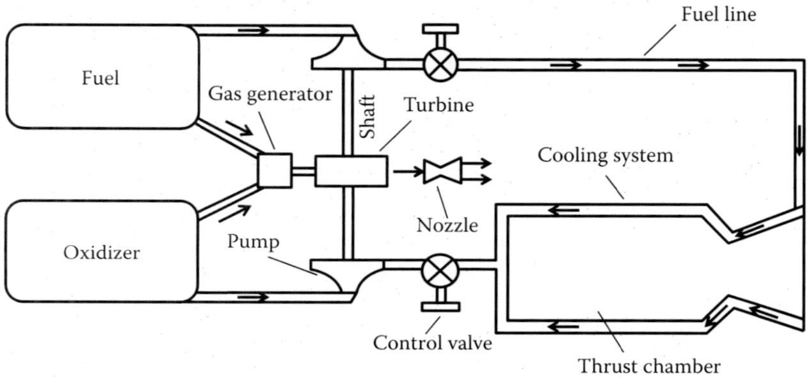

FIGURE 8.3 Schematic of bipropellant LPR engine.

in terms of specific impulse and offers higher versatility in terms of restarting, variable thrust, a wider range of operations, and so on. However, it has several disadvantages as it has higher failure modes, complexities, and price tags as compared to monopropellant rocket engine. However, this kind of engine is used for launch vehicles, missiles, and other applications extensively, due to its higher performance levels. As mentioned earlier, a typical pump-fed bipropellant liquid rocket engine, shown in Figure 8.3, has thrust chamber, injection system, cooling system, propellant feed system, nozzle, and so on. Generally, fuel and oxidizers as liquid propellants are fed through respective atomizers to convert bulk liquid into spray, which mix with each other and subsequently get vaporized and react with each other on ignition to produce high-temperature and high-pressure gas. The resultant gas is expanded through the convergent–divergent nozzle to produce thrust.

Based on the nature of ignition, the bipropellant liquid rocket engines can be further divided into two categories: (1) hypergolic and (2) nonhypergolic. In case of hypergolic propellant rocket engines, when liquid fuel and oxidizer come in contact, they ignite without the aid of any external ignition energy, leading to combustion while undergoing exothermic chemical reactions. As mentioned earlier, several fuels like aniline, triethylamine, hydrazine, MMDH, and UDMH are hypergolic when they react with oxidizers like white fuming nitric acid (4%–6% nitrogen tetroxide) and red fuming nitric acid (10%–14% nitrogen tetroxide). Note that these propellants can be stored under normal pressure and temperature without any special arrangements. However, hypergolic propellants like liquid fluorine–liquid hydrogen are to be stored in cryogenic conditions. In case of nonhypergolic rocket engines, ignition energy must be supplied externally to initiate combustion. As discussed earlier in Chapter 6, several fuels like kerosene, hydrocarbon, alcohol, and so on, can be used as nonhypergolic propellants. There is another class of most preferred nonhypergolic bipropellant rocket engines, due to its higher specific impulse, in which liquid hydrogen and liquid oxygen propellants are being used and for which cryogenic storage system is required. Semicryogenic nonhypergolic systems, namely, liquid kerosene and liquid oxygen, are being used for rocket engines. There is some unusual combination, namely, ammonia and liquid oxygen nonhypergolic, being used in some rocket engines like X-15 research aircraft.

Example 8.1

A monopropellant rocket engine is to be designed to produce thrust of 55 N with specific impulse of 169 s using hydrazine fuel. If the mass flux through the hydrazine catalytic bed is 220 kg/m^{2}s, determine the diameter of this bed.

Solution

The mass flow rate of hydrazine propellant can be determined as

$$\dot{m} = \frac{F}{I_{sp} g} = \frac{55}{169 \times 9.81} = 0.572 \text{ kg/s}$$

The cross-sectional area A_b of the catalyst bed is determined as

$$A_b = \frac{\dot{m}}{\dot{m}''} = \frac{0.572}{220} = 0.0026 \text{ m}^2$$

The diameter of the catalyst bed is evaluated as

$$D_b = \sqrt{\frac{4 \times 0.0026}{3.14}} = 0.056 \text{ m}$$

8.4 COMBUSTION OF LIQUID PROPELLANTS

The processes involved during the combustion of liquid propellant in a rocket engine are quite complex and have not been understood completely, because the physical processes such as vaporization, diffusion of species, mixing, and heat transfer take place along with heat release due to overall exothermic chemical reactions during combustion. The liquid fuel, being injected into the combustion chamber at a high velocity (20–100 m/s), gets converted into spray, as shown in Figure 8.4, during which breaking of liquid stream into ligaments/lobes accompanied by secondary atomization of ligaments into fine droplets also takes place [3]. The vaporization of droplets occurs due to heat transfer from the combustion zone and hot wall through all three heat transfer modes, namely, conduction, convection, and radiation. Apart from liquid phase mixing, intimate gas phase

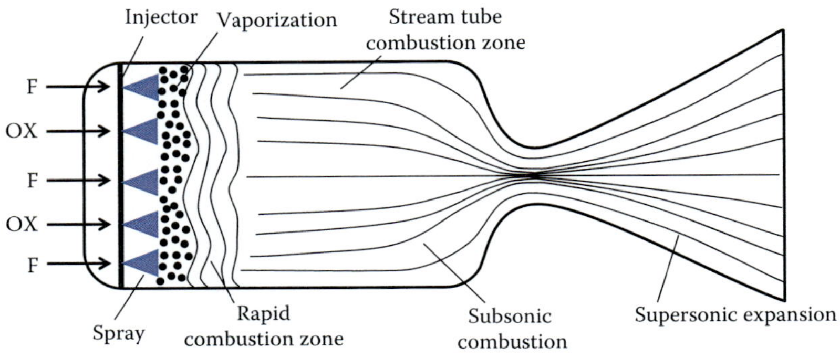

FIGURE 8.4 Schematic representation of various processes in the combustion chamber.

mixing of fuel and oxidizer does take place in a narrow zone of combustion chamber as shown in Figure 8.4. The gaseous or liquid or, both, gaseous and liquid propellants undergo complex chemical reactions with several intermediate products, liberating large amount of heat within a narrow zone of combustion chamber. Depending on the droplet size of the spray, several liquid droplets move with gas across entire combustion zone even till the end of combustion chamber, making the flow to be inherently two-phased in nature. During the early stage of the two-phase flow, droplets moving at velocity of 50–100 m/s can cause drag to gas flow, but subsequently accelerate the gas flow due to both vaporization of droplets and heat release caused by burning of fuel. Besides this, turbulent diffusion of active species such as H, OH, and H_2 makes the flow more complex and difficult to handle analytically. The occurrence of recirculation zone and turbulent vortices of gas around the injector augments the mixing and heat transfer, which helps vaporization and combustion. Note that the processes are so complex that they would not be occurring sequentially as described. Rather, some of these processes occur simultaneously. Besides this, it is difficult to decide which process has the greatest influence on the combustion of propellants.

The combustion in the liquid rocket engine is quite dynamic because the flame front moves across the entire mass, which might be caused due to intense turbulence level in the gases of the combustion chamber. The residence time of both fuel and oxidizer is quite small, that is, even less than 10 ms. The heat release rate per unit volume in typical rocket engine is

TABLE 8.1 Comparison of Various Parameters between Typical Liquid-Propellant Rocket Engine Combustor and Aerogas Turbine Engine Combustor

Sl. No.	Parameter	Aerogas Turbine Engine Combustor	Liquid-Propellant Rocket Engine Combustor
1.	Peak temperature	2300 K	3400 K
2.	Chamber pressure	5–40 atm	50–200 atm
3.	Liquid flow rate	2.5 kg/s	600 kg/s
4.	Droplet size	20–60 μm	20–150 μm
5.	Mode of spray	Dilute	Dense
6.	Recirculation zone	Significant	Very little
7.	Heat release rate density	10 MW/m^3	1000 MW/m^3

1000 MW/m^3, which is even 100 times compared to typical aerogas turbine combustor in aircraft, because the higher rate of heat release during the combustion of propellant takes place at high-pressure and high-temperature conditions (see Table 8.1) prevailing in rocket engine. The droplet size range for rocket engines is higher as compared to aerogas turbine combustor. Hence, in the case of rocket engine, diameter and length of combustion chamber is chosen judiciously such that most of the droplets are at least vaporized in the combustion chamber itself. In contrast, complete combustion of droplets takes place within the initial portion of combustion chamber length of aerogas turbine engine as later portion of combustion chamber is used to ensure uniform temperature at its exit.

We will now explore how different phases take place in the combustion chamber for two types of liquid propellants, namely, (1) hypergolic and (2) nonhypergolic. In case of hypergolic propellant combustion, injectors are designed such that both fuel and oxidizer are mixed in the liquid phase itself as they can react easily in the liquid phase to produce certain amount of heat while undergoing exothermic chemical reactions. For example, certain hypergolic propellants like triethylamine (fuel) and nitric acid (oxidizer) or xylidine and nitric acid can be mixed in liquid phase to form stable solution even at low temperature and pressure. The heat liberated during liquid phase exothermic reactions is utilized to vaporize the liquid fuel–oxidizer mixture and hot gases are produced quickly, which helps in initiating chemical reactions in gaseous phase without any aid of external ignition energy. Hence, actual combustion between both fuel and oxidizer will take place leading to formation of burnt gas that mixes with other vapor. This intense mixing of fuel and oxidizer in gaseous phase can form turbulent premixed flame/flamelets in the combustion chamber to produce final combustion products. Apart from this, there is a chance that droplets are formed during the atomization of both fuel and oxidizer propellant. These droplets can undergo combustion, either forming a single diffusion flame around a single droplet or diffusion flame around a group/cluster of droplets. These processes involved during hypergolic combustion are shown schematically in Figure 8.5a. It can be observed that there is another path in which both fuel and oxidizer droplets formed during atomization process can undergo vaporization and mixing to form gaseous mixture that can lead to the formation of premixed flame. There might be several other permutations and combinations of processes that can occur during combustion of hypergolic propellants in a liquid-propellant rocket engine.

In case of nonhypergolic propellant, both fuel and oxidizer do not have chemical affinity for each other in liquid phase. In other words, there would not be any chemical reaction between fuel and oxidizer in liquid phase.

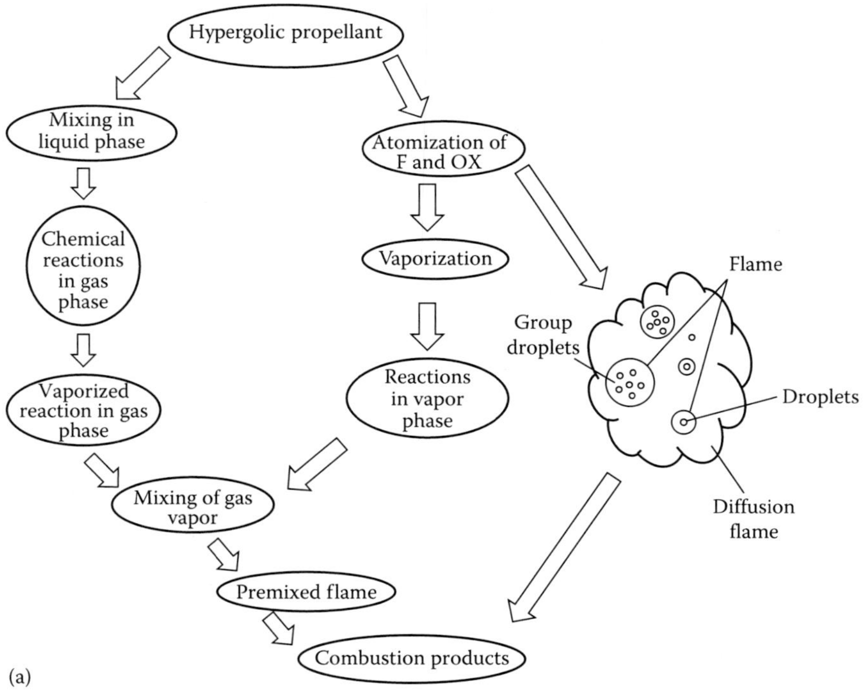

FIGURE 8.5 Schematic of combustion processes in liquid-propellant rocket engine combustor for (a) hypergolic. (*Continued*)

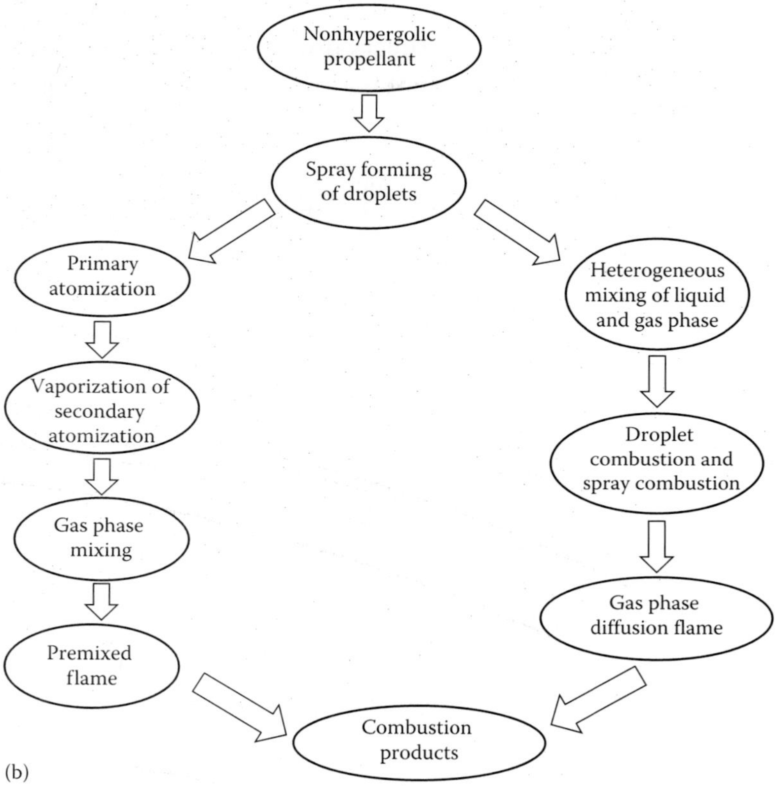

FIGURE 8.5 (*Continued*) Schematic of combustion processes in liquid-propellant rocket engine combustor for (b) nonhypergolic [3].

Hence, exothermic reactions that occur during the gas phase reactions are used to vaporize the liquid propellants such that they can be converted into gaseous phase before being ignited for sustenance of combustion process. That means, the combustion processes will be influenced by the heat release rate and physical properties of liquid propellants. Although there might be liquid phase mixing as shown in Figure 8.5b during atomization process, but there would not be any chemical reactions in the liquid phase. During this process, some of the liquid-propellant droplets can get vaporized into gas phase. This vaporization process can be influenced by heat release from both homogeneous and heterogeneous combustion processes, as some of the liquid fuel droplets can be floated as single droplet or cluster of droplets or cluster of group droplets. When vaporized fuel from a single or group of droplets comes in contact with gaseous oxidizer, diffusion flame is formed while undergoing heterogeneous combustion to produce combustion products. Besides these, primary droplets may undergo secondary atomization, which accelerates the vaporization processes, although with the help of heat transfer from combustion zone by both convection and radiation heat transfer. The fuel, oxidizer, and combustion products in gaseous phase are mixed by molecular diffusion and turbulent flow to form premixed mixture, which undergoes chemical reactions to form premixed flame. The processes involved during combustion of nonhypergolic propellants are shown schematically in Figure 8.5b. It can be noted that these complex processes are tentative in nature, which need to be developed with further research with the help of advanced optical diagnostic and numerical tools in future.

The processes described provide a qualitative picture of the complex phenomena describing various stages during combustion of liquid propellants. It is quite difficult and cumbersome to determine the exact time required for each process. In order to design combustion chamber length, we need to determine the time required for certain partial or complete processes that ensure complete combustion. Let us then consider the change of specific volume v, from the time of propellant entry to the formation of combustion products in the form of hot gas. We will consider two cases: (1) hypergolic and (2) nonhypergolic propellant combustion.

8.4.1 Hypergolic Propellant Combustion

For hypergolic propellant combustion, it can be assumed that exothermic chemical reaction can be initiated as soon as both fuel and oxidizer propellants are injected from a suitable injection system. As a result, specific volume starts increasing from the point of injection, as shown in Figure 8.6. Subsequently, specific volume increases continuously till the end of complete burning of propellants, as gaseous phase combustion continues both in the form of premixed and diffusion flames that results in an increase in specific volume. It can be noted that it would not be possible to differentiate between liquid-phase and gas-phase chemical reactions. However, if we can assume the first-order chemical reaction to be occurring during liquid phase chemical reaction, then its time constant can be obtained easily, assuming a theoretical profile for specific volume with time, as shown in Figure 8.6a. The time for liquid-phase reaction t_{lc} can be determined as the reciprocal of specific reaction rate. This can be defined by the tangent to the exponential curve at time $t = 0$ as described in Figure 8.6a. The time for the reactions to take place in the combustion chamber t_{rg} is equal to the

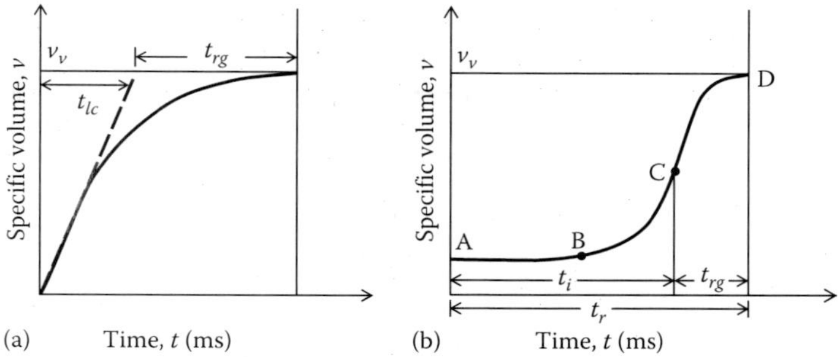

FIGURE 8.6 Variation of specific volume with time for (a) hypergolic and (b) nonhypergolic combustion.

residence time of the gaseous flow of propellant. Hence, two characteristic times, namely, liquid-phase combustion time t_{lc} and residence time for gaseous phase combustion t_{rg} can be used for the determination of the length of combustion chamber.

8.4.1.1 Nonhypergolic Propellant Combustion

In this case, liquid-phase combustion does not take place. Hence, when propellant enters into the combustion chamber through injector, the propellant does not burn immediately; rather, it takes certain interval of time known as ignition time t_i, during which fuel and oxidizer propellant are mixed and vaporized. As a result, specific volume remains almost constant during this time, as shown in Figure 8.6b, and the specific volume starts increasing at B due to vaporization of liquid propellant caused by atomization and heat transfer from the combustion zone. Of course, this increase in specific volume occurs at a slower rate till point C. The time duration from the starting of propellant injection till point C is known as ignition delay time t_i. Subsequently, the specific volume gets accelerated beyond point C till point D as the exothermic chemical reactions are initiated. The gas generated due to burning of propellant has to stay for certain time in the combustion chamber to ensure complete combustion. Hence, the residence time duration from the initiation of combustion till its completion, during which propellants and their combustion products will be in gas phase, is termed t_{rg}. Hence, the total residence time t_r is the sum of ignition delay time and residence time in gas phase, as given in the following:

$$t_r = t_i + t_{rg} \tag{8.3}$$

8.5 COMBUSTION CHAMBER GEOMETRY

In order to ensure complete combustion in a liquid-propellant rocket engine, we need to provide minimum volume of the combustion chamber such that liquid propellant must have enough residence time to stay inside the combustion chamber. The volume of thrust chamber V_c can be related to the characteristics length L^* and throat area A_t, as given in the following:

$$V_c = L^* A_t; \quad V_c = L_c A_c \tag{8.4}$$

where

L_c is the combustor chamber length
A_c is the combustion chamber cross-sectional area

The characteristics length L^* can be related to the total residence time as [3]

$$\frac{L^*}{C^*} = f_1\left(t_i\right) + f_2\left(t_{rg}\right) \tag{8.5}$$

where C^* is the characteristic velocity. Hence, in order to determine the volume, we need to know the $f_1(t_i)$ and $f_2(t_{rg})$ along with characteristic velocity and throat area. For simplicity, the ignition time t_i can be neglected as it is small as compared to residence time in gas phase for certain cases. As the residence time in gas phase t_{rg} is the ratio of mass of the gas m_g in the combustion chamber to the mass flow rate of propellant being injected into the combustion chamber $\dot{m}$, which is expressed as

$$t_{rg} = \frac{m_g}{\dot{m}} \tag{8.6}$$

For steady-state conditions, let us assume that entire combustion chamber volume V_c has average density $\bar{\rho}_g$, which is equal to the density of the burnt gases. Then by using the definition of the characteristic velocity C^*, the expression (Equation 8.6) for the residence time in gas phase t_{rg} can be restated as

$$t_{rg} = \frac{m_g}{\dot{m}} = \frac{\bar{\rho}_g V}{\dot{m}} = \frac{\bar{\rho}_g L^* C^*}{P_c} \tag{8.7}$$

By invoking Equation 8.4, we can express $f_2(t_{rg})$ as

$$f_2\left(t_{rg}\right) = \frac{P_c}{\bar{\rho}_g C^{*2}} \cdot t_{rg} \tag{8.8}$$

By assuming the specific heat ratio γ to remain constant across the entire combustion zone up to the throat of the nozzle, we can have

$$\frac{P_c}{\bar{\rho}_g C^{*2}} = \Gamma^2 \tag{8.9}$$

Then, we can have a simple expression for the characteristic length L^*, which is given by

$$\frac{L^*}{C^*} = \Gamma^2 t_{rg} \tag{8.10}$$

Similarly, we have an expression for combustion chamber volume V_c, which is given by

$$V_c = L^* A_t = \Gamma^2 t_{rg} A_t \tag{8.11}$$

Note that for design of liquid-propellant rocket engine, the specific heat ratio γ can be considered as 1.25; then we can evaluate the combustion chamber volume V_c easily by knowing the residence time in gas phase t_{rg}. For most cases, experimental studies indicate that the residence time in gas phase t_{rg} varies from 2×10^{-3} to 7×10^{-3} s. But in order to evaluate the volume accurately, we need to evaluate $f_1(t_i)$. But this residence time function $f_1(t_i)$ is dependent on the injection system. In other words, it is dependent on the injection pressure drop across fuel ΔP_F and oxidizer ΔP_{ox}, and chamber pressure, turbulence level, type of injector and its arrangement, and type of propellants used in liquid-propellant rocket engine. It is quite difficult to develop a general expression for $f_1(t_i)$, and this is not yet possible. However, experimental data can be used for design purposes. Besides this, the characteristic length L^* can be obtained experimentally. As mentioned earlier, characteristic length L^* is dependent on the nature of propellant, as evident from the experimental data shown in Table 8.2.

TABLE 8.2 Typical Range of Characteristic Length L^* for Different Liquid-Propellant Systems

Propellant System	Characteristics Length L^*(m)
Liquid fluorine–hydrazine	0.60–0.71
Liquid fluorine (LF)–liquid hydrogen (LH_2)	0.63–0.76
Nitric acid–hydrazine	0.76–089
Liquid oxygen–ammonia	0.76–1.0
Liquid oxygen (LOX)–liquid hydrogen (LH_2)	0.8–1.0
Liquid oxygen–kerosene (RP1)	1.01–1.27
Nitric acid–UDMH	1.5–2.0
Liquid oxygen–ethyl alcohol	2.5–3
Nitric acid–hydrocarbon	2–3

Sources: Barrere, M. et al., *Rocket Propulsion*, Elsevier Publishing Company, New York, 1960; Huzel, D.K. and Huang, D.H., *Modern Engineering for Design of Liquid Propellant Rocket Engines*, Vol. 147, AIAA Publication, Washington, DC, 1992.

Apart from this, characteristic length L^* is not only dependent on the injection system but also on the cross-sectional area ratio $\varepsilon_c = A_c/A_t$. This represents the optimum cross section required for the minimum chamber length. In order to determine the cross-sectional area ratio $\varepsilon_c = A_c/A_t = A_2/A^*$. Note that this cross-sectional ratio ε_c is dependent on the size of the rocket motor, which decreases as the thrust level increases. For example, the American designers use $\varepsilon_c = 1.2–2$ for 100 tons of thrust, $\varepsilon_c = 2–3$ for 10 tons of thrust, $\varepsilon_c = 3–4$ for 1 ton of thrust. It can be observed that the American designers prefer the cross-sectional area ratio ε_c in the range of 1.2–4.0, while the German designer uses the cross-sectional area ratio ε_c in the range of 4–15. Then, the German-designed rocket engines will be smaller in length as compared to the American design for the same thrust level and propellant system. But a lower value of ε_c can incur losses due to heat transfer from its wall and higher pressure drop in cooling system, which result in lower overall efficiency.

After choosing proper value of the cross-sectional area ratio $\varepsilon_c = A_c/A_t = A_2/A^*$, we need to determine the pressure losses in the combustion chamber that are incurred due to heat releases rate, increase in flow Mach number along combustion chamber and frictional heat losses. The actual determination of pressure drop across even the geometrically simple combustion chamber is quite difficult and complex as the flow is three-dimensional in nature accompanied with higher level of heat release. For this purpose, let us consider a constant area combustion chamber for which we can derive the static pressure ratio with assumption of Rayleigh flow [6,7], given as follows:

$$\frac{P_2}{P_1} = \frac{1+\gamma M_1^2}{1+\gamma M_2^2} \tag{8.12}$$

Then the total pressure ratio between stations (1) and (2) can be derived easily as

$$\frac{P_{t2}}{P_{t1}} = \frac{1+\gamma M_1^2}{1+\gamma M_2^2}\left[\frac{1+\dfrac{\gamma-1}{2}M_2^2}{1+\dfrac{\gamma-1}{2}M_1^2}\right]^{\frac{\gamma}{\gamma-1}} \tag{8.13}$$

We can assume the Mach number M_1 at the end of injection to be negligibly small. Then, for $M_1 \simeq 0$, Equations 8.12 and 8.13 become

$$\frac{P_2}{P_1} = \frac{1}{1+\gamma M_2^2}; \quad \frac{P_{t2}}{P_{t1}} = \frac{\left[1+\dfrac{\gamma-1}{2}M_2^2\right]^{\frac{\gamma}{\gamma-1}}}{1+\gamma M_2^2} \tag{8.14}$$

We need to determine the Mach number at station (2) to determine the pressure losses across the combustor. By assuming the flow to be isentropic in the exhaust nozzle of the rocket engine, we can derive an expression for the cross-sectional area ratio $\varepsilon_c = A_c/A_t = A_2/A^*$ as follows:

$$\frac{A_c}{A_t} = \frac{A_2}{A^*} = \frac{1}{M_2}\left[\frac{2}{\gamma+1}\left(1+\frac{\gamma-1}{2}M_2^2\right)\right]^{\frac{\gamma+1}{2(\gamma-1)}} \tag{8.15}$$

Assuming the specific heat ratio γ to be around 1.25 for typical liquid-propellant combustion, we can determine the exit Mach number M_2 for typical set of the cross-sectional area ratio $\varepsilon_c = A_2/A_t$, as shown in Figure 8.7. It can be noted that Mach number decreases with increase in the cross-sectional area ratio ε_c. In contrast, both static and total pressure ratios increase with the cross-sectional area ratio ε_c. Although it is prudent to use the cross-sectional area ratio $\varepsilon_c = A_2/A_t = 1$ corresponding to straight nozzle, the cross-sectional area (contraction) ratio ε_c between 1.2 and 4 is used routinely.

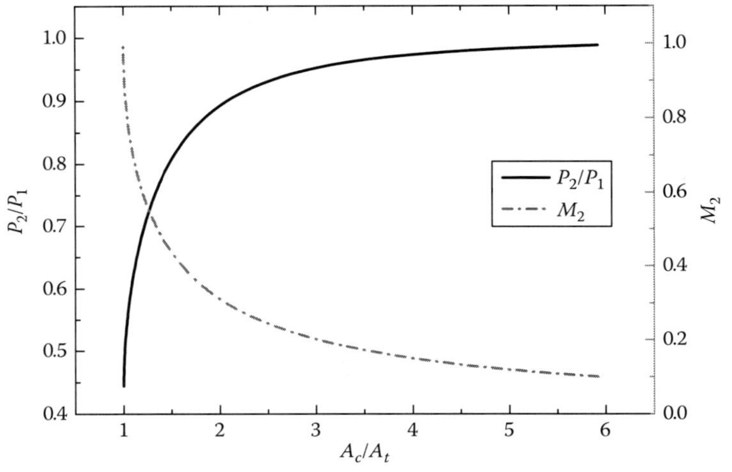

FIGURE 8.7 Variation of Mach number, M_2, static and total pressure ratios with cross-sectional area ratio ε_c for $\gamma = 1.2$.

Before leaving this section, let us discuss the options of choosing proper shape of the combustion chamber available to the designer. From Equation 8.13, it can be inferred that designer can choose any shape of the combustion chamber as the residence time is independent of the combustion chamber volume. But in the real situation, the choice of proper shape of combustion chamber is quite important in determining the performance of the liquid rocket engine. We know that a long combustion chamber with a smaller cross-sectional area can incur high pressure losses due to higher frictional and gas dynamic losses. In contrast, shorter combustion chamber with higher cross-sectional area can have higher zone of atomization and vaporization, leading to poorer combustion due to shorter mixing and combustion time. Besides these factors, other factors such as heat transfer, combustion instability, weight, cost, reliability, and reignition capability are to be considered while deciding about the shape of combustion chamber. Three distinct shapes of combustion chamber that have been employed in practice are shown in Figure 8.8. Ideally, the spherical shape of the combustion chamber shown in Figure 8.8a is considered to be the best among all three shapes as it has the highest surface-to-volume ratio for the same material strength and chamber pressure. Besides this, it can have minimum wall thickness required for same level of chamber pressure, which happens to be almost half of that of a cylindrical shape chamber. It has added advantages of lesser cooling surface and weight. But it is quite

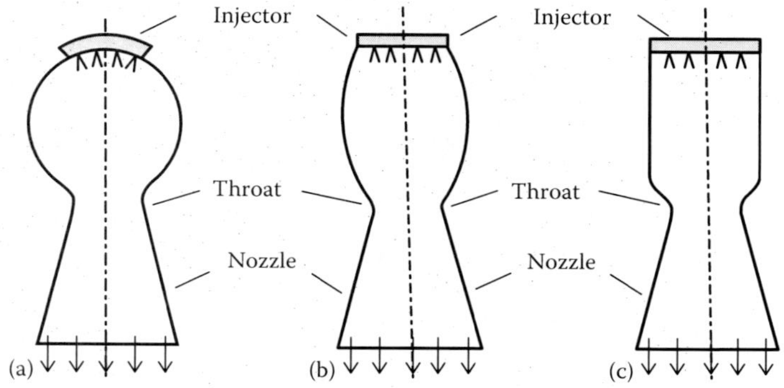

FIGURE 8.8 Various shapes of combustion chamber: (a) spherical, (b) near spherical, and (c) cylindrical.

complex to manufacture and incurs loss in overall performance of combustion due to difficulties encountered during atomization, vaporization, and combustion processes. Early European designers preferred this kind of shape for liquid rocket engine. Subsequently, in order to overcome the problems of manufacturing, they designed and developed near-spherical shape combustion chamber, as shown in Figure 8.8b. But, American engineers prefer cylindrical-shaped combustion chamber, as shown in Figure 8.8c, which is currently being used in most parts of the globe as it is quite easy to manufacture and maintain.

Example 8.2

In a liquid-propellant rocket engine, propellant is injected into its combustion chamber at pressure of 6 MPa and temperature of 3800 K. If the residence time happens to be 0.7 ms, determine the length of cylindrical combustion chamber. Assume that instantaneous combustion occurs, which produces flow with Mach number of 0.3. Take $\gamma = 1.2$ and MW of product gas = 25.

Solution

The theoretical characteristic velocity of this engine can be determined as

$$C^* = \frac{\sqrt{RT_c}}{\Gamma} = \frac{\sqrt{(8314/25)\times 3800}}{0.649} = 1732 \text{ m/s}$$

as $\Gamma = \sqrt{\gamma\left(\frac{2}{(\gamma+1)}\right)^{\frac{\gamma+1}{\gamma-1}}} = \sqrt{1.2\left(\frac{2}{(1.2+1)}\right)^{\frac{1.2+1}{1.2-1}}} = 0.649$

By using Equations 8.8 and 8.5, we can evaluate

$$\frac{L^*}{C^*} = \frac{P_c}{\bar{\rho}_g C^{*2}} \cdot t_{rg} = \frac{6\times10^6 \times 0.7\times10^{-3}}{4.75\times1732^2} = 0.29\times10^{-3}$$

Assuming ideal gas law, we can have

$$\bar{\rho}_g = \frac{P_c}{RT_c} = \frac{6\times10^6\times25}{8314\times3800} = 4.75\ \text{kg/m}^3$$

Let us now evaluate the A_c/A_t by using Equation 8.15:

$$\frac{A_c}{A_t} = \frac{1}{M_2}\left[\frac{2}{\gamma+1}\left(1+\frac{\gamma-1}{2}M_2^2\right)\right]^{\frac{\gamma+1}{2(\gamma-1)}}$$
$$= \frac{1}{0.3}\left[\frac{2}{1.2+1}\left(1+\frac{1.2-1}{2}(0.3)^2\right)\right]^{\frac{1.2+1}{2(1.2-1)}} = 2.07$$

The length of combustor L_c can be determined as

$$L_c = \frac{V_c}{A_c} = \frac{V_c}{A_t}\frac{A_t}{A_c} = L^*\frac{A_t}{A_c} = \frac{L^*}{C^*}C^*\frac{A_t}{A_c}$$
$$= 0.29\times10^{-3}\times1732\times2.07 = 1.04\ \text{m}$$

8.6 COMBUSTION INSTABILITIES IN LPRE

The combustion process in a liquid-propellant rocket engine occurs inherently under unsteady conditions. Generally, certain level of unsteadiness occurred within pipe line can lead to low-amplitude, random flow fluctuations to prevail during the operations of liquid rocket engines without affecting its performance significantly. But in certain situations, liquid-propellant rocket engine does encounter certain violent unsteady combustion, resulting in irregular pressure fluctuations of large amplitude in the combustion chamber and nozzle flow. These fluctuations in pressure are in

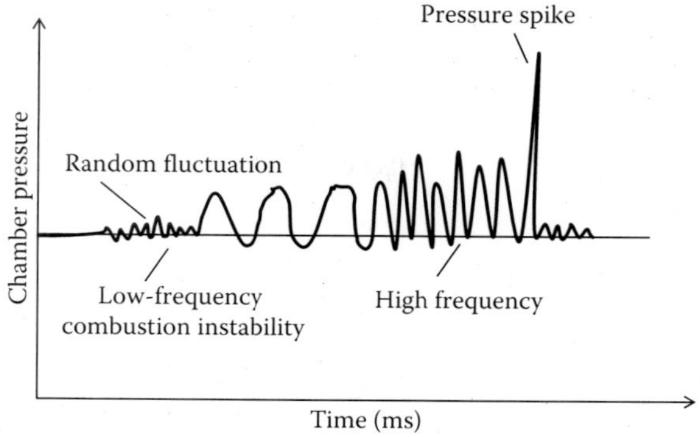

FIGURE 8.9 Chamber pressure variation time indicating combustion instabilities in LPRE.

the acoustic or below acoustic range, caused due to the positive feedback between the heat release rate during the process of propellant combustion and flow field disturbance. Figure 8.9 indicates a typical variation of chamber pressure with time of a liquid rocket engine. A liquid rocket engine is considered to be operated under stable condition, although there will be random fluctuations in pressure, as shown in Figure 8.9, however, with lower limits around its mean value. There might be some instances when a single large amplitude, as shown in Figure 8.9, known as pressure spike occurs during its operation. If this pressure fluctuation happens to interact with the natural frequency of the feed system or the acoustic frequency of the combustion chamber, it may lead to the formation of periodic pressure oscillations at certain characteristic frequency. These oscillations may be amplified or maintained or dampened out by the combustion process. These sustained oscillations during combustion process in the combustion chamber are often termed as combustion instability. By carrying out the fast Fourier transform (FFT) analysis of pressure time history, it can be found out during combustion instability phase that the pressure fluctuations have several definite frequencies. In other words, the combustion instabilities have more natural acoustic frequencies, which can cause several detrimental effects. As discussed in Chapter 9, depending on the range of frequency, combustion instability can be broadly classified into three kinds: (1) LFI, (2) IFI, and (3) HFI. When combustion instability occurs below 400 Hz, then it is termed as low frequency instability (LFI). This is also known as chugging/system instability, which mainly results due to coupling between the combustion and propellant feed system. Intermediate frequency instability

(IFI) is likely to occur in the frequency range of 400–1000 Hz, which is also known as buzzing/entropy wave. This frequency is believed to be caused due to pressure fluctuations in injector and manifold caused by flow eddies, mechanical vibrations of engine, and propellant feed system resonance. Beyond 1000 Hz, it is known as high frequency instability (HFI), which is attributed to the coupling between the pressure oscillation and heat release rate. This is often termed as screaming/screeching. Recall that based on the predominant acoustic mode, combustion instabilities can be classified into various types. Generally, three predominant modes, namely, longitudinal (axial), transverse, and radial modes, and their combinations in different frequency ranges can occur during the combustion processes in the liquid rocket engine. We will discuss these modes in liquid rocket engine in subsequent sections. Note that sizable combustion oscillations, either periodic or random, are considered to be detrimental for the overall performance and life of the liquid rocket engine, because pressure oscillation of 10% about its mean value can cause a thrust oscillation of 10%–100% of the mean thrust. Such oscillations can have a coupling with other portions of vehicle and can develop abnormal stresses on engine components and thus spoil the normal operation of other onboard systems. The form of self-excited combustion instability that is linked to the structure of rocket engine is often termed as POGO, which occurs mainly in large vehicles in the range of a few hertz to 1000 Hz, particularly close to burnout time. That is why, pressure oscillations in liquid rocket engines greater than its mean value by 5% is considered to be a serious concern to the designer, which must be avoided at any cost.

We have learnt that low-frequency combustion instability is not deadly but need to be attended to as it affects the performance of the rocket engine and is accompanied with chugging sound. The combustion efficiency decreases by as much as 20% due to the occurrence of low-frequency instability (LFI). This is mainly caused by the phase difference between feed system response and combustion chamber pressure oscillation. Hence, it can be controlled by isolating the pressure oscillation in the combustion chamber such that it cannot be sensed by the feed system. One way of accomplishing this is to increase the pressure drop across the injector. The frequency and the amplitude of this low-frequency instabilities are dependent on the characteristic length L^*, the chamber pressure P_c, and the pressure difference across injector ΔP_i. Experimentally, it has been observed that both amplitude and frequency decrease with increase in the characteristic length L^*. But with an increase in chamber pressure, amplitude of instability decreases accompanied with an enhancement in its frequency.

Interestingly, with increase in the pressure difference across injector ΔP_i, although the frequency of instability increases marginally in initial phase, stability condition is recovered subsequently.

8.6.1 Analysis of Bulk Mode Combustion Instability

The pressure oscillation can occur either over entire volume of combustion chamber or with different amplitude and phases with different locations in combustion chamber. We will be restricting our discussion to the bulk mode of combustion instability, during which bulk gas oscillates as lumped mass in the entire combustion chamber [2]. Let us consider a simplified liquid-propellant engine in which propellant with density ρ is injected with constant injection pressure P_i through injection orifice of A_i. Due to combustion of liquid propellant, gaseous mass is generated in the combustion chamber. We know that the chamber pressure of rocket engine depends on the rate of gaseous mass generation during burning of propellant, mass flow rate expelled through nozzle, and accumulation of gas in the combustion chamber. By using the continuity equation, we can have

$$\frac{dm}{dt} = \dot{m}_g - \dot{m}_n \tag{8.16}$$

where the first term dm/dt in Equation 8.16 represents the rate of mass accumulation in combustion chamber, $\dot{m}_g$ is the mass generation due to propellant burning, and $\dot{m}_n$ is the mass flow rate leaving through the nozzle. Recall that certain residence time t_r is required before complete combustion of propellant. Hence, the gas mass flow rate $\dot{m}_g$ generated during combustion corresponds to the mass flow rate of propellant $\dot{m}_p$ that is injected before time $(t - t_r)$. Hence, Equation 8.16 can be written as

$$\frac{dm}{dt} = \dot{m}_{p,(t-t_r)} - \dot{m}_n \tag{8.17}$$

The mass of propellant injected through the injector with cross-sectional area A_i and discharge coefficient C_d is given by

$$\dot{m}_p = C_d A_i \sqrt{2\rho\left(P_i - P_c\right)} \tag{8.18}$$

where

P_i is the injection pressure
P_c is the chamber pressure

The mass flow rate through the nozzle can be expressed as

$$\dot{m}_n = \frac{A_t P_c}{C^*} \tag{8.19}$$

where
P_c is the chamber pressure
A_t is the throat area
C^* is the characteristic velocity

By combining Equations 8.16 through 8.19, we can get

$$\frac{dm}{dt} = \frac{d\left(P_c V_c / RT_c\right)}{dt} = \left[C_d A_i \sqrt{2\rho\left(P_i - P_c\right)}\right]_{t-t_r} - \frac{A_t P_c}{C^*} \tag{8.20}$$

We know that chamber pressure fluctuates due to combustion instability. The chamber pressure P_c can be decomposed into the steady state ($\bar{P}_c$) and fluctuating component (P_c') of chamber pressure $\left(P_c = \bar{P}_c + P_c'\right)$. By substituting this in Equation 8.20, we have

$$\frac{dP_c}{dt} = \frac{RT_c}{V_c}\left[C_d A_i \sqrt{2\rho\left\{P_i - \left(\bar{P}_c + P_c'\right)\right\}}\right]_{t-t_r} - \frac{\Gamma^2 C^*\left(\bar{P}_c + P_c'\right)}{L^*} \quad \text{as} \quad \frac{RT_c A_t}{V_c C^*} = \frac{\Gamma^2 C^*}{L^*} \tag{8.21}$$

The first term of Equation 8.21 can be simplified as

$$\frac{RT_c}{V_c}\left[C_d A_i \sqrt{2\rho\left\{P_i - \left(\bar{P}_c + P_c'\right)\right\}}\right]_{t-t_r} = \frac{RT_c}{V_c}\frac{\bar{P}_c A_t}{C^*}\sqrt{\left[1 - \frac{P_c'}{P_i - \bar{P}_c}\right]_{t-t_r}}$$

Note that $C_d A_i \sqrt{2\rho\left(P_i - \bar{P}_c\right)} = \bar{P}_c A_t / C^*$ for steady state flow condition.

(8.22)

By using Equation 8.22, we can rewrite Equation 8.21 as

$$\frac{dP_c}{dt} = \frac{\Gamma^2 C^*}{L^*}\left(\bar{P}_c \sqrt{\left[1 - \frac{P_c'}{P_i - \bar{P}_c}\right]_{t-t_r}} - P_c\right) \tag{8.23}$$

Note that $L^*/(\Gamma^2 C^*)$ is the nondimensional time, which can be considered as residence time t_r of propellant in the combustion chamber. Using the residence time and expanding root term in Equation 8.23, we can get

$$\frac{dP_c}{dt} = \frac{\bar{P}_c}{t_r}\left[1 - \frac{1}{2}\left(\frac{P_c'}{P_i - \bar{P}_c}\right)\right]_{t-t_r} - \frac{P_c}{t_r} \tag{8.24}$$

By defining the nondimensional pressure perturbation as $\psi = P_c'/\bar{P}_c$, Equation 8.24 can be expressed as

$$\frac{dP_c}{dt} = \frac{\bar{P}_c}{t_r}\left[1 - \beta\psi\right]_{t-t_r} - \frac{\bar{P}_c(1+\beta)}{t_r}; \quad \beta = \frac{\bar{P}_c}{2\left(P_i - \bar{P}_c\right)} \tag{8.25}$$

By simplifying, this expression can be expressed in terms of nondimensional pressure perturbation ψ as follows:

$$\frac{d\psi}{dt} + \frac{\psi}{t_r} + \frac{\beta\psi_{t-t_c}}{t_r} = 0; \quad \text{as } P_c = \bar{P}_c + P_c' = \bar{P}_c(1+\psi) \tag{8.26}$$

Let us express the nondimensional pressure perturbation ψ in terms of growth of perturbation α and angular velocity ω, as follows:

$$\psi = Ae^{(\alpha + i\omega)t} \tag{8.27}$$

Note that the growth constant α is less than zero, pressure perturbation decreases with time. In contrast for $\alpha > 0$, the amplitude of pressure oscillations grows with time. Of course for $\alpha = 0$ the amplitude of pressure oscillations remains constant with time.

By solving Equation 8.27, we can derive expression for combustion delay time t_C and residence time t_r corresponding to the stable condition ($\alpha = 0$) as

$$t_C = \frac{\pi - \tan^{-1}\sqrt{\beta^2 - 1}}{\omega} \tag{8.28}$$

$$t_r = \frac{\sqrt{\beta^2 - 1}}{\omega} \tag{8.29}$$

By using Equations 8.27 and 8.28,

$$\frac{t_C}{t_r} = \frac{\left(\pi - \tan^{-1}\sqrt{\beta^2 - 1}\right)}{\sqrt{\beta^2 - 1}} \tag{8.30}$$

It can be noted that for large values of β, combustion instability is likely to occur at lower combustion delay or larger residence time. In contrast, combustion instability occurs for large combustion delay or small residence time when β is greater than unity. The condition for unstable combustion is given by

$$\frac{t_C}{t_r} > \frac{\left(\pi - \tan^{-1}\sqrt{\beta^2 - 1}\right)}{\sqrt{\beta^2 - 1}} \tag{8.31}$$

In other words, for this condition the combustion instability can occur when $\left(\Delta P_i / \bar{P}_c\right) < 1/2$. Note that this injection pressure drop, ΔP_i, depends on the chamber pressure and feed line pressure. In order to avoid the bulk mode of combustion instability, pressure drop across injector ΔP_i must be less than half of the average chamber pressure. This is known as the Summerfield criterion for combustion stability which can be rewritten for fuel and oxidizer streams of bipropellant rocket engine as

$$\frac{\left(\Delta P_i\right)_F}{\bar{P}} > \frac{1}{1+MR}; \quad \frac{\left(\Delta P_i\right)_{ox}}{\bar{P}} > \frac{MR}{1+MR}; \tag{8.32}$$

where MR is the mixture ratio, which is defined as the ratio of the oxidizer flow rate $\dot{m}_{Ox}$ to the fuel flow rate $\dot{m}_F$. A smaller combustion chamber (smaller residence time) that uses less reactive propellant (large delay time) is less prone to this bulk mode of combustion instability.

Example 8.3

In a liquid-propellant rocket engine, propellant is injected into its combustion chamber of 0.5 m length at pressure of 1.2 MPa with pressure drop of 0.4 MPa. If the combustion delay time and residence time happen to be around 12 and 4.5 ms, respectively, determine its frequency of oscillation. Is it subjected to combustion instability or not?

Solution

We know that for stable combustion bulk mode, the criterion is given by Equation 8.30:

$$\frac{t_C}{t_r} = \frac{\left(\pi - \tan^{-1}\sqrt{\beta^2 - 1}\right)}{\sqrt{\beta^2 - 1}} = \frac{\left(\pi - \tan^{-1}(1.118\right)}{1.118} = 2.06$$

But we can evaluate β as

$$\beta = \frac{\bar{P}_c}{2(P_i - P_c)} = \frac{1.2}{2 \times 0.4} = 1.5; \quad \sqrt{\beta^2 - 1} = \sqrt{1.5^2 - 1} = 1.118;$$

The ratio of combustion delay and residence time can be evaluated as

$$\frac{t_C}{t_r} = \frac{12}{4.5} = 2.67$$

Hence, the combustion instability will occur for this condition.

The frequency of combustion oscillation is determined by using Equation 8.28, as given in the following:

$$\omega = \frac{\pi - \tan^{-1}\sqrt{\beta^2 - 1}}{t_C} = 191.58 \text{ rad/s}; \quad f = \frac{\omega}{2\pi} = 30.51 \text{ Hz}$$

8.6.2 Control of Combustion Instability

In the last section, we have learnt about the analytical method by which probability of certain modes of combustion that can occur in liquid-propellant combustion chamber can be predicted to some extent. But we need to make certain physical changes to avoid the occurrence of combustion instability during its entire operating range. Based on physical changes, the methodologies of controlling combustion instabilities can be broadly divided into three categories: (a) chemical, (b) aerodynamic, and (c) mechanical.

8.6.2.1 Chemical Method

In this method, chemical additives are used to dampen the heat release profile in the combustion chamber by a factor of 2 or more. The type of additives is specific to the propellant. Besides this, this chemical additive can augment the droplet shattering, and thus alters the heat release profile to dampen the combustion oscillations. It has an advantage that no changes in hardware are called for to control combustion instabilities, unlike in other methods.

8.6.2.2 Aerodynamic Method

The combustion instability can be minimized by changing the prevailing aerodynamic of the combustion chamber, which can be carried out by the introduction of tangential velocity, nonuniform distribution of droplets, changes in location of injectors and the angle and orifice sizes of the injection elements, and so on. Besides this, a change in the injector design can affect the prevailing pattern of atomization that can lead to altering the heat release profile such that it can prevent the occurrence of combustion instability. The designer can use larger diameter orifice for injector for same mass flow rate such that the injection velocity can get reduced such that atomization takes place longer distance to complete to avoid pressure oscillations. Another way is to reduce the pressure drop across the injector such that larger droplets are being formed to reduce the heat release rate and this dampens the combustion instability. The designer has to move the velocity gradient point downstream or to increase A_c/A_t such that vaporization rate can be slowed down so that heat release occurs over longer axial distance than that of concentrated region, which prevents occurrence of combustion instability.

8.6.2.3 Mechanical Method

In this method, several mechanical fixtures, namely, baffles, acoustic dampers, ablative liners, and so on are devised to prevent high-frequency oscillations in liquid-propellant combustion chamber. Generally, baffles at the injector head, as shown in Figure 8.10, are provided to dampen the destructive mode of combustion oscillations. Besides this, they can prevent the higher mode of oscillations to derive energy from the mean flow to sustain their further growth. These are mainly designed on the assumptions

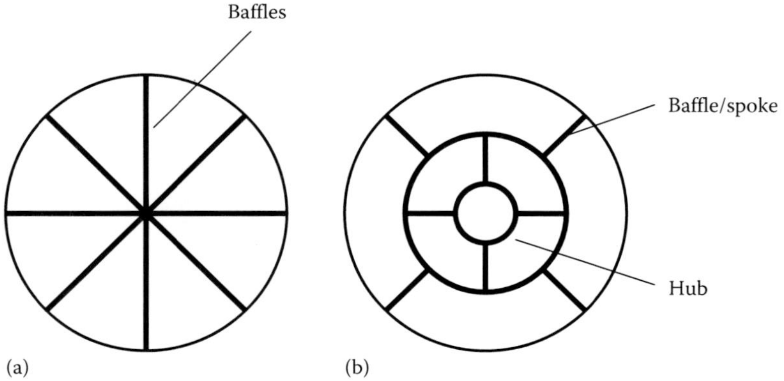

FIGURE 8.10 Types of baffles: (a) radial and (b) hub and spoke.

that highest heat release rate (chemical reactions zone) that drives the combustion oscillations occurs near the injector face. It is believed that it minimizes the coupling between the heat release rate and acoustic oscillations and thus reduces the amplification of gas dynamic forces within combustion chamber. Hence, the depth of the baffles must be selected such that they can cover the heat release zone. As they are placed in the highest heat release zone, they must be strong enough to resist the thermal load even under pressure oscillations conditions. But the depth of the baffles should be such that individual compartments of combustion zone must not be created with their own acoustic characteristics. Note that odd number of baffles is being used by most designers as even number of baffles can enhance standing modes of combustion instability. Generally, baffles are more effective for both radial and tangential modes of oscillations. The transverse acoustic modes of oscillations can be minimized by providing certain number of radial baffles with sufficient depth along the axis of the combustion chamber. During the transverse mode of combustion oscillations, acoustic energy is being lost due to high-velocity turbulent flow at the tips of baffles. Based on empirical design, several possible baffles, shown in Figure 8.10, have been evolved, which are being deployed in rocket engines. In order to avoid radial modes of oscillations, circular baffles with radial blades can be employed such that standing waves can be interrupted easily. Based on experimental studies, certain guidelines for baffle design have been developed particularly for number of baffles and their depths. It has been observed that the spinning tangential mode of oscillation can be eliminated using a two-bladed baffle, while first and second tangential modes can be avoided using three-blade baffle design. A periodically arranged four-blade baffle can be deployed to reduce second tangential mode of oscillations. The blade depth of 15%–30% of the chamber diameter with lower sizes with large diameter LPR engine is preferred to dampen oscillations.

Another way of dampening the combustion oscillation is to absorb the energy in the combustion chamber. Generally, artificial acoustic absorbers such as acoustic liners and discrete cavities are devised, which can be used along the wall of the combustor or near the injector. It can be noted that cavities are being used routinely particularly at the corner of the injector face as pressure antinode exists for almost all resonant modes of vibration, including longitudinal, tangential, radial, and combination of these oscillations and velocity oscillations attains a minimum value.

8.7 IGNITION SYSTEMS

The combustion of liquid propellants in the thrust chamber can be initiated by a suitable ignition system. Several ignition systems for both thrust chamber and gas generator have been designed and developed over the years for successful applications in rocket engines. Proper selection of ignition system depends on the nature and phase of propellants, need for restart, system safety, altitude relight capability, weight, and space considerations. For example, with regard to hypergolic propellants, better mixing of propellants can be good enough to ignite them. But in case of bipropellant, rapid, reliable ignition of incoming propellants must be ensured before propellants are accumulated in the combustion chamber. Otherwise, if ignition delay is too large, ignition of accumulated propellant can lead to explosion. Note that proper ignition of liquid propellants in the thrust chamber is dependent on proper selection of the ignition method, quality of design, and integrity of ignition system.

As mentioned earlier, the main function of igniter is to supply requisite but sufficient amount of heat energy to initiate the chemical reactions in the main propellant. Generally, igniter draws its energy from the limited stored energy onboard with the rocket engine to initiate combustion. Besides, it is essential to have flow of propellant during ignition. Generally, the starting propellant flow rate is kept lower than the full flow rate, which not only smoothens the ignition process but also prevents an excessive accumulation of unignited propellants. As a result, a quite different fuel/oxidizer mixture ratio around stoichiometric condition is maintained for successful ignition of propellant mixture, because initial vaporization and mixing of propellants happen to be poor due to low starting injection velocity of propellant. It is essential that once propellant is ignited, it must remain on combustion mode as it acts as pilot ignition source for fresh propellant entering into the thrust chamber. Although several igniters have been devised and used in various liquid-propellant rocket engines, we will restrict our discussion to the following five types of igniters: (1) catalytic igniter, (2) hypergolic igniters, (3) spark plug and spark-torch igniters, and (4) resonance igniters.

1. *Catalytic igniter*: In rocket engines, catalysts are used mainly to initiate and sustain the combustion of monopropellants. For example, potassium permanganate/potassium solution was used as a catalyst for initiating combustion of hydrogen peroxide in V2 ballistic missile during the Second World War. Subsequently, these liquid catalysts

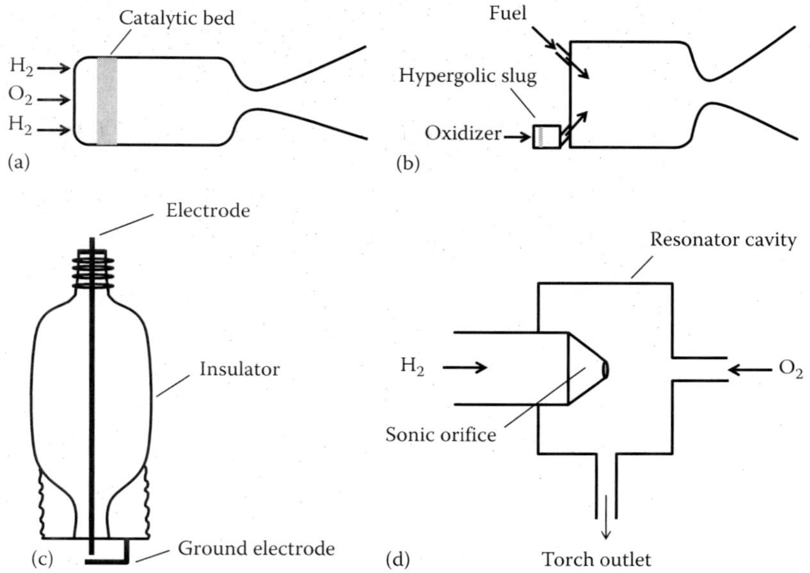

FIGURE 8.11 Types of igniter: (a) catalytic, (b) hypergolic, (c) spark plug, and (d) resonance.

were abandoned due to complicated valve timing and interlocking devices. Rather, these are replaced by solid catalysts; liquid monopropellant is passed through solid catalyst bed for initiation of combustion. Since hydrogen peroxide has lower I_{sp}, it is used along with solid catalyst for igniting other liquid propellants. In recent times, catalyst igniter is emerging as an alternative to the augmented spark ignition system, even for hydrogen and oxygen systems. A typical catalyst ignition system for hydrogen–oxygen mixture is shown in Figure 8.11a, in which gaseous mixture of hydrogen and oxygen will have to pass through properly designed catalyst bed for igniting this mixture. For this purpose, any of three reliable and durable catalysts, namely, platinum, iridium, and palladium, can be used in igniting hydrogen/oxygen mixture. Efforts are being made to develop better catalyst, which must have a longer life with least cost.

2. *Hypergolic igniters*: Recall that the term "hypergolic" means occurrence of spontaneous initiation of overall exothermic chemical reaction when two hypergolic propellants come in contact with each other. On successful ignition, a signal is sent to actuate remotely controlled valves for propellant feed line for supplying the main

propellants to combustion chamber to get requisite thrust. But it suffers from problems of clumsiness, frequent clogging of feed lines, and need to eject a large amount of solid materials. In order to overcome this problem, a small amount of fluid that is hypergolic with one of the propellants is stored in a cylindrical cartridge, as shown in Figure 8.11b. The diaphragm of this cartridge gets ruptured when requisite pressure is applied to it for the initiation of ignition. Generally, this loaded cartridge is placed as a bypass to the high-pressure main feed line. Note that a fuel that is hypergolic to the oxidizer is preferred over the opposite combination in actual practice. On the initiation of oxidizer valve, the pump supplies oxidizer, raising its pressure, which ruptures the diaphragm of the hypergolic propellant cartridge leading to spontaneous ignition. Subsequently, main fuel propellant is fed into combustion chamber to sustain ignition flame to reach main stage level. Triethylaluminum, a room-temperature storable liquid, is preferred in rocket engine as it is easily hypergolic with liquid oxygen with higher ignition delay characteristics. But it produces tenacious residue, which affects subsequent combustion. Hence, it is mixed with another liquid known as triethylboron, which has relatively lower ignition delay characteristics. It has been established that 10%–15% of triethylaluminum by weight is mixed with triethylboron to obtain a satisfactory ignition delay characteristic with liquid oxygen while producing an acceptable residue. Organometallic hypergol slug is being used routinely for the ignition of main propellants RP1/LOX, which are being used in several spacecraft, namely, Atlas, Delta, F1, and so on. Organometallic liquids are not only hypergolic but also hypophoric with liquid oxygen. Besides these, reactive oxidizers, namely, gaseous fluorine and chlorine tri-fluoride, have been found to be very effective ignition systems particularly in research engines. The quantity of hypergol required for successful ignition of liquid propellants is dependent on the feed system, start sequence, and level of thrust delivered by rocket engine. The selection of proper hypergolic ignition system is dependent on the safety requirement, level of thrust delivered by rocket engine, ignition reliability, cost, packaging, and ease of handling.

3. *Spark plug and Spark-torch igniters*: Spark ignition system can be easily used in liquid-propellant rocket engines in which initial energy required for ignition is provided by producing spark with help of

high-tension and low-tension capacitive discharge system. A typical spark plug ignition system with integral ignition exciter is shown in Figure 8.11c. It consists of dual gap spark plug, ceramic metal seals, high-voltage transformer, storage capacitor, and electronic circuit module. Generally, proper gap between electrodes has to be maintained for successful discharging of electrical energy for proper ignition to occur. A properly designed capacitor-discharge ignition system can manage to ignite a chilled combustion chamber in space even at chamber pressure of 2.5 MPa. A typical spark igniter exciter can deliver energy around 350 mJ per spark when high-voltage order of 20 kV or more is applied to its electrodes. But around 100 mJ per spark is produced at the spark plug tip. Note that electrical power for this system is generally supplied from onboard battery. Generally, spark plug, ignition exciter, and high-voltage components are packed in a high-pressure-sealed enclosure so that it can operate even in near-vacuum condition, especially in space. But it has limitation of igniting large rocket engines due to heat losses. In order to overcome this problem, the spark-torch igniter has been designed and developed, in which small amount of engine propellants are fed into spark-torch igniter. Of course, combustion is initiated using spark plug. The flame with copious amount of high-temperature and high-pressure gases is ejected out to ignite the main propellants in the thrust chamber. In order to augment ignition capability, an augmented spark igniter, as shown Figure 8.11c, has been designed and developed. In this, fuel is injected through multiple tangentially aligned orifices for ignition combustion. Of course, spark plugs are placed upstream of fuel injection orifices around recirculation zone for ensuring successful and rapid ignition. The combustion chamber of this igniter can have a convergent or straight duct that can supply high-temperature gas to the main thrust chamber. The torch igniter size varies from 10 to 35 mm diameter, depending on the thrust level of rocket engines. This kind of igniter can be placed as assembly onto injector end or any other place. Generally, it is mounted on the centerline of the injector as it provides better performance.

The torch igniter can provide a large number of ignitions even in high altitude as compared to spark plug. It can also operate satisfactorily over a wide range of propellant flow rates and mixture ratios. In recent times, spark torch has been miniaturized, which can be employed in smaller rocket engines.

4. *Resonance igniters*: In this igniter, resonance heating created from the flow of high-pressure gas is used to ignite liquid propellant. A typical sketch of resonance igniter is shown in Figure 8.11d, which consists of propellant supply line, sonic orifice, mixing chamber, and resonator cavity. In this igniter, high gas (hydrogen) is introduced through sonic nozzle that is expanded in mixing chamber and subsequently directed to the resonator cavity in which it is cyclically compressed and expanded. As result, the temperature of this gas is raised at the close end of this cavity. This high-temperature gas can be mixed with spilled gas and is exhausted through the torch outlet. Hydrogen gas temperature as high as 1400 K could be generated by this method within 60 ms. After fuel has reached certain threshold high temperature, oxidizer is introduced to the hot zone, causing ignition of this mixture. Note that this concept can not only be used for H_2/LOX system but also for other premixed propellant combinations. Several types of resonance igniters have been designed and developed successfully.

8.8 COOLING SYSTEMS

We know that heat release in the case of rocket engine is quite high, in the order of 30,000 MW/m^3. As a result, temperature in the range of 3000–3500 K prevails in the combustion chamber of liquid-propellant engine, which is much higher than the melting temperature of most available metal alloys used for rocket engines. Recall that strength of wall materials decreases at high temperature, which is likely to fail structurally under high temperature and pressure condition. Of course one can use thicker wall to circumvent the problem of heating which is allowed in rocket design as it is essential to minimize the mass of vehicle at particular operating condition. Hence, cooling system is used to lower the wall temperature of combustion chamber and nozzle. The most vulnerable part of the rocket engine is the throat of the nozzle where heat transfer rates are one order of magnitude higher than the exit portion of exhaust nozzle. Several methods such as (1) regenerative cooling, (2) film/sweat cooling, and (3) ablative cooling have been adopted for cooling of rocket engines, which are discussed in the following.

8.8.1 Regenerative Cooling

This is the most widely used cooling system for liquid-propellant rocket engine. In this case, liquid fuel/oxidizer is employed as a coolant, which

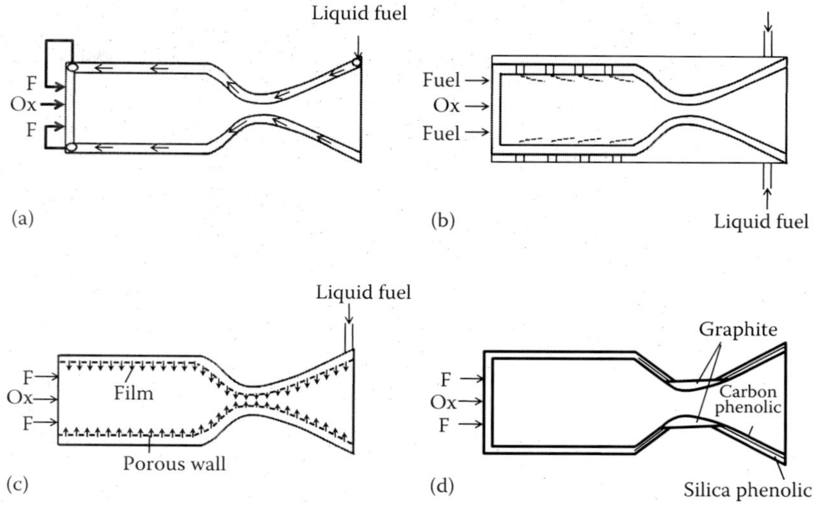

FIGURE 8.12 Schematic of cooling system: (a) regenerative cooling, (b) film cooling, (c) transpiration cooling, and (d) ablative cooling.

is allowed to pass through the passages placed outside of the combustion chamber and nozzle, as shown in Figure 8.12a, before being fed into thrust chamber. As a result, heat is transferred to the incoming propellant from the combustion hot gases and enhances its enthalpy. Since the heat from hot gases are reused as in regenerative cycle of power plant, this kind of cooling system is known as regenerative cooling system. This method of cooling is used in several rocket engines such as Saturn vehicle and Apollo missions. Generally, in this system, array of suitably shaped tubes are brazed to the walls of the thrust chamber of liquid-propellant rocket engine that are supported by steel bands. In case of cryogenic engine, hydrogen is used as a coolant for regenerative cooling system, because the pressure of the hydrogen is much above the critical pressure of boiling and avoids boiling during regenerative cooling. As a result, liquid hydrogen is continuously converted into hydrogen gas, which can enhance heat transfer rate due to its higher thermal diffusivity.

8.8.2 Film/Sweat Cooling

In this film cooling process, a thin film of coolant is formed on the inner side wall of the thrust chamber, as shown in Figure 8.12b, which acts as an insulating layer. Note that the coolant on the film absorbs heat from the hot gases due to both sensible and latent heat absorption. As a result, the temperature of film is much lower than that of the hot combustion products. This is an effective way to protect the wall of the thrust chamber from hot gases by directing the cooling air into boundary layer to provide a protective cool film along the surface. In certain practical systems, injectors are designed such that fuel-rich mixture layer prevails near the inner wall of the thrust chamber. The temperature of this fuel-rich mixture layer is much lower than that of the oxidizer-rich mixture in the core of thrust chamber. Besides this, fuel-rich mixture as film layer is preferred over oxidizer-rich mixture as oxidation of metal can be reduced drastically.

The effect of film cooling is dependent on the coolant mass flow rate, velocity of coolant, number of rows of cooling holes, and so on. If too much coolant is injected into boundary layer or its velocity is too high, then the coolant may penetrate into the boundary layer, defeating the main purpose of using film cooling. In order to overcome this problem, sweat/transpiration cooling method, shown in Figure 8.12c, can be used, in which the cooling air is forced through porous walls of thrust chamber into boundary layer to form relatively cool, insulating film. This is considered to be the most efficient cooling technique, but it is not being used in practice due to nonavailability of suitable porous materials that can withstand high-temperature and high-pressure gases in the thrust chamber. Note that the pores should be small to enhance cooling rate. However, this can lead to pore blockage due to soot particles or foreign material. It is also more economical as only 1.5%–2% of the total coolant (fuel/oxidizer) mass flow rate can reduce, the blade temperature in the range of 200°C–300°C.

8.8.3 Ablative Cooling

In this cooling system, insulating material that melts and vaporizes at high temperature is used on the combustion gas side walls of the thrust chamber. As a result, lower wall temperature as compared to combustion product gas temperature is maintained, as its temperature remains almost constant corresponding to melting point. This kind of cooling, in which thermal erosion is used to maintain low temperature, is known as ablative cooling (see Figure 8.12d). Besides this, ablative material being good thermal insulator allows low rate of heat transfer to the outer structure. Some ablative composite materials from epoxies, unsaturated polyester, phenolic resin along with silica and carbon fibers are produced for use in rocket engines. Two popular composites, namely, carbon phenolic and silica phenolic, which ablate around 1400 K, are being used in rocket engines. However, silica phenolic is being preferred in current times due to its higher thermal insulating characteristics and higher oxidation resistance. The inner wall of the thrust chamber is lined with a layer of ablative composites, as shown in Figure 8.12d, which protects it from high-temperature hot gases in the thrust chamber. The ablative cooling system is being used in solid propellant and low chamber pressure liquid propellant short-duration rocket engines.

8.9 HEAT TRANSFER ANALYSIS FOR COOLING SYSTEMS

We need to determine heat transfer from combustion product hot gas in either combustion chamber or nozzle to its respective wall. However, it is quite difficult and complex to estimate accurately the amount of heat transfer as it is affected by boundary layer thickness. Recall that boundary layer thickness is dependent on the type of flow, fluid, wall curvature, pressure gradient, temperature gradient, and so on. In case of nozzle, the flow will be accelerating across it. The boundary layer thickness decreases along flow direction to a minimum value at the throat of nozzle and subsequently increases marginally in divergent portion of nozzle. Hence, maximum heat transfer occurs at the throat of nozzle. In order to estimate the convective heat transfer, one has to solve the Navier–Stokes equation along with energy equation, which, being nonlinear and coupled in nature, is quite cumbersome to solve. Besides this, one has to invoke conjugate heat transfer equations for getting accurate prediction. However, we will resort to quasi-one-dimensional heat transfer analysis, as depicted in Figure 8.13,

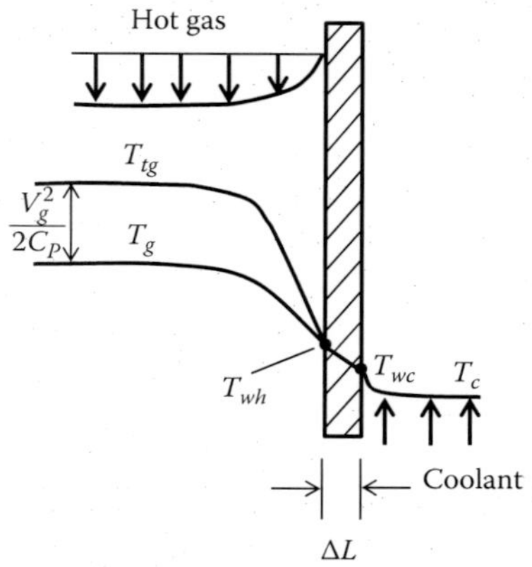

FIGURE 8.13 Schematic of one-dimensional steady heat transfer model.

which indicates steady-state heat transfer through the combustion chamber wall with coolant flowing in the ducts placed on outer side. The hot gas with velocity V_g and temperature T_g is flowing over the inner wall of thrust chamber whose velocity and temperature profiles are shown in Figure 8.13. If the wall happens to be adiabatic, the adiabatic wall temperature T_{wa} is expected to be equal to total gas temperature T_{tg}. But if the gas flow velocity happens to be supersonic, the slowing down of gas near wall would not be adiabatic. Note that there will be heat transfer from the gas near the wall region with higher static temperature to the adjacent static temperature gas away from the wall. As a result, the temperature of gas at the wall will be lower than T_{tg}, as shown in Figure 8.14. Besides this, T_{tg} will increase slightly away from the adiabatic wall, as shown in Figure 8.14, to satisfy steady flow energy equation. In order to relate the T_{wa} to T_{tg}, we can define recovery factor RF as

$$RF = \frac{T_{wa} - T_g}{T_{tg} - T_g} \tag{8.33}$$

For subsonic flow, RF is equal to 1.0, while for supersonic flow, RF will be less than 1.0. Note the supersonic flow prevails in the exhaust nozzle for rocket engine. For Mach number 4.0, RF is found out to be 0.91 [4]. At the throat of rocket engine, sonic condition prevails, as flow is considered to be

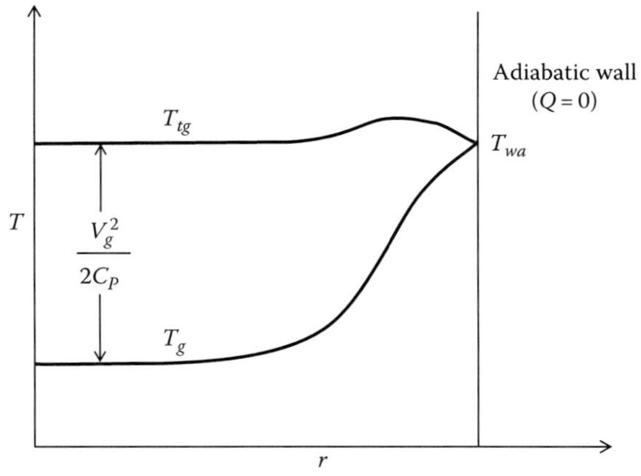

FIGURE 8.14 Schematic of adiabatic wall temperature for defining recovery factor.

choked. By considering γ of hot gas to be 1.25, we can determine T_g^*, given as follows:

$$T_g^* = T_{tg}\left(\frac{2}{\gamma_g + 1}\right) = 3000\left(\frac{2}{1.25+1}\right) = 2666\ \text{K} \tag{8.34}$$

For $T_{tg} = 3000$ K, $RF = 0.91$, we can determine T_{wa} as follows:

$$T_{wa} = T_g^* + RF\left(T_{tg} - T_g^*\right) = 2970\ \text{K} \tag{8.35}$$

The difference between T_{wa} and T_{tg} happens to be within 1% error, which falls within experimental error. Hence, we can assume T_{wa} to be equal to T_{tg} at the throat of nozzle.

By considering one-dimensional steady flow heat transfer, we can determine convection heat transfer rate per unit as $\dot{H}_g''$, given as follows:

$$\dot{H}_g'' = h_g\left(T_{wa} - T_{wh}\right) \tag{8.36}$$

where

h_g is the heat transfer coefficient of hot gas
T_{wa} is the wall temperature under adiabatic condition
T_{wh} is the well temperature on hot gas side

Besides, heat transfer from the hot gases to the well can also take place due to radiative heat transfer mode. Note that same amount of heat transferred from hot gas to the wall due to both radiative convection heat transfer modes will be transferred to cold side of wall due to heat conduction under steady-state condition. By considering one-dimensional heat transfer due to conduction through the wall, we can determine $\dot{H}_w''$ as

$$\dot{H}_w'' = -k_w\frac{dT}{dx} = \frac{k_w\left(T_{wh} - T_{wc}\right)}{\Delta L} \tag{8.37}$$

where

k_w is the thermal conductivity of wall
T_{wc} is the wall temperature on the coolant side
ΔL is the wall thickness

As the coolant flows through the coolant passage, heat has to transfer from wall to the coolant due to convection. This convective heat transfer rate per unit area due to convection of coolant $\dot{H}_c''$ can be determined as

$$\dot{H}_c'' = h_c \left(T_{wc} - T_c\right) \tag{8.38}$$

where

T_c is the coolant (liquid) temperature on free stream side
h_c is the convective heat transfer coefficient of coolant

It is quite cumbersome to determine radiative heat flux $\dot{H}_r''$ due to complexities in geometric and various modes, namely, absorption, reflection, and radiation mechanisms of heat transfer. Besides this, nonuniform temperature condition prevails both in combustion chamber and nozzle. The combustion gas contains several product gases, namely, CO, CO_2, H_2O, and so on, which can radiate heat in discrete spectrum. Hence, determination of radiative heat transfer rate per unit area is quite complex. But we will resort to a simple model, for which total radiative heat flux rate can be estimated as

$$\dot{H}_r'' = \varepsilon_g \sigma T_g^4 \tag{8.39}$$

where

ε_g is the emissivity of gas considering complexities of geometry and gas properties
σ is the Stefan–Boltzmann constant (5.687×10^{-8} W/m$^2 \cdot$K^4)
T_g is the gas temperature

The wall is assumed to be a black body without any emission of radiation. Note that hot gases in rocket engine do contain solid/liquid particles that contribute appreciable amount of radiative energy to the wall. The contribution of the radiative heat transfer happens to be in the range of 5%–40%, depending on the gas temperature and composition.

By considering steady one-dimensional heat transfer for regenerative cooling, we can have a heat balance across combustion chamber wall with regenerative cooling passage as

$$\dot{H}'' = \dot{H}_g'' + \dot{H}_r'' = \dot{H}_w'' = \dot{H}_c'' \tag{8.40}$$

By using Equation 8.36, we can recast Equation 8.40 as follows:

$$\dot{H}'' = h_g\left(T_{wa} - T_{wh}\right) + \dot{H}_r'' \tag{8.41}$$

By using Equations 8.37, 8.38 and 8.40, we can have

$$T_{wh} = \dot{H}_w'' \frac{\Delta L}{k_w} + T_{wc} = \dot{H}'' \frac{\Delta L}{k_w} + T_{wc} = \dot{H}'' \frac{\Delta L}{k_w} + \frac{\dot{H}''}{h_c} + T_c \tag{8.42}$$

By using Equations 8.41 and 8.42, we can express

$$\dot{H}'' = \frac{\left(T_{wa} - T_c\right) + \dot{H}_r''/h_g}{\dfrac{1}{h_g} + \dfrac{\Delta L}{k_w} + \dfrac{1}{h_c}} \tag{8.43}$$

This one-dimensional steady analysis can be extended for complex configuration by considering effective area on both hot gas and coolant side, which can be higher than actual area. Hence, Equation 8.40 can be modified as follows:

$$\dot{H}'' A_g = \left(\dot{H}_g'' + \dot{H}_r''\right) A_g = \dot{H}_w'' A_w = \dot{H}_c'' A_c \tag{8.44}$$

where

A_g is the gas side area
A_w is the effective wall area
A_c is the effective coolant side area

This equation can be rewritten as

$$\dot{H}'' = \frac{\left(T_{wa} - T_c\right) + \dot{H}_r''/h_g}{\dfrac{1}{h_g} + \dfrac{A_g \Delta L}{A_w k_w} + \dfrac{A_g}{A_c h_c}} \tag{8.45}$$

We need to determine the heat transfer coefficient for both hot gas and coolant flow. The empirical relationship for Nusselt number *Nu* from the heat transfer literature that can be used for hot gas flow can be expressed in terms of Reynolds number *Re* and Prandtl number *Pr* as follows:

$$Nu_g = \frac{h_g D_g}{k_g} = 0.026 Re_g^{0.8} Pr_g^{0.4} = 0.026\left(\frac{\rho VD}{\mu}\right)_g^{0.8}\left(\frac{\mu C_p}{k}\right)_g^{0.4} \tag{8.46}$$

where

- D is the local diameter of combustion chamber/nozzle
- k is the thermal conductivity
- μ is the viscosity
- ρ is the density
- V is the flow velocity
- C_p is the specific heat corresponding to gas

Note that fluid properties are dependent on the gas temperature, which is generally evaluated at average film temperature T_f between T_g and T_{wh} ($T_f = 0.5(T_g + T_{wh})$). The coolant side heat transfer coefficient can be evaluated by using empirical relation, as given in the following:

$$Nu_c = \frac{h_c D_c}{k_c} = 0.023 Re_c^{0.8} Pr_c^{0.33} = 0.023\left(\frac{\rho VD}{\mu}\right)_c^{0.8}\left(\frac{\mu C_p}{k}\right)_c^{0.33} \tag{8.47}$$

where

- D is the local hydraulic diameter of combustion chamber/nozzle
- k is the thermal conductivity
- μ is the viscosity
- ρ is the density
- V is the flow velocity
- C_p is the specific heat corresponding to coolant

From Equation 8.47, we can express heat transfer coefficient h_c in terms of coolant mass flux and specific heat as

$$\frac{h_c}{(\rho V C_p)_c} = 0.023\left(\frac{\rho VD}{\mu}\right)_c^{-0.2}\left(\frac{\mu C_p}{k}\right)_c^{-0.67} = 0.023(Re)_c^{-0.2}(Pr)_c^{-0.67} \tag{8.48}$$

It can be observed from this expression that heat transfer coefficient of coolant increases with coolant mass flux and specific heat for a fixed value of *Re* and *Pr*. Hence, designer can choose coolant with higher specific heat to have maximum heat transfer. Generally, hydrogen is chosen in rocket engine as coolant as it has higher specific heat. Recall that maximum heat transfer rate per unit area occurs at the throat of the nozzle; hence, higher coolant flux can be used to augment the heat transfer coefficient by choosing proper cross-sectional area of the coolant passage. The density of coolant being constant, coolant velocity is to be increased by decreasing cross-sectional area of the coolant passage at the throat region. Generally, coolant velocity in the range of 15–25 m/s is used in throat region.

Example 8.4

The combustion chamber with 0.45 m diameter of a liquid propellant engine is cooled by regenerative cooling to maintain its outer wall temperature at 300 K. The temperature and the pressure of combustion chamber are maintained constant at 3000 K and 0.75 MPa, respectively. The heat loss due to gas radiation happened to be 25% of total heat loss. If the wall thickness happens to be 5.2 mm with thermal conductivity of 21 W/m · K, determine inner wall temperature under steady-state condition. Consider $Re = 10^6$, $Pr = 0.73$, $k_g = 0.17$ W/m · K.

Solution

We know by Equation 8.40, the heat flux through the wall can be expressed as

$$\dot{H}_w = \dot{H}_r + h_g \left(T_{tg} - T_{wh}\right) \tag{A}$$

As $\dot{H}_r = \dfrac{\dot{H}_w}{4}$, then from Equation A we can have

$$\dot{H}_w = \frac{4}{3} h_g \left(T_{tg} - T_{wh}\right) \tag{B}$$

$$\dot{H}_w = \frac{T_{tg} - T_{wh}}{\left[1 + \dfrac{\Delta L}{k_w} \dfrac{4}{3} h_g\right]} \tag{C}$$

We evaluate the h_g from semiempirical relationship (Equation 8.46) as

$$h_g = \frac{k_g}{D} 0.026 \left(Re\right)^{0.8} \left(Pr\right)^{0.4} = \frac{0.17}{0.45} 0.026 \left(10^6\right)^{0.8} \left(0.73\right)^{0.4}$$
$$= 546.44 \text{ W/m}^2 \cdot \text{s}$$

Dividing Equation B by (C), we can get

$$\frac{T_{tg} - T_{wh}}{T_{tg} - T_{wC}} = \frac{\frac{3}{4h_g}}{\left[\frac{3}{4h_g} + \frac{\Delta L}{k_w}\right]} = \frac{\left(0.75/546.44\right)}{\left[\frac{0.75}{546.44} + \frac{5.2 \times 10^{-3}}{21}\right]} = 0.85 \quad \text{(D)}$$

From Equation D, we can determine T_{wh} as

$$T_{wh} = 3000 - 0.85\left(3000 - 250\right) = 705 \text{ K}$$

REVIEW QUESTIONS

1. What are the differences between liquid-propellant and solid-propellant rocket engines?
2. What are the advantages of liquid-propellant engine over solid-propellant rocket engine?
3. What are the disadvantages of liquid-propellant engine over solid-propellant rocket engine?
4. What is a monopropellant rocket engine? How is it different from bipropellant rocket engine? What are its applications?
5. What is hypergolic propellant combustion? How is it different from nonhypergolic propellant combustion?
6. Define combustion delay time? How is it different from chemical delay time?
7. Define chemical delay time? How can it be determined for liquid propellant?
8. What do you mean by characteristic length L^*? How is it different from combustion chamber length?

9. What are the shapes of combustion chamber that can be used in LPRE? Which one is preferred and why?
10. What do you mean by POGO instability?
11. What do you mean by Chuff? Why does it occur in liquid-propellant rocket engine (LPRE)?
12. If you observe chugging sound in LPRE, what kind of frequency of the combustion oscillation can it experience? What are the ways of alleviating this instability problem?
13. What are the methods used for suppression of combustion instability in LPRE?
14. Why are baffles used for suppression of combustion instability in LPRE?
15. What are the types of igniters used for liquid-propellant rocket engine?
16. What is a pyrotechnic igniter? How is it different from catalytic igniter?
17. What is a resonance igniter? How is it different from spark plug igniter?
18. Why is the cooling of nozzle required for liquid-propellant rocket engine?
19. What are the types of cooling methods adopted for liquid rocket engine?
20. What do you mean by regenerative cooling? Explain it with a neat sketch.
21. What do you mean by ablative cooling? Can it be preferred for liquid rocket engine? If so, under what condition can it be used?
22. What is the difference between film and transpiration cooling system? Which is preferred, and why?
23. What do you mean by recovery factor? Why is it required at all?
24. What is Nusselt number? How is it different from Biot number?
25. What do you mean by Prandtl number? What does it indicate?

26. How does the heat flux vary along the CD nozzle?

27. Why is cooling of nozzle throat critical for thermal failure? How can this problem be overcome?

PROBLEMS

8.1 A small monopropellant liquid hydrazine rocket engine is designed to produce thrust of 25 N and I_{sp} of 1985, which is to be operated for 5 h. Two layers of alumina granules are used for the catalyst bed with blade loading B_L of 3.75 $\text{g/cm}^2 \cdot \text{s}$. Determine diameter of catalyst bed and amount of hydrazine required for this operation.

8.2 A liquid-propellant rocket engine is used to develop a thrust of 1.5 kN with a characteristic velocity of 1750 m/s at chamber pressure of 4.5 MPa. If its thrust coefficient C_F happens to be 1.2, determine (1) throat of nozzle, (2) total mass flow rate of both oxidizer and fuel.

8.3 In a liquid-propellant rocket engine, propellant is injected into its combustion chamber at pressure of 1.5 MPa and temperature of 3500 K. If the residence time happens to be 1.2 ms, determine the length of cylindrical combustion chamber. Assume that instantaneous combustion occurs that produces flow with Mach number of 0.3. Take $\gamma = 1.25$ and MW of product gas = 23.

8.4 The propellant is injected into its combustion chamber of 0.65 m length at pressure of 1.8 MPa with pressure drop of 0.5 MPa. The combustion delay time and residence time happen to be around 8 ms and 3.5 ms, respectively. Is it subjected to combustion instability or not? If so, determine its frequency of oscillation.

8.5 In a liquid-propellant rocket engine, propellant is injected into its cylindrical combustion chamber with a diameter of 0.5 m at pressure of 25 MPa and temperature of 3800 K. If the residence time happens to be 1.2 ms, determine the length of cylindrical combustion chamber. Assume that instantaneous combustion occurs that produces flow with Mach number of 0.3 at its exit. Take $\gamma = 1.25$ and MW of product gas = 25. Determine the fundamental frequency of instability along longitudinal, radial, and tangential direction, assuming it to be a closed end cavity.

8.6 The combustion chamber with 0.5 m diameter of a liquid propellant engine is cooled by regenerative cooling to maintain its outer wall temperature at 300 K. The temperature and the pressure of combustion chamber are maintained constant at 3000 K and 0.75 MPa, respectively. The heat loss due to gas radiation happens to be 30% of total heat loss. If the wall thickness happens to be 2.5 mm with thermal conductivity of 25 W/m · K, determine inner wall temperature under steady-state condition. Consider $Re = 5 \times 10^6$, $Pr = 0.83$, $k_g = 0.21$ W/m · K.

8.7 A rocket engine is operated at 2.8 MPa and 2800 K, when coolant temperature (T_L) is maintained at 300 K. The hot wall at throat with thickness (ΔL) 3.5 mm is maintained at 1050 K. The thermal conductivity of throat wall (k_w) and heat transfer coefficient of coolant (h_c) are 23 W/m · K and 0.78 MW/m³ · K, respectively. Assume that ($\Delta L/k_w$), T_c and h_c and fraction of radiative heat transfer remains constant, and determine the chamber pressure when T_{wh} is increased to 1275 K with using of new material at the throat. Use semiempirical relationship for Nusselt number $Nu = 0.023\, Re^{0.8} Pr^{0.4}$; where Re = Reynolds number, Pr = Prandtl number.

REFERENCES

1. Sutton, G.P. and Biblarz, O., *Rocket Propulsion Elements*, 7th edn., John Wiley & Sons Inc., New York, 2001.
2. Ramamurthi, K., *Rocket Propulsion*, MacMillan Published India Ltd, New Delhi, India, 2010.
3. Barrere, M., Jaumotte, A., and Vandenkerckhove, J., *Rocket Propulsion*, Elsevier Publishing Company, New York, 1960.
4. Huzel, D.K. and Huang, D.H., *Modern Engineering for Design of Liquid Propellant Rocket Engines*, Vol. 147, AIAA Publication, Washington, DC, 1992.
5. Humble, R.W., Hennery, G.N., and Larson, W.J., *Space Propulsion Analysis and Design*, McGraw Hill, New York, 1995.
6. Mishra, D.P., *Gas Turbine Propulsion*, MV Learning, London, U.K., 2015.
7. Hill, P.G. and Peterson, C., *Mechanics and Thermodynamics of Propulsion*, 2nd edn., Addision-Wesley Publishing Company, Reading, MA, 1999.

CHAPTER 9

Hybrid Propellant Rocket Engine

Power does not corrupt a man but it only reveals the corrupted mind of a person.

D.P. MISHRA

9.1 INTRODUCTION

A hybrid rocket engine is one in which both liquid and solid propellants can be used simultaneously. For example, oxidizer can be in liquid phase, while fuel can be in solid phase or vice versa. Several combinations of solid/liquid fuels and liquid/solid oxidizers have been used by researchers for the design and development of hybrid rocket engines. However, liquid oxidizer and solid fuel is preferred in hybrid rocket engine design. The other combination of solid oxidizer and liquid fuel is not preferred as most of the solid oxidizers are available in crystal form, which cannot be cast in a specific form of propellant grain.

The concept of hybrid combustion goes back to the history of the first forest fire. But in 1946, Bartel and Rannie [1] were the first to work on the hybrid propellant concept for ramjet engines using solid fuels. Later on, around 1962, several people started working again on the concept of hybrid rocket engine [2,3]. It is interesting to note that hybrid rocket engine is the only chemical rocket that was designed and developed for the first time based on sound fundamental knowledge base.

Hybrid propellant rocket engine (HPRE) has certain characteristics that are different from solid-propellant rocket engine (SPRE) and liquid-propellant rocket engine (LPRE). The specific impulse at vacuum in the range of 300–380 s can be achieved particularly for high-energy hybrid propellants. A combustion efficiency of 96% of theoretical values can be achieved easily. Some of the characteristic features of HPRE as compared with SPRE and LPRE are given as follows:

- The specific impulse range lies above that of SPRE but below that of maximum value of LPRE.
- It has higher-density specific impulse than that of the LPRE. But it has lower-density specific impulse than that of the LPRE.
- Its specific impulse varies even during steady-state operation due to the inherent variation in the mixture ratio.
- It has the ability to change thrust level smoothly over a wide range.
- Its operation can be switched off or on easily during flight unlike SPRE.
- It has relatively lower system cost.
- It has a higher safety envelope during storage than LPRE and during operation as compared with SPRE.
- The mass fraction of HPRE is lower than that of SPRE as certain slivers of fuel propellant remain underutilized at the end of operation.

HPREs find applications where command shutdown, restarting, and throttling are essential for a certain mission. Besides this, HPREs are well suited for long-range mission, where storage of propellants, manufacturing, and launch are to be tackled easily. Hence, this kind of rocket engine can find applications in booster and upper stages of launch vehicle and satellite maneuvering systems due to its superior performance level. Besides this, it can be used for low cost tactical and target missile systems. It is contemplated to be used even for aircraft application as it is quite safe to use as compared with other chemical rocket engines.

9.2 COMBUSTION CHAMBER

Let us consider a typical liquid oxidizer and solid fuel hybrid rocket engine as shown in Figure 9.1, which consists of a combustion chamber, a nozzle, a liquid propellant tank, a propellant feed system, an atomizer,

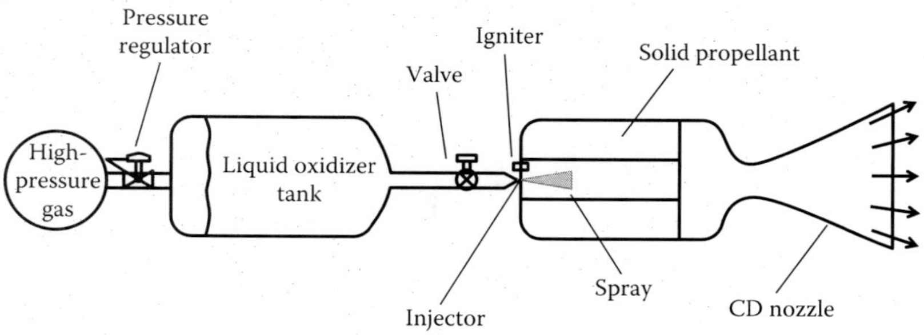

FIGURE 9.1 Schematic of a typical hybrid rocket engine.

an igniter, etc. The combustion chamber, along with the nozzle, is commonly known as the thrust chamber. It houses injectors and the igniter, which atomizes the liquid propellants, mixes, and ignites, leading to combustion of the liquid propellants. As discussed earlier, the main function of the combustion chamber is to produce high-temperature and high-pressure gas by burning the solid propellants. In this case, a liquid oxidizer is injected through the atomizer into the precombustion chamber upstream of solid fuel grain to produce a fine spray. Generally, solid fuel grains contain several axial combustion ports in which fuel vapors are produced to react with liquid vapor. Note that both fuel and oxidizer must be vaporized and mixed well for combustion to take place within the shorter zone of the combustion chamber. In order to have better-quality spray, requisite average droplet size, and distribution, the propellant must be fed into the injectors at high pressure. Of course, liquid propellant must be stored in a high-pressure tank, particularly in small rocket engines, as in the case of LPREs. However, turbo-pumps are used to feed high-pressure propellants, as in larger LPREs. Generally, a rear mixing chamber is provided at the end of the solid fuel grain to ensure complete mixing and burning of fuel and oxidizer before being ejected through the exhaust nozzle. The function of the ignition system is to provide a certain amount of initial ignition energy for the initiation of combustion. The high-pressure and high-temperature gas is expanded in convergent and divergent nozzles to produce thrust.

9.3 PROPELLANTS FOR HPRE

As mentioned earlier, several combinations of solid fuel and liquid oxidizer can be used as propellants. Some of the liquid oxidizers that can be used in HPRE are red fuming nitric acid (RFNA), white fuming nitric acid

(WFNA), nitrogen tetraoxide (N_2O_4), hydrogen peroxide (H_2O_2), liquid oxygen (LOX), etc. Certain hydrocarbon fuel grains, namely, hydroxyl terminated polybutadine (HTPB), cyclo terminated polybutadine (CTPB), poly methyl methacrylate (PMM: plexiglass), polyethylene, etc., can be used in HPRE. Hydrogen peroxide has been tested with polyethylene fuel, whose average propellant density is higher than that of the equivalent liquid-propellant system. One can get relatively higher I_{sp} of around 228 s at chamber pressure of 20 bar as compared with the I_{sp} (136 s) developed during monopropellant-mode burning of hydrogen peroxide. The most common solid-propellant fuels HTPB and CTPB, along with hydrogen peroxide, are being used routinely for upper-stage spacecraft, although they have lower energy content and lower combustion efficiency. Recall that in an SPRE, HTPB is used as binder for combining aluminum fuel and ammonium perchlorate (NH_4ClO_4). However, in hybrid rocket engine, HTPB can be used as a stand-alone fuel due to its low cost, processability, and safety reason. But for larger rocket engines, LOX, along with HTPB/CTPB, is preferred as a better oxidizer as compared with hydrogen peroxide due to its safety consideration and higher performance level.

Some of the high-energy oxidizers, namely, fluorine/LOX mixture and chlorine/fluorine compounds such as ClF_3 and ClF_5, can be used along with high-energy fuels such as hydrides of metals, namely, aluminum, beryllium, lithium, etc., of course, combined with suitable proportions of polymeric binder. Although this combination of high-energy hybrid propellants can produce rocket engines of higher specific impulse, I_{sp}, in the range of 350–400 s, they have not been used in actual flight rocket vehicles.

9.4 GRAIN CONFIGURATION

We know that grain configuration plays an important role in deciding the level of thrust and thrust law, apart from the composition of propellants. In the case of SPRE, both single-port and multi-port grains of various shapes and sizes can be used depending on the mission requirement. Unfortunately, in the case of hybrid rocket engine, single-port grain configuration cannot be used as the fuel regression rate of a typical hybrid propellant system is lower than that of the equivalent composite solid propellant almost by a factor of one-third. Hence, it is quite difficult to design a grain configuration for hybrid rocket engine whose regression rate can be comparable with that of SPRE. As a result, it is advisable to use a multiport solid fuel grain to meet the high-thrust requirement of the mission,

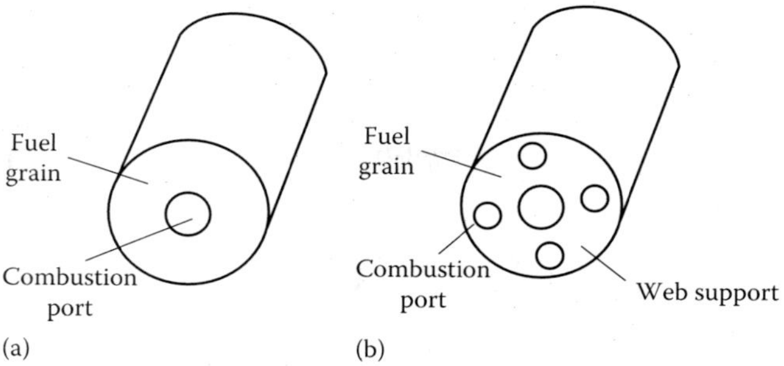

FIGURE 9.2 Schematic of two grain configurations for HRE: (a) single port and (b) multi-port.

particularly for HPRE design. It must be noted that the performance of hybrid rocket engine is dependent on the extent of the mixing of fuel and oxidizer, leading to higher combustion efficiency, which is caused due to multiple combustion ports and an adequate-sized mixing chamber provided at the rear end of the fuel grain. Both single- and multi-port grain configurations are shown in Figure 9.2, which can be used for HPRE. Of course, the number of ports in a fuel grain will be dependent on the desired level of thrust required for a certain specific mission, motor length, diameter of the fuel grain, fuel–oxidizer combination, etc. Besides this, the oxidizer/fuel ratio plays an important role in deciding the regression rate, which is dictated eventually by the characteristic velocity across the port. In other words, the regression rate of the solid fuel is dependent on the oxidizer flow rate through the port of grain.

9.5 COMBUSTION OF HYBRID PROPELLANTS

The combustion process in hybrid rocket engine is quite different from that of solid propellant combustion due to the fact that the fuel grain of a hybrid rocket engine does not contain any oxidizer. The process involved during the burning of hybrid propellant is quite complex, which has not been understood completely till date, because the physical processes such as vaporization, diffusion of species, mixing, and heat transfer not only for solid fuel but also for liquid oxidizer take place, along with heat release, due to overall exothermic chemical reactions during combustion. The liquid fuel being injected into the combustion chamber just upstream of the fuel grain gets converted into a spray and, subsequently, gets vaporized due to heat transfer from the combustion zone through all three heat transfer modes, namely, conduction, convection, and radiation. In a similar way, due to heat transfer from the flame and hot combustion products, the solid fuel undergoes pyrolysis and gasification processes and gets converted into gaseous form. Both oxidizer and solid fuel undergo complex chemical reactions with several intermediate products, liberating a large amount of heat within a narrow zone of the combustion chamber. Besides this, diffusion of active species such as H, OH, H_2, etc., makes the flow more complex and difficult to handle analytically, being a turbulent flow, along with chemical reactions involving mass transfer. The occurrence of the recirculation zone and turbulent vortices of gas along the solid fuel bed although augments the mixing and heat transfer, leading to an enhancement of vaporization and combustion, but also makes the flow complex to analyze using analytical tools. Note that the processes are so complex that they would not be occurring sequentially as described earlier. Rather some of these processes occur simultaneously in a complex manner. Besides this, it is quite difficult to ascertain which process has the greatest influence on the combustion of propellants.

Let us consider a simplified model as shown in Figure 9.3 for the combustion of hybrid propellant in an HPRE in which a nonmetalized fuel grain is used. We know that the solid fuel gets vaporized due to heat transfer from the flame and hot product gases. The solid fuel grain is decomposed thermally and is gasified at the solid surface. The extent of heat transfer to the solid surface is dependent on the extent of convection, radiation, the flame position, and composition of the fuel. In a

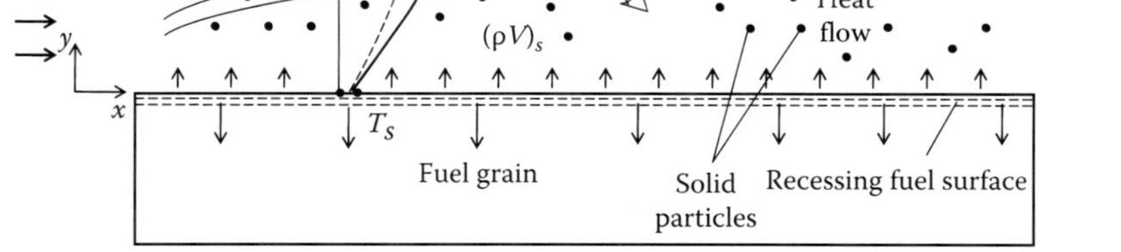

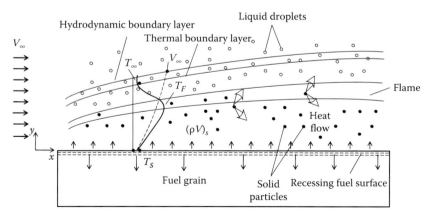

FIGURE 9.3 Schematic of a simplified model for the combustion process in a typical hybrid rocket engine.

metalized fuel grain, the radiation mode of heat transfer plays an influential role as compared with convective heat transfer. However, in the case of a nonmetalized fuel grain, the radiation mode of heat transfer is insignificant as compared with the convective mode, and hence can be neglected easily. The heat received by the surface of the solid fuel is utilized in converting the fuel from solid to vapor phase, which is known as the heat of vaporization. But for polymeric fuels, apart from the heat of vaporization, heat is to be provided to break polymer chains, which is known as the heat of depolymerization. The fuel vapors get transported away from the solid fuel grain upward and meet the gaseous oxidizer diffusing from the core in a relatively narrow region closer to the solid surface. As a result, a flame is formed near the fuel grain due to exothermic chemical reactions inside the hydrodynamic boundary layer, where the stoichiometric fuel/oxidizer ratio exists ideally. But in actual situation, combustion takes place on the fuel-rich side of stoichiometry, forming a flame whose thickness is around 10% of the boundary layer thickness. Typical velocity and temperature profiles on the solid fuel surface during the combustion process are shown in Figure 9.3, in which a thin sheet of flame is likely to occur inside the hydrodynamic turbulent boundary layer. It can be noted that the boundary layer grows along the flow direction until it attains a fully developed flow. The height of the flame from the surface of the solid fuel grain is dependent on the regression rate of the fuel grain, the oxidizer flow velocity, the diffusion rate of the vaporized fuel, and the oxidizer. But the thickness of the flame is dependent on the reaction rate, which is influenced by the chamber pressure, temperature of the solid fuel grain, and also the free-stream gas and ingredients of the fuel. The rate of chemical reaction in the flame is dependent on the rate of heat transfer from the flame to the fuel grain surface as per Equation 9.2, rather than the chemical kinetics, of course, except at lower pressures. Hence, the rate of combustion is limited by the transport processes rather than by chemical kinetics as in diffusion flame and hence the combustion in HPRE is diffusion controlled. We know that the regression rate $\dot{r}$ (mm/s) of the solid fuel is dependent on the extent of heat transfer from the flame surface and hot stream due to both convection and radiation modes. In other words, the regression rate $\dot{r}$ (mm/s) can be related to free-stream propellant mass flux (flow rate per unit cross-sectional area) G, which varies both with time and axial distance along the fuel grain. We can derive an expression for the regression rate $\dot{r}$ (mm/s) by striking a balance between the heat transfer to the solid fuel surface and the heat transported away by the decomposed fuel under steady-state condition as

$$\dot{Q}_w = \dot{m}_F A_b \Delta H_v = \rho_F \dot{r} A_b \Delta H_v \tag{9.1}$$

where

$\dot{Q}_w$ is the total heat transfer, including convective $\dot{Q}_c$ and radiative $\dot{Q}_r$ heat transfer

$\dot{m}_F$ is the mass-burning rate of the fuel

A_b is the burning surface area of the fuel grain

$\rho_F \dot{r}$ is the product of solid fuel density and regression/burning rate, which is basically the mass of fuel burnt per unit surface area

ΔH_v is the heat of phase change from solid to gas, which is required to raise the temperature of the solid fuel from ambient to the surface temperature T_s

Note that the heat absorbed by the solid fuel can be supplied as heats of melting, liquification, and vaporization, sensible heat to raise its temperature from liquid to its boiling point, and from vapor to surface temperature. Besides this, the heat can be utilized in the degradation of the polymer in the case of polymer-based fuel grains. Generally, the surface temperature attained by the fuel grain varies from 500°C to 600°C. We can neglect the radiation heat transfer for simplicity, particularly in the case of nonmetalized solid fuel grains. The convective heat transfer $\dot{Q}_c$ can be considered to be dependent on the prevailing enthalpy difference between the flame surface and the solid fuel surface and free-stream propellant mass flux G. Of course, this assumption can be valid only if the flame is considered to be a thin one as compared with the boundary layer thickness. Hence, the convective heat transfer $\dot{Q}_c$ can be determined as

$$\dot{Q}_c = h_c \rho_g V_g A_b \Delta h_T \tag{9.2}$$

where

Δh_T is the enthalpy difference between the flame and fuel surface

h_c is the convective heat transfer coefficient

$\rho_g V_g$ is the mass flux of hot gases, which can be evaluated using the Reynolds analogy [2,3,6]

Note that the heat transfer coefficient h_c is related to the skin friction factor C_f, which is defined as [2]

$$C_f = \frac{\tau_w}{0.5\rho_g V_g^2} \tag{9.3}$$

where

τ_w is the shear stress on the solid fuel surface
$0.5\rho_g V_g^2$ is the kinetic energy of the free stream

According to the Reynolds analogy, the heat transfer coefficient h_c is related to the skin friction factor C_f as

$$h_c = \frac{C_f}{2} \tag{9.4}$$

We can derive an expression for the regression rate $\dot{r}$ in terms of the fluid properties from the boundary layer theory by using Equations 9.1 through 9.4 as

$$\rho_F \dot{r} = \frac{C_f \rho_g V_g \Delta h_T}{2\Delta H_v} = \frac{C_f \rho_g V_g}{2} B \tag{9.5}$$

where B is the mass transfer number based on thermochemical parameters, which can be expressed as

$$B = \frac{\Delta h_T}{\Delta H_v} = \frac{2\rho_F \dot{r}}{C_f \rho_g V_g} \tag{9.6}$$

Note that the mass transfer number B is physically a ratio of enthalpy difference Δh_T between the flame and fuel surface to the enthalpy of vaporization, ΔH_v. As discussed earlier, the enthalpy difference Δh_T dictates the heat being transferred per unit area from the flame surface to the fuel grain surface, while the enthalpy of vaporization ΔH_v is the heat required to gasify unit fuel mass. It can be noted that this transfer number B is proportional to the regression rate $\dot{r}$. We can use the well-known empirical relationship for the skin friction factor C_f as [3]

$$\frac{C_{f0}}{2} = 0.03\left(\frac{\rho_g V_g x}{\mu_g}\right)^{-0.2} = 0.03\left(Re_x\right)^{-0.2} \tag{9.7}$$

It must be kept in mind that this analysis is valid only for solid surfaces, which would not incur any mass addition from the surface to the boundary layer. However, during hybrid propellant combustion, solid fuels get regressed, whose vapor is added to the boundary layer and thus the previous relation would not be valid, which must be modified adequately to be applied for the present problem of hybrid propellant combustion. We know that the mass injection from the solid surface due to the regression of the solid propellant into the boundary layer affects the extent of convective heat transfer from the flame surface. In other words, the mass injection from the solid propellant into the boundary layer alters the velocity profile of the boundary layer, causing the axial velocity to grow more slowly as compared to the nonblowing (no mass injection) case. Thus, there will be a reduction in convective heat transfer, which will reflect in a reduction in C_f. That means that we need to modify the term C_f in Equation 9.5, which must be expressed in terms of C_{f0} of Equation 9.7 for the nonblowing (no mass injection) case. Hence, Equation 9.5 for the nonblowing case can be re-expressed as

$$\rho_F \dot{r} = B\left[\frac{\rho_g V_g}{2}\left(\frac{C_f}{C_{f0}}\right)C_{f0}\right] \tag{9.8}$$

where B is the mass transfer number, which is dependent on thermochemical properties, and the terms within square brackets is dependent on the fluid dynamics parameter. The skin friction ratio C_f/C_{f0} indicates the reduction in heat transfer caused by the mass addition to the boundary layer. We will determine C_f by using the same definition as given in Equation 9.3

$$C_f = \frac{\tau_w}{0.5\rho_g V_g^2} \tag{9.9}$$

In this case, the turbulent wall shear stress τ_w in fluid flow can be defined as

$$\tau_w = \left(V_y + \rho\varepsilon\right)\left(\frac{\partial V_y}{\partial y}\right)_s \tag{9.10}$$

where

- ε is the turbulent diffusivity
- V_y is the velocity of gas along the y-direction from the solid propellant wall

The analysis of Marxman et al. [2,3] for the turbulent flow, which is quite involved, can provide a sufficiently accurate relationship for C_f/C_{f0}

$$\frac{C_f}{C_{f0}} = 1.27B^{-0.77} \quad 5 \le B \le 100 \tag{9.11}$$

We can definitely apply this expression for hybrid propellant combustion as the mass transfer number B for this turns out to be around 10. By using Equations 9.7 and 9.11 in Equation 9.8, we get,

$$\rho_F \dot{r} = 0.038 G Re_x^{-0.2} B^{0.23} = 0.038 G^{0.8} \left(\frac{x}{\mu_g} \right)^{-0.2} B^{0.23} \tag{9.12}$$

Note that G is the combined mass flow rate per unit area contributed by both oxidizer and fuel. At any location x along the grain surface, the mass flux, G_x, can be evaluated by considering the total mass flux passing through the port divided by its area. It must be noted that the mass flux, G_x, increases with the length of the grain, as more mass due to the vaporization of the solid propellant from the grain surface is added to the core stream. The increase in the mass flux, G_x, along the length of the grain is compensated to some extent by the $x^{-0.2}$ term in Equation 9.12. Note that the regression rate increases along with the length of the grain, as it is related to the mass flux, G, as per Equation 9.12 ($\dot{r} \propto G^{0.8}$). Besides this, as the port area increases with time due to the regression of the solid propellant, the mass flux, G, decreases with time. The decrease in the regression rate, $\dot{r}$, with time is compensated, to some extent, due to the increase in the burning surface area. However, the overall effect is that the regression rate, $\dot{r}$, would be affected by the smaller grain length, particularly for $L/D < 25$. It can be noted from Equation 9.12 that the regression rate, $\dot{r}$, is also weakly dependent on the mass transfer number B ($\dot{r} \propto B^{0.23}$). Any large increase in the enthalpy difference Δh_T between the flame and the fuel surface or any large decrease in the enthalpy of vaporization ΔH_v of the solid propellant would produce a significant change in the regression rate. Let us understand it by analyzing the physical processes during the regression process of a solid propellant. If there is an increase in the regression rate, $\dot{r}$, there will be an increase in the mass injection into the boundary layer, which eventually blocks the heat to be transferred to the solid propellant surface. As a result, this decrease in convective heat transfer to the propellant surface will definitely reduce the evaporation rate, and thus the regression rate is reduced. That is the reason why the regression rate of a hybrid propellant is found to be lower than that of a composite solid propellant while both are being operated at the same chamber pressure.

Generally, it has been observed that the regression rate, $\dot{r}$, of hybrid propellant combustion does not get affected significantly by the chamber pressure, particularly at higher pressures ($P_c > 20$ atm)

$$\dot{r} = CG_{Ox}^{n} \tag{9.13}$$

where

$\dot{r}$ is the regression rate (m/s)
G_{Ox} is the oxidizer mass flux (kg/m^2 s)
C is the regression rate constant,
n is the mass flux index

Typical experimental values of n span from 0.5 to 0.8. However the regression rate, $\dot{r}$, does get affected significantly at lower pressures, as the kinetics plays a very important role due to the slower overall chemical reaction rate. It has been found that at low-pressure regimes, the regression rate, $\dot{r}$, can be related to the chamber pressure as

$$\dot{r} = CG^{n} P_{C}^{m} \tag{9.14}$$

where

C is the regression rate constant
Both pressure index, m, and the mass flux index, n, must be less than unity ($m < n < 1.0$)

We know that solid propellants contain a certain amount of nonvolatile particles, which get released into the core stream during gasification of the propellant. We know that metal particles on the propellant grain are added to enhance the properties of propellant grains. These particles are generally converted into their respective oxides during gasification and combustion. For example, aluminum particles (30–40 μm) meant to enhance the heat of reaction of the solid propellant grain get burnt on the surface of the grain to produce Al_2O_3, which eventually absorbs a certain amount of gaseous oxygen. These oxide particles are generally much smaller than actual metal particles, which vary between 0.1 and 10 μm. These particles will affect the regression rate of the solid propellant grain due to several reasons, namely, (1) the heat of gasification of the propellant (volatile component) gets enhanced depending on the extent of nonvolatile particles loading in the propellant grain, as nonvolatile particles need to be heated from ambient temperature to the surface temperature of the propellant grain. (2) The heat flux from nonvolatile particles in the combustion chamber is radiated back to the grain surface. (3) As the nonvolatile particles lag behind the fluid flow velocity and temperature during expansion in the rocket nozzle, there will be a decrease in thrust and I_{sp}. Of course, the relative volume of these nonvolatile particles as compared with that of volatile fuel vapor is quite small, even less than 0.1%. As a result, these nonvolatile particles do not affect the convective heat transfer process and also the blocking effect, which is dependent on the gas blowing rate. Of course, they may affect the radiative heat transfer to the grain surface. Note that we are not considering the radiative heat transfer in our analysis. Even then, these nonvolatile particles will affect the regression rate as a certain amount of heat reached at the grain surface is utilized for inert heating of these metal particles. Hence, to take account of this effect, we can modify the expression for the regression rate (Equation 9.12) as

$$\rho_g \dot{r} = \rho_F (1-K)\dot{r} = 0.036G^{0.8}\left(\frac{x}{\mu_g}\right)^{-0.2} B^{0.23} \tag{9.15}$$

where

- K is the mass loading of metal particles
- Hence, $\rho_F(1-K)$ is the vaporizing component of the fuel

Note that the heat of gasification defined in Equation 9.1 can be considered as a sum of the heat required to gasify the volatile component and the heat required to bring the temperature of the nonvolatile component to the surface temperature of the grain. This is known as the effective heat of gasification of the propellant, which can be expressed as

$$\Delta H_{v,eff} = \Delta H_{vb} + \frac{K}{K-1} C_{nv}\left(T_s - T_i\right) \tag{9.16}$$

where

- ΔH_{vb} is the binder heat of vaporization
- K is the mass loading of nonvolatile particles
- C_{nv} is the specific heat of nonvolatile particles
- T_s is the surface temperature
- T_i is the surface temperature of the grain

Note that the value of ΔH_{vb} is used to evaluate the mass transfer number B for solid propellant grains containing nonvolatile particles. As most metal additives have relatively low heat capacities and low temperature differences, the effective heat of gasification does not differ from the value of ΔH_{vb}. Hence, the regression rate can be considered to be almost inversely proportional to $K/K-1$.

9.5.1 Effects of Thermal Radiation on Hybrid Propellant Combustion

In the case of solid fuel grains with a high metal loading or a heavily sooting fuel, heat transfer to the fuel grain surface due to thermal radiation plays a very important role that affects the regression rate of the fuel grain. The aspects of thermal radiations must be considered particularly when combustion gases contain large amounts of solid particles (>10^7 particle/cc). It can be noted that with the increase in radiative heat transfer $\dot{Q}_r$, there will be an increase in the regression rate of the fuel grain, which eventually inhibits the convective heat transfer $\dot{Q}_c$. By considering both convective and radiative heat transfer, Marxman et al. [3] have derived an expression for the regression rate

$$\rho_g \dot{r} = \frac{\dot{Q}_c}{\Delta H_{v,eff}}\left(e^{-(\dot{Q}_r/\dot{Q}_c)} + \dot{Q}_r/\dot{Q}_c\right) \tag{9.17}$$

and

$$\dot{Q}_r = \sigma\varepsilon_s\left(\varepsilon_g T_F^4 - \alpha_g T_s^4\right) \tag{9.18}$$

where

- T_F is the flame temperature
- T_s is the surface temperature
- σ is the Stefan–Boltzmann constant
- ε_s is the emissivity of the grain surface
- ε_g is the emissivity of the gas
- α_g is the absorptivity of the gas phase

The radiative heat transfer from the grain surface to the gas phase can be neglected easily, as the surface temperature is around 15% of the flame temperature. As a result, Equation 9.18 becomes

$$\dot{Q}_r = \sigma\varepsilon_s\varepsilon_g T_F^4 \tag{9.19}$$

The emissivity of gas, ε_g, is dependent on the nature of the gas and its contents. In other words, the value of emissivity of gas, ε_g, depends on the constituents of the gas, and particle size and its distribution, etc. For gas-phase radiation, the emissivity of gas, ε_g, is related to the pressure, *P*, and optical path length, *Z*, as

$$\varepsilon_g = 1 - e^{-C_g PZ} \tag{9.20}$$

When the hot gas is laden with several particles, the radiation from the particles will be predominant over gas-phase radiation and thus can be neglected. Note that the value of *Z* can be approximated to the interior diameter of the grain. The emissivity from particles, ε_g, is related to the number density of particles, *N*, as

$$\varepsilon_g = 1 - e^{-C_s N} \tag{9.21}$$

The number density of particles, *N*, can be determined from the burning test for a particular propellant, which can also provide data about particle diameter and vaporized mass density, *G*. It must be noted that the particle number density, *N*, increases along with the axial distance as more particles are released. The value of constants C_g and C_s can be determined empirically from the experimental data, which varies with propellant type. Now combining the Equations 9.19 through 9.21, we can have an expression for radiative heat transfer $\dot{Q}_r$ as

$$\dot{Q}_r = \sigma \varepsilon_s T_F^4 \left(1 - e^{-\left(C_g PZ + C_s N\right)}\right) \tag{9.22}$$

where T_F is the effective flame temperature, which is taken care of by both convective and radiative heat transfer caused by both gas-phase and solid-phase thermal radiation.

It has been reported in the literature that the regression rate of the solid fuel grain is dependent on the type of propellant and its metal loading, fuel port size, oxidizer atomizer design, flow structure in fuel port, and property variations in fluid properties in the boundary layer. Some of these effects on the regression rate are illustrated qualitatively in Figure 9.4. Three distinct regimes can be observed in this figure for the regression

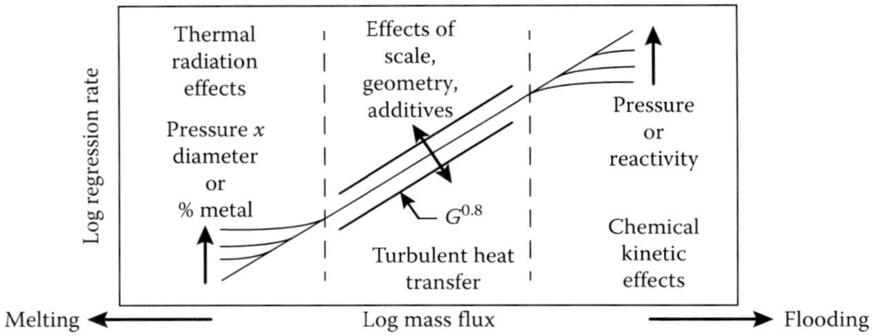

FIGURE 9.4 Various regimes of regression rate dependency [4]. (From Chiaverini, M.J. and Kuo, K.K. (Eds.), *Fundamentals of Hybrid Rocket Engine Combustion and Propulsion*, Vol. 218, AIAA Progress in Astronautics and Aeronautics, Academic Press, New York, 2007.)

rate. In the low mass flux regime, the thermal radiation plays an important role in influencing the regression rate as heat flux due to convection is quite insignificant. Radiation heat transfer becomes more important at a given mass flux condition when the product of port diameter (*PD*) and chamber pressure increases, as it enhances the optical density. The regression rate gets enhanced even beyond classical theory prediction with an increase in the *PD* value as gas emits thermal radiation more efficiently. For metalized fuel grains, there will also be an increase in the regression rate as its metal loading is enhanced. If the mass flux decreases below certain threshold values, the fuel gets melted, leading to the formation of char. In certain situations, subsurface decomposing of the solid fuel grain can occur due to the penetration of thermal waves deep into the subsurface region. As a result, it will be difficult to establish stable combustion of the solid fuel. This threshold limit is known as the cooking limit in literature. Such kind of situation must be avoided during fuel grain operation to avert the mishap.

As the mass flux increases further to a certain intermediate level, there will be an increase in the regression rate, as this regime is mainly predominant by turbulent flow heat transfer, as indicated in Figure 9.4. In this regime, the regression rate and the mass flux relation is influenced significantly neither by radiation nor by chemical kinetics, although these effects cannot be avoided altogether. When the mass flux increases further to a high level, the increase in the regression rate is more influenced by both heterogeneous and gas-phase reaction rates. In other words, the chemical kinetics plays an important role in influencing the regression rate. In this regime of mass flux, the chamber pressure also affects the regression rate. There is a certain threshold value of chamber pressure for a given mass flux and a certain fuel/oxidizer system, where a lower regression rate is obtained as compared with the predicted value of the regression rate by classical hybrid boundary layer theory with a finite-rate chemistry model, which takes care of both gas-phase reactions and heterogeneous reactions. With further increase in the mass flux, the flooding limit is reached, which depends on the chamber pressure and the reactivity of the fuel/oxidizer system. The flooding of hybrid propellants can likely occur in the case of highly reactive fuel/oxidizer combinations, namely, lithium hydride and fluorine, particularly at higher mass flux situations. In contrast, at the same condition of mass flux, hydrocarbon/nitrous oxide is less likely to attain the flooding limit due to its lower reactivity.

9.6 IGNITION OF HYBRID PROPELLANTS

The ignition of hybrid propellants can be accomplished by several devices, namely, a solid squib, park ignition system, hypergolic slug, etc. As discussed earlier, a solid propellant charge is being ignited by an electrical squib, which produces hot gases laden with solid and hot particles for initiating the ignition of hybrid propellants by the heating of the oxidizer. In a similar fashion, a spark plug is used to ignite gaseous fuel and oxidizer mixtures. In the case of a hypergolic slug system, hypergolic propellants are used to produce high-temperature gases, which act as a heat source for ignition of hybrid propellants. In all cases, the heating of the fuel in the case of HPRE takes place with the help of these ignition systems, which act as a short-duration heat source. During the initial phase of the ignition process of hybrid propellants, ignition of the aft portion of the fuel grain is initiated in the beginning and, subsequently, the flame spreads upstream to cover the entire grain length, thus ensuring a steady-state burning of the propellant. Toward the end of this ignition transient, combustion gases leaving the rocket engine attain a steady-state oxidizer/fuel ratio, which is generally rich in the oxidizer. In contrast, in the case of SPRE, the oxidizer–fuel ratio remains almost constant. The time required for ignition to occur in the case of HPRE happens to be quite higher as compared with the SPRE, as longer time is taken for the establishment of the combustion boundary layer over the solid fuel grain. Hence, the time lag to reach steady-state combustion in the HPRE is quite higher as compared with SPRE.

9.7 COMBUSTION INSTABILITY IN HPRE

Combustion instability is an important and difficult phenomenon that can also occur in that case of hybrid rocket engine. As discussed earlier, a certain level of unsteadiness can be manifested due to transient phenomena within the pipeline, combustion chamber, nozzle, etc., that can lead to irregular pressure fluctuations in the combustion chamber and nozzle flow. When the oscillation is sustained during the combustion process of hybrid rocket engine, this phenomenon is often termed as combustion instability. Generally, in the case of HPRE, the amplitude of the pressure oscillation is restricted to a maximum of 50%–60% of the mean pressure value. As a result, it would not cause the disastrous failure of the combustion chamber as in solid- or liquid-propellant rocket engines. Of course, the combustion chamber pressure oscillations must be restricted to 2%–3% of the mean chamber pressure. In the case of HPRE, although rougher pressure variation with time is observed as compared with either solid or liquid rocket engines, unbounded growth of pressure oscillations has not been reported till date. Hence, oscillatory combustion occurring during the instability phase of HPRE needs to be looked at seriously, as this could result in major thrust oscillation, apart from introducing excess structural/thermal loads due to the oscillatory regression rate of the solid fuel. The thrust oscillations must be minimized, as an excess of its variation can lead to the failure of the mission while affecting spacecraft structure and payload. Interestingly, pressure oscillations have been always observed to grow to their limiting amplitude, which is dependent on the fuel grain size and shape, oxidizer velocity, injector characteristics, and mean chamber pressure.

As in a liquid-propellant rocket engine, combustion instabilities in the case of HPRE can be classified into (1) low frequency, (2) medium frequency, and (3) high frequency [5,6]. Generally, low-frequency pressure oscillations are likely to occur in HPREs, and hence, we will be focusing our discussion on low-frequency combustion instabilities, which can be divided into three categories, namely, (1) feed system coupled instabilities, (2) chuffing, and (3) intrinsic low-frequency instabilities (ILFLs).

9.7.1 Feed System–Coupled Instabilities

We have already learnt in the previous chapter that instability arises due to feed system large amplitude and low frequency oscillations in feed system, commonly known as chugging, which generally occurs in the case of cryogenic oxidizer systems. Such kind of instability is likely to be triggered due to the presence of a two-phase flow under high pressure. This kind of combustion instabilities can be minimized by eliminating the propagation of pressure disturbances from the combustion chamber upto the feed system, which can be accomplished by increasing the pressure drop across atomizers. Note that the root cause of such kind of instabilities is the coupling of feed system with combustion processes in the thrust chamber, triggered due to vaporization lag between the feed line and combustion chamber oscillations. The existence of coupling between the combustion chamber and the feed system is a necessary but not sufficient condition for this kind of instabilities. A reasonably longer vaporization delay may be required for the occurrence of pressure oscillations. It is quite easy to identify this kind of instabilities due their specific characteristics, namely, (1) they are likely to be experienced in liquid feed systems, (2) these oscillations are quite regular in nature in comparison with low-frequency instabilities, and (3) they occur within a narrow bandwidth of frequency. Generally, the fundamental frequency does not change and its intensity remains almost constant over the entire period of operation. The transient behavior is predominantly affected by the feed system–coupling instability, while other low-frequency instabilities are likely to stay inactive during its run.

9.7.2 Chuffing

During the operation of hybrid rocket engine, pressure oscillations in the range of 1–5 Hz are likely to occur, which is commonly known as chuffing. Generally, chuffing instability is caused due to the accumulation and breakoff of the char/molten layer from the fuel surface as the fuel surface regresses at a slower rate. Its frequency is controlled by the thermal lag time of the solid fuel. Hence, the chuffing frequency can be scaled with thermal lag time of solid fuel, which is related as $t_T = \alpha / \dot{r}^2$, where α is the thermal diffusivity (m^2/s) and $\dot{r}$ is the regression rate (m/s) of the fuel. Chuffing is likely to occur only at very low mass flux of oxidizers, during which the thermal layer thickness in the solid fuel grows significantly due to the melting of the solid fuel underneath the port surface area. Such kind of removal of soft layers of the solid fuel and its sudden burning give rise to a pressure surge in the chamber pressure. The repeated occurrence of such phenomenon can lead to oscillations in the chamber pressure, thus resulting in chuffing in rocket engines.

9.7.3 Intrinsic Low-Frequency Instabilities

This type of combustion instability is quite predominant in HPRE over a wide range of its operating conditions. The nature of such kind of instabilities is more or less similar in most hybrid rocket engines. Let us consider an example of a hybrid rocket engine developed at the NASA Ames Research Center, in which a fast-burning paraffin-based fuel and a gaseous oxidizer are used. The variation in the chamber pressure with time for a typical case in this rocket engine is shown in Figure 9.5a, along with feed line pressure. It is interesting to note that there is no oscillation in the feed line pressure–time trace, indicating the absence of coupling between the feed line and chamber pressure oscillation. In other words, the feed system and the chamber pressure are decoupled from each other. The fast Fourier transform (FFT) of the chamber pressure–time trace is shown in Figure 9.5b, in which there are three dominant broad peaks, namely, the intrinsic low-frequency instability (ILFI), the Helmholtz mode, and the first longitudinal mode. It can be noted that the ILFI has lower frequency and higher amplitude as compared with the other two modes of oscillations. Of course, the first longitudinal mode exhibits the highest frequency among all three modes of oscillations.

It is a common belief that low-frequency instability is generated by a linear mechanism, but its amplitude is governed by nonlinear effects. Some of the potential driving mechanisms responsible for low-frequency instability as reported in open literature are as follows: (1) atomization (2) mass flux coupling, (3) chuffing, (4) pressure coupling, vortex shedding, etc. It has been claimed by researchers that vaporization lag due to the atomization system can couple the combustion process with the acoustic pressure field, leading to ILFI. By increasing the residence time of the droplets

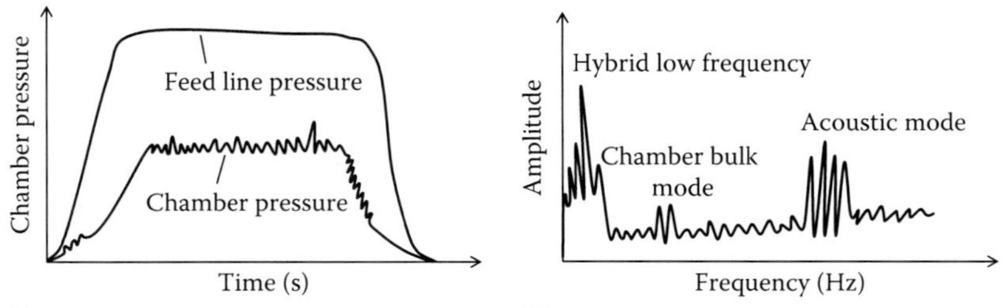

FIGURE 9.5 (a) Schematic of chamber pressure variation with time and (b) its FFT plot.

in the precombustion chamber, such type of instability can be minimized, particularly for liquid oxidizer hybrid rocket engine. Of course, it may not work well for systems with gaseous oxidizer, and thus calls for a comprehensive model that can explain the phenomenon of low-frequency combustion instability. Another mechanism of low-frequency instability is the mass flux coupling, which is considered to supply energy feedback to instability as the regression rate of the fuel is strongly dependent on the mass flux of the oxidizer. Chuffing, as discussed earlier, is the kind of instability that is associated with the formation and sudden breakup of the char/molten layer of the fuel, which occurs, of course, at very low oxidizer mass flux levels. In two extreme operating regimes, pressure-coupling combustion instability is likely to occur, as the regression rate is dependent on the chamber pressure, not the mass flux rate. When the oxidizer mass flux rate is quite high, chemical kinetics dictates the rate of fuel grain regression, then pressure-coupling combustion instability is likely to occur. In the case of a low oxidizer mass flux rate, pressure coupling would provide energy feedback mechanism, leading to combustion instability, as radiative heat transfer will be the dominant mode, not convective heat transfer. Sometimes, pressure-dependent combustion can be coupled with longitudinal acoustic modes to produce medium-frequency instability. Some researchers have argued that low-frequency combustion instability can be caused due to vortex shedding in a rapid expansion region, particularly in the mixing chamber of rocket engines. On several occasions, this mechanism may drive combustion oscillation to medium frequency in a steady low-frequency range. The various aspects of the mechanism of combustion instability in the case of HPRE are yet to be matured and considered as one of the active research areas in the field.

In order to suppress the problems of combustion instabilities in HPREs, several methods can be devised, which are as follows:

1. Isolating acoustic communication between the combustion chamber and the oxidizer feed line system.
2. Designing a proper injector that can produce a proper oxidizer spray pattern compatible with the fuel grain in the precombustion chamber.
3. Providing flame holding on solid fuel surfaces near the precombustion chamber to eliminate flame instabilities so that oscillation in the pyrolysis front upstream of the fuel grain can be avoided.

4. Providing compartments at the fore end of the fuel grain and solid fuel layers in the precombustion chamber to minimize the transverse mode of instabilities.

5. Selecting proper fuel grain geometry that can suppress combustion instability.

REVIEW QUESTIONS

1. How is a hybrid rocket different from a liquid-propellant rocket engine?

2. What are the advantages of a hybrid rocket over a liquid-propellant rocket engine?

3. What are the advantages of a hybrid rocket over a solid-propellant rocket engine?

4. What are the oxidizers commonly used for a hybrid rocket engine?

5. What are the fuels commonly used for a hybrid rocket engine?

6. Draw the schematic of a typical hybrid rocket engine and explain each component?

7. Why is a multi-port grain preferred over a single-port grain?

8. Draw the schematic of a simplified model for the combustion process in a hybrid rocket engine?

9. Derive a relationship for the regression rate of a propellant in a hybrid rocket engine.

10. What do you mean by mass flux rate? Why is it used for regression rate expression?

PROBLEMS

9.1 A hybrid propellant (HTPB + O_2) rocket engine is used to develop a thrust of 115 kN with an I_{sp} of 340 s and a chamber pressure 3.5 MPa. The cylindrical HTPB fuel grain with an outer diameter of 120 cm undergoes combustion with a mixture ratio of 1.5 and produces an initial chamber pressure of 3.5 MPa. If five ports with an initial diameter of 50 mm are used, determine (1) mass flow rate of the gaseous fuel and liquid oxidizer, and (2) regression rate and length of grain.

Use semiempirical relations for the regression rate as $\dot{r} = CG^n P_c^m$, where pressure index, m, is 0.5; mass flux index, n, is 0.68; and regression rate, C, is 4×10^{-7} for SI units. Take the density of solid HTPB = 1280 kg/m^3, $\gamma = 1.23$, molecular weight of HTPB = 25 g/mol.

9.2 A hybrid propellant rocket engine consisting of HTPB and a liquid oxidizer is used to develop a thrust of 1.2 kN for a period of 150 s. A cylindrical HTPB fuel grain undergoes combustion with a mixture ratio of 2.1 and produces a chamber pressure of 4.5 MPa at 3408 K using a nozzle with an expansion ratio of 7.5 with an initial mass flux rate of 39.2 kg/s m^2. If its thrust coefficient, C_F, happens to be 1.05 and characteristic velocity is equal to 2750 m/s, determine (1) throat diameter of the nozzle and (2) mass flow rate of the gaseous fuel and liquid oxidizer. By considering pressure index, m, to be 0.3, mass flux index, n, to be 0.75; and regression rate, C, to be 12.89×10^{-7}, determine the initial port diameter. Take ρ of HTPB = 1280 kg/m^3, $\gamma = 1.23$, molecular weight of HTPB = 25 g/mol.

REFERENCES

1. Bartel, H.R. and Rannie, W.D., Solid fuels combustion ramjet, Progress Report 3-12, Jet Propulsion Laboratory, California Institute of Technology, Pasadena, CA, 1946.
2. Marxman, G.A., Combustion in turbulent boundary layer on a vaporizing surface, *10th International Symposium on Combustion*, The Combustion Institute, Pittsburgh, PA, 1965, pp. 1337–1349.
3. Marxman, G.A., Wooldrige, C.E., and Muzzy, R.J., *Fundamental of Hybrid Boundary Layer Combustion, Heterogeneous Combustion*, Green, L., Jr. (Ed.), Vol. 15, AIAA Progress in Astronautics and Aeronautics, Academic Press, New York, 1964, pp. 485–521.
4. Chiaverini, M.J. and Kuo, K.K. (Eds.), *Fundamental of Hybrid Rocket Engine Combustion and Propulsion*, Vol. 218, AIAA Progress in Astronautics and Aeronautics, Academic Press, New York, 2007.
5. Karabeyoglu, M.A., Zilliac, G., Cantwell, B.J., DeZilwa, S., and Castellucci, P., Scale-up tests of high regression rate paraffin based hybrid rocket fuels, *Journal of Propulsion and Power*, 20(6), 1037–1045, November–December 2004.
6. Sutton, G.P. and Ross D.M., *Rocket Propulsion Elements*, 5th edn., John Wiley & Sons, New York, 1975.

CHAPTER 10

Liquid-Propellant Injection System

Excellence can only be achieved through ceaseless energetic actions.

D.P. MISHRA

10.1 INTRODUCTION

We learnt in Chapters 8 and 9 that the injection system located on the rear portion of the thrust chamber is an important part of both liquid and hybrid propellant engines as it dictates their performance significantly. The main function of an injection system is to supply the requisite quantity of liquid propellant that ensures proper mixing of fuel and oxidizer in the combustion chamber such that smooth and stable combustion can be ensured during its entire operating range. Besides this, its proper design is essential for the combustion system to have (1) high combustion efficiency, (2) shorter combustor length, and (3) wall temperature within metallurgical limits. The injector system consists of two components: (1) distributor (see Figure 10.1) and injector head. The injector head carries several injectors, as shown in Figure 10.1. The main function of a distributor is to ensure homogeneous supply of propellant to the individual injectors. In order to achieve this, a much lower velocity of liquid propellant in the distributer must be maintained such that it can act as settling chamber. Generally, the minimum area ratio between settling chamber and injector orifices must be greater than 16 for the distributor to act as a settling chamber. In addition, a perfect sealing must be maintained between fuel

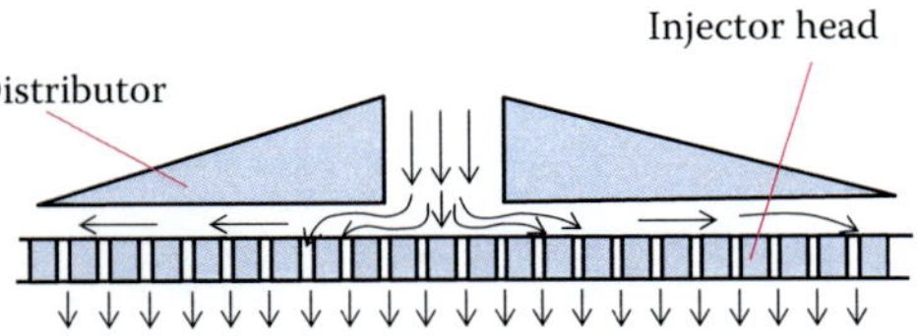

FIGURE 10.1 Schematic of a typical injection system.

and oxidizer distributors for a safe operation. The feed lines of the propellant must be located symmetrically to the axis of the chamber. In order to avoid short-circuiting problems in propellant supply to individual orifices of injectors, the diameter of distributor must long enough, while maintaining the compactness of the thrust chamber.

10.2 ATOMIZATION PROCESS

In order to accomplish combustion of the liquid fuel, it must be converted into fine droplets so that evaporation and mixing of liquid fuel with oxidizer can be achieved easily in the least possible time. This process of converting bulk liquid fuel into an array of several smaller droplets is known as *atomization* [1,2]. As discussed earlier, the main function of a fuel injector/atomizer is to produce a high surface-to-mass ratio in the liquid phase such that a high evaporation rate can be accomplished. It must also produce a proper fuel/oxidizer mixture ratio, which must be distributed uniformly in the combustion zone for ensuring a stable combustion over the whole range of engine operating conditions.

We know that during the process of atomization, bulk liquid is converted initially into a liquid sheet or jet [1,2]. Basically, this liquid sheet or jet gets disrupted subsequently due to various forces such as internal and external forces and surface tension force. Of course, in the absence of any other disruptive forces, the surface tension force tends to pull the liquid in the form of sphere as it has minimum surface energy. However, the viscosity of the liquid affects the atomization process adversely as it opposes any change in the shape and geometry of the liquid. In contrast, the aerodynamic force helps in disrupting the liquid ligaments into droplets. In other words, it can be stated that when the sum of disruptive forces exceeds the overall surface tension force, the liquid is disintegrated into ligaments and subsequently gets converted into smaller droplets. Note that this disintegration of liquid jet into droplet is governed by a nondimensional number known as *Weber number* We_j, which is given by

$$We_j = \left(\frac{\text{Inertia force}}{\text{Surface tension force}} \right) = \left(\frac{\rho V_j^2 d_j}{\sigma} \right) \quad (10.1)$$

where

V_j is the velocity
d_j is the diameter of undisturbed liquid jet
σ is the surface tension of the liquid in contact with the surrounding gas

A liquid jet breaks up into droplets only when inertial force is much greater than that of the surface tension force. Therefore, the Weber number must be larger than unity for droplet to be formed from bulk liquid. The jet Reynolds number that plays an important role in the formation of droplets is defined as

$$Re_j = \left(\frac{\text{Inertia force}}{\text{Viscous force}} \right) = \left(\frac{\rho V_j d_j}{\mu} \right) \quad (10.2)$$

The jet disintegration can be described comprehensively by another nondimensional number known as *Ohnesorge number* Oh_j, which can be defined as

$$Oh_j = \left(\frac{\mu_l}{\sqrt{\rho \sigma d_j^2}} \right) = \frac{\left[\rho_l / \rho_g We_j \right]^2}{Re_j} \quad (10.3)$$

This nondimensional number is sometimes known as stability number, which helps in describing the various mechanisms of jet breakup. Let us consider the variation of Ohnesorge number with the jet Reynolds number, as shown in Figure 10.2. Four regimes can be observed in this figure. Regime I is known as the *Rayleigh zone*, in which jet breakup is due to the effects of surface tension forces since the jet Reynolds number is low. The jet structure in this zone is predominantly varicose, in which the jet breakup occurs only when jet disturbance wavelength is about 4.6 d_j, as predicted by Rayleigh. With further increase in Re_j, the disintegration of jet is initiated by the jet oscillations with respect to the jet axis. Such kind of mechanism is found to occur in regime II, in which sinuous waves in the jet are observed prior to its breakup into ligaments. This zone marks the beginning of the influence of the ambient gas and is also known as *wind-induced regime*, which is divided further into first and second wind-induced

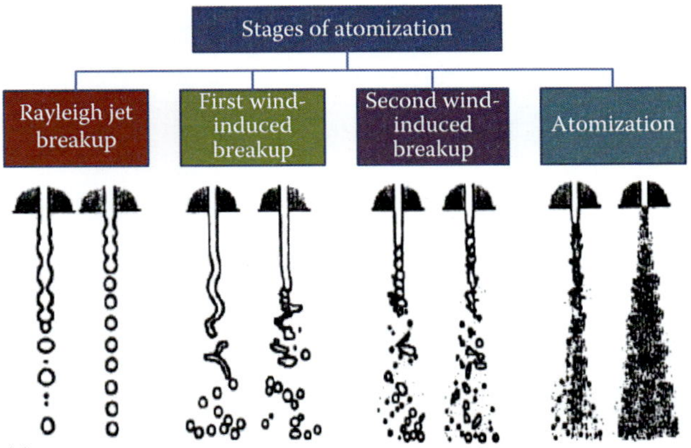

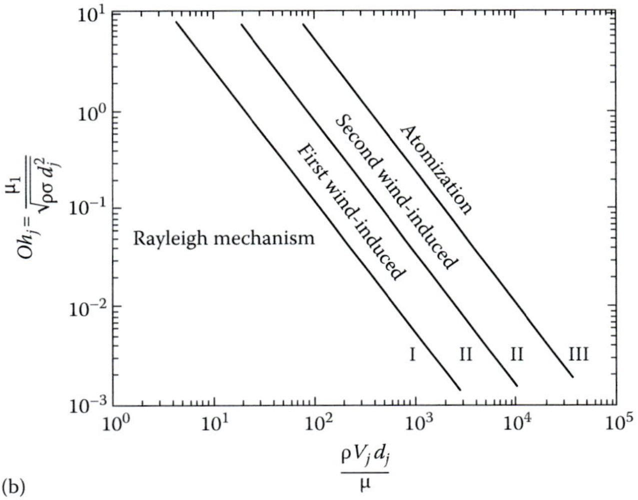

FIGURE 10.2 Four stages of liquid fuel disintegration in a simple orifice: (a) schematic, (b) *Oh* vs. $Re = \rho V_j d_j/\mu$ plot. (From Lefebvr, A.H., *Atomization and Spray*, Taylor & Francis, New York, 1989.)

regimes. At higher Reynolds number, liquid jet can be disintegrated within a short distance from the orifice, which is known as *atomization regime*. In this case, the jet breakup is caused due to the interactions of jet with ambient gas, combined with aerodynamic drag and flow turbulence. It has been observed that the size of ligaments decreases with increase in relative air speed, resulting in breakup into smaller droplets. As a result, average drop diameters in this zone are found to be much less than jet diameters.

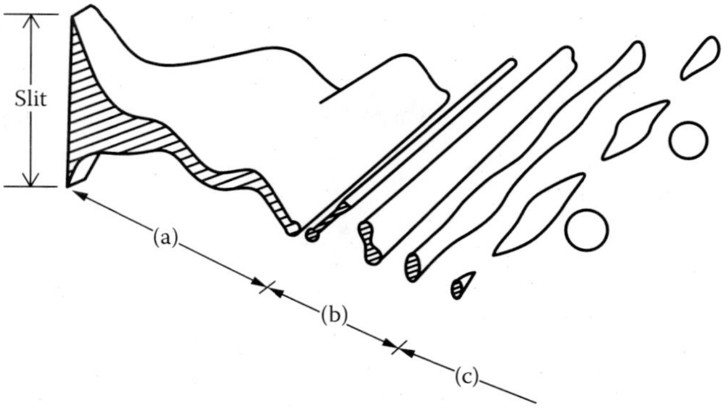

FIGURE 10.3 Processes involved in fuel atomization: (a) growth of waves in sheet, (b) formation of ligaments, (c) break-off of ligaments and droplet formation. (From Lefebvr, A.H., *Atomization and Spray*, Taylor & Francis, New York, 1989.)

Let us ask ourselves how the liquid is converted into droplets. Let us understand the mechanism of droplet formation in a qualitative manner. Let the liquid be forced through a small slit, as shown in Figure 10.3. Note that pressure energy is used to impart a high velocity to the liquid fuel. As a result, a liquid sheet/jet is created at the exit of orifice, which quickly becomes unstable and subsequently breaks up into fine droplets. During droplet formation, the following processes, as shown in Figure 10.3, may occur in sequence:

1. Stretching of fuel into liquid jet/sheet
2. Appearance of ripples and protuberances
3. Disintegration of liquid jet/sheet by atomizing forces (surface tension, inertia, and aerodynamic forces)
4. Formation of ligaments or holes in the sheet/jet
5. Breaking up of ligaments into droplets leading to spray formation
6. Agglomeration or shredding of droplets

10.3 INJECTOR ELEMENTS

As discussed earlier, the injector element is the most delicate and important part of the liquid-propellant rocket engine as it dictates the atomization, distribution, and mixing of the propellant in the combustion chamber. The specific injector configuration that is used in the rocket engine is called the injector element. Hence, we will use the terms "injector element," "injector," or "atomizer" interchangeably throughout this text. The desirable characteristics of injector elements are as follows:

1. Deliver the requisite flow rate of propellant (fuel and oxidizer).
2. Ensure good atomization and mixing in the combustion chamber.
3. Maintain proper local mixture ratio of fuel and oxidizer in the combustion chamber.
4. Maximize characteristic velocity for a given chamber length.
5. Provide cooling to the combustion chamber wall and injector face.
6. To have higher turn-down ratio of the combustion chamber.
7. Easier to manufacture and maintain.

10.3.1 Types of Injectors

Several types of injectors have been devised for rocket engines. They can be broadly divided into two categories: (1) nonimpinging and (2) impinging. Among nonimpinging types of injectors, three types, namely, (a) shower-head injector, (b) coaxial injector, and (c) swirl atomizers, are used in liquid-propellant rocket engines. The impinging injectors are broadly classified into two: (a) unlike-impinging and (b) like-impinging injectors. All these injectors are discussed in detail in the following.

10.3.1.1 Nonimpinging Injectors

1. *Shower-head injector*: This is one of the oldest nonimpinging injectors, in which fuel and oxidizer are ejected from injector head normal to its face, as shown in Figure 10.4a(A), which is similar to water shower. Note that it was used in the V2 rocket engine by German during the Second World War. In this case, axial streams of fuel and oxidizer form the spray cones/sheets. The sprays from individual elements interact with each other, which promotes atomization and mixing. They are atomized and mixed due to turbulence and diffusion. As a result, this incurs inefficient atomization and ineffective mixing. Hence, it requires longer length of combustor for complete combustion. However, it is quite effective in cooling the combustor

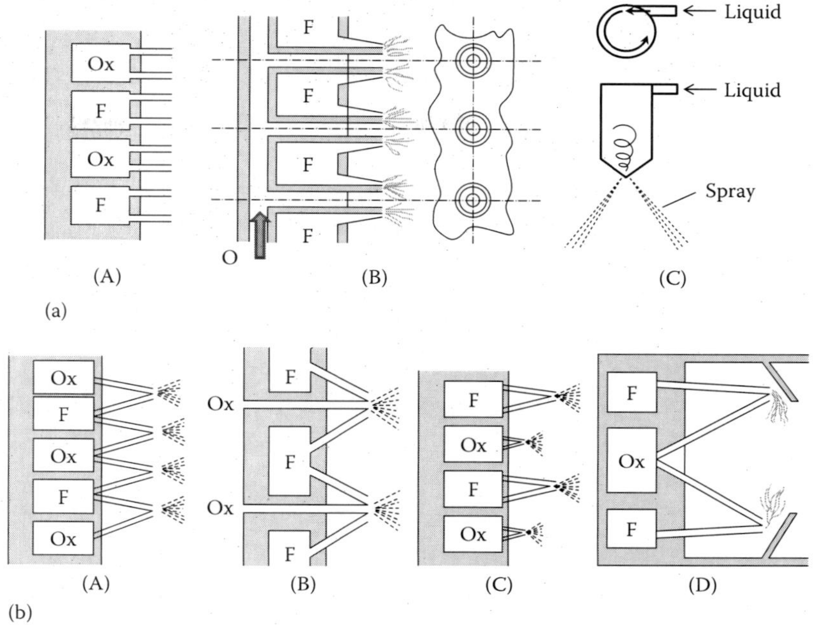

FIGURE 10.4 Types of injector elements: (a) nonimpinging: (A) shower head, (B) co-axial injector, (C) swirl injector and (b) impinging: (A) unlike doublet, (B) unlike triplet, (C) like doublet, (D) splash plate (Ox, oxidizer; F, Fuel).

wall as axial flow remains intact and thus prevents heat transfer from combustion zone to the wall. Besides this, the rocket engine can be throttled easily by varying the width spray cone/sheet with the help of axially movable sleeves without incurring excessive injection pressure drops. By using this concept, it has been demonstrated in lunar excursion module that throttling over 10:1 flow rate can be achieved easily without causing much change in mixture ratio.

2. *Coaxial injector*: It is the most common type of nonimpinging injector that is used mainly for nonhypergolic propellant. Particularly, it is preferred for semicryogenic liquid-propellant rocket engines. It was first developed by NASA during early experiments of cryogenic liquid-propellant engines. The schematic of a typical coaxial injector is shown in Figure 10.4a(B), which consists of two concentric tubes with a recessing length. Generally, liquid propellant (oxygen) passes through the central tube at a slower velocity (less than 30 m/s),

while gaseous fuel passes through the co-centric outer tube at a much higher velocity (more than 300 m/s).

The liquid is injected at a lower velocity intentionally, to a reduced velocity into the recess region. In contrast, the gaseous fuel is injected at high velocity into recess region, whose shearing action breaks the liquid surface into ligaments and subsequently into fine droplets. This also ensures better atomization, and mixing of fuel and oxidizers. Hence, it is considered to be a high performance and stable injector, and is being used profusely for gaseous fuel and liquid oxygen as in semicryogenic rocket engines. In this case, the oxidizer is surrounded by the fuel, which tends to shield the combustion process and thus combustion instability is less likely to occur. Besides this, this oxygen stream surrounded by fuel restricts heat transfer to the combustion wall from the combustion zone. Note that the performance of coaxial injector deteriorates drastically for two liquid streams or as the optimal momentum flux ratio required for better performance for atomization is not being obtained easily.

3. *Swirl injector*: In this case, liquid enters into the injection chamber through tangential entry relative to its axis, as shown in Figure 10.4a(C), and thus a hollow cone of liquid is formed with cone angle ranging from 40° to 100°. This liquid conical sheet will break into ligaments and subsequently into droplets. Note that with increase in swirl components, liquid sheet with large cone angle is formed, which results in uniform distribution of liquid droplets. This kind of atomizer is preferred for such nonhypergolic bipropellants that require rapid vaporization and adequate mixing of fuel and oxidizer in gas phase for successful combustion.

10.3.1.2 Impinging Injectors

In impinging type of injectors, two/three streams of propellant jets are impinge on each other to break bulk liquid jet/sheet into spray. The majority of rocket engines use impinging injectors because of their better performance, simplicity, and lower cost. Generally, impinging injectors are preferred for nonhypergolic propellant. But, it has been used for hypergolic propellants as well. As mentioned earlier, the impinging injectors are broadly divided into two categories: (1) unlike-impinging and (2) like-impinging injectors.

1. *Unlike-impinging injector*: In this case, two/three different liquid streams impinge on each other when they are issued from two/three angled orifices. Several types of unlike-impinging injectors have been developed for rocket engine applications. Some of them are (a) unlike-impinging doublets, (b) unlike-impinging triplets, which are discussed here:

   a. *Unlike-impinging doublets*: In the case of unlike doublet injector, two streams of fuel and oxidizer propellant jets impinge on each other, as shown in Figure 10.4b(A). The impact of two streams at impinging points produces a fan-shaped spray consisting of the mixture of two different liquids. During this process of atomization, waves are formed that help convert bulk liquid jet/sheet into ligaments and subsequently undergo fragmentation into smaller droplets. Impingement of liquids helps in enhancing the atomization process and fuel–oxidizer distribution in the combustion chamber. The processes of disintegration of two impinging jets are shown in Figure 10.5. It can be observed that spherical wave emanated from the impinging point propagates outwardly, which converts the jets into ligaments. The extent of liquid jet disintegration is dependent on jet diameter, momentum, injection pressure drop, chamber pressure, and angle of impingement. If two impinging streams are different, then it is known as unlike-impinging doublet (see Figure 10.4b(i)). As mentioned, the two impinging jets produce a two-dimensional fan-shaped spray in

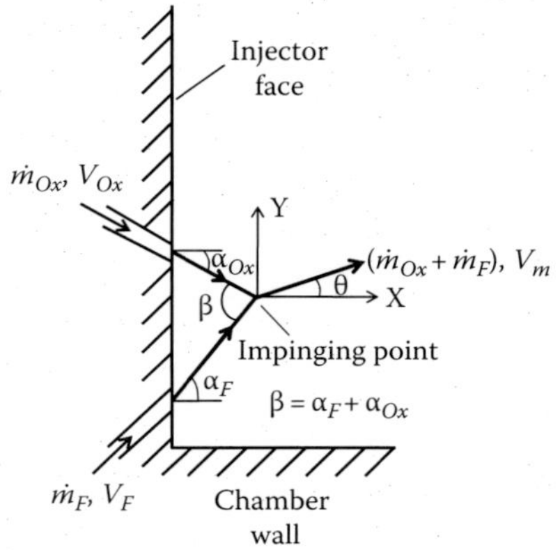

FIGURE 10.5 Simple model for doublet impinging injector element.

a plane consisting of a mixture of two different impinging liquids, provided no chemical reactions take place in liquid phase. The performance of impinging injectors in terms of droplet size and its distribution, mixing, and so on, can be affected by any of these parameters, namely, different momentum of liquid streams, stream-diameter mismatch of impinging liquids, and so on. The mis-impingement of two streams can distort the shape and size of the spray fan, leading to poor atomization and mixing. Besides this, when fast chemical reactions, as in the case of hypergolic propellants, are likely to occur simultaneously during atomization, it affects the performance of atomization and mixing as reactive-stream separation (blow apart) and reactive demixing are likely to occur. As hypergolic propellants have very short ignition delay, it produces copious amount of gases even before completion of hydrodynamic impact of two liquid streams, leading to separation of reacting surfaces. In the case of nonhypergolic propellant system, namely, liquid kerosene–liquid oxygen, similar phenomena are likely to occur particularly at high pressure, as observed experimentally. The occurrence of combustion during atomization of hypergolic propellants can modify the mixing and mass distribution of the injected spray. Besides this, due to hot gas cross-flow and higher turbulence near the injector face, radial winds can cause deformation in the spray pattern by stripping off the most rapidly atomizing portion of the injected propellants.

b. *Unlike-impinging triplets*: As discussed earlier, in doublet impinging injector, spray may be swayed away from the actual axis of desired impinging point due to mismatching of stream size and momentum between the fuel and oxidizer propellant streams. In this situation, the distorted spray fan results in poor atomization and mixing. This problem can be easily overcome by use of a symmetrical axial central stream of one propellant surrounded by two symmetrical impinging streams of other propellant streams, which is known as unlike triplet injector, as shown in Figure 10.4b(B). This kind of unlike triplet injector element can have either two fuel streams impinging on oxidizer stream (F-Ox-F) or the two oxidizer streams impinging on fuel stream (Ox-F-Ox). The latter is more likely to be preferred as total oxidizer area will be more as compared to fuel being operated at fuel lean conditions, which ensures better mixing. But it may lead to oxidizer-rich streaks, which must be avoided near the combustion wall. The advantage of unlike triplet injector is higher level of mixing resulting in higher combustion efficiency as compared to unlike doublet injector. However, this kind of injector is quite prone to the combustion instability problems. In the similar line, several possible combinations of unlike streams can be used to produce spray, although with increasing complexities. Although, properly designed and fabricated quadlets or pentads or hexa can produce higher level of mixing but result in poorer mass distribution. Hence, unlike-impinging injectors beyond pentads are not being used in practice. Note that this multiple-impinging injector is more prone to combustion instabilities. However, this kind of injector with more number of streams is preferred for high propellant flow rate applications.

2. *Like-impinging injectors*: In this case, two/three or more same liquid streams are impinged on each other when they are issued from their respective angled orifices. This kind of injectors is also known as self-impinging injectors. Several types of like-impinging injectors have been developed for rocket engine applications. Some of them, namely, (a) like-impinging doublet and (b) like-impinging triplets, are in the following:

   a. *Like-impinging doublets*: In this case, two streams of same propellant jets are impinging on each other, as shown in Figure 10.4b(C), leading to formation of spray. The impact of a fan-shaped spray consisting of droplets of the same liquid is similar to unlike-impinging doublets. During this process of atomization, waves are formed on the two-dimensional fans along the direction of resultant momentum vectors, leading to dissipation of energy, which helps convert bulk liquid jet/sheet into ligaments and subsequently undergo fragmentation into smaller droplets. But there would not be any mixing within this fan like in unlike-impinging doublets due to the presence of one type of liquid in both streams. The extent of mixing of two propellant streams is influenced by the orientation of initial fans for secondary impingements and overlapping of sprays. Note that this interaction between doublets spray dictates the performance of this kind of injectors. Like-impinging injectors are generally used for rocket engines that use nonhypergolic liquid–liquid propellants as it circumvents the problems of reactive demixing of unlike-impinging injectors. As a result, it can ensure higher combustion stability margin as compared to unlike-impinging doublets injector. Although this kind of like-impinging doublets injectors result in lower mixing level as compared to unlike-impinging doublets injectors, higher combustion efficiency can be easily obtained by adopting better design concept.

b. *Like-impinging triplets*: In order to avoid the problems of undesired shifting of impinging point due to mismatching of stream size and momentum between the two propellant streams like in unlike triplets, three streams of the same propellant can be allowed to impinge on a single point, as shown in Figure 10.4. Generally, like-impinging triplets injector results in a narrower spray fan leading to larger droplets as compared to equivalent doublets injector, which can effect an overall loss in its performance. Besides this, smaller size orifices must be used as more number of triplets injectors are to be accommodated in the same manifold surface area. Similar problems can be encountered in other multiple-streams self-impinging injectors, namely, quadlets, pentads, and so on.

10.3.1.3 Other Types of Injectors

Several other types of injectors have been tried during the development of rocket engines. Two of them, namely, (1) splash plate and (2) premixing injectors, shown in Figure 10.4, are discussed:

1. *Splash plate injector*: In this case, the principle of impinging along with a splash plate, as shown in Figure 10.4b(D), is used to promote breaking of liquid jets/sheets, leading to better mixing of propellants in liquid state itself. As liquid streams are impinged against splash plate and thus the problem of misalignment of impinging points of steams encountered in doublets can be avoided easily, this enhances

its performance over a wide range of operating conditions. Generally, these injection methods have been used for certain storable propellants successfully.

2. *Premixing injector*: In this case, both liquid fuel and oxidizer are mixed before being sprayed into the combustion chamber. The dimension of premixing chamber is influenced by time delay of the reactions and residence time of propellant streams in the premixing chamber. It is important to avoid the occurrence of explosion of premixing propellants inside its chamber during its operation. There is greater chance of flame traveling to premixing chamber from combustion chamber particularly leading to explosion, especially when rocket engines are being operated under high-pressure and high-mass-flux conditions. In some cases, swirls are imparted in liquid stream to avoid explosion. Due to these problems, it is very uncommon in rocket engine and however it can be also used in combustion chamber of other engines. However, it has been tried out for nonhypergolic propellants, although it causes excessive thermal loads on its structure as precombustion is likely to occur in premixing chamber. Hence, its use is preferred only to overcome certain special problems of injection.

10.4 DESIGN OF INJECTOR ELEMENTS

Recall that proper design of injector elements plays an important role in determining the performance of rocket engine design. Before embarking on the design of injector elements that are generally placed on the injector heads, it is important and essential to understand the flow characteristics of each individual injector element, because these flow characteristics dictate the desired injection pressure, mass flow rate, and mixture ratio in injector elements. We know that for a given thrust level, certain specific effective exhaust velocity is to be achieved corresponding to a certain mass flow rate of propellants, which can be obtained by $\dot{m} = F/V_e$. The mixture ratio MR is defined as the ratio of the oxidizer flow rate $\dot{m}_{Ox}$ and fuel flow rate $\dot{m}_F$ and can be expressed in terms of total mass flow rate of propellant as

$$\dot{m}_{Ox} = \frac{MR}{1+MR}\dot{m}; \quad \dot{m}_F = \frac{1}{1+MR}\dot{m} \tag{10.4}$$

By applying Bernoulli equation, for incompressible flow through circular holed orifices, we can express velocity, volumetric and mass flow rate of liquid propellant in terms of injection pressure drop as

$$V = C_d\sqrt{\frac{2\Delta P}{\rho}}; \quad \dot{Q} = C_d A_i\sqrt{\frac{2\Delta P}{\rho}}; \quad \dot{m} = C_d A_i\sqrt{2\rho\Delta P} \tag{10.5}$$

where
- V is the propellant velocity through orifice
- $\dot{Q}$ and $\dot{m}$ are the volumetric and mass flow rate of propellants, respectively
- C_d is the discharge coefficient
- A_i is the injection area
- ρ is the density of liquid propellant
- ΔP is the pressure drop across the injector

Note that the above relationships can be applied not only to the injector but also to the propellant feed system and so on. Generally, the value of this discharge coefficient C_d can be determined experimentally. Note that it does not remain constant over a wide range of operating conditions. Rather, it varies with physical properties of propellant, geometry of injector, pressure drop across the injector ΔP, and chamber pressure P_c. We know that a cylindrical orifice can be described geometrically by length of cylindrical part l and diameter of orifice d, which can be expressed well in terms of l/d. The performance of injector made of round orifice depends not only on the l/d ratio, but also on the nature of orifice entry [3–5]. The entry of the orifice can be sharp-edged, chamfered or smooth-rounded, sharp-edged cone, with swirler. However, sharp-edged orifices are preferred over others as they have better flow distribution due to well-defined vena contracta, particularly for $l/d > 5$. It has been reported that the discharge coefficient C_d, varies from 0.6 to 0.75 when l/d is between 0.5 and 1. By using Equation 10.5 we can determine easily the injection area and orifice diameter, as given in the following:

$$A_i = \frac{\dot{m}}{C_d\sqrt{2\rho\Delta P}}; \quad d_i = \sqrt{\frac{4}{\pi}\frac{\dot{m}}{NC_d\sqrt{2\rho\Delta P}}} \tag{10.6}$$

where N is the number of orifices used in an injection head. The total injection area is quite easy to compute provided one can choose the pressure drop across the injector and discharge coefficient for a particular mass flow rate. For estimation of total injection area using the Equation 10.6, we need to choose pressure drop across each injector. A good starting point would be to consider 20% of the chamber pressure P_c, which is considered to be a thumb rule for injector design for rocket engine. For example, if the chamber pressure happens to be 10 MPa, then pressure drop across the injector would be around 2 MPa, which must be provided for the injection pressure resistance to have a better spray. This differential pressure drop across injector orifice causes the propellant to flow with a higher velocity and thus provides requisite kinetic energy for atomization, leading to controlled mass distribution and better mixing between fuel and oxidizer propellants in the combustion chamber. Besides this, the injection pressure drop provides isolation between chamber pressure oscillation and propellant flow rate. If such coupling problems of chamber pressure oscillation and propellant flow rate happen to be rampant, injection pressure drop can be much higher than 20% of chamber pressure.

The orifice-sizing process is quite complex as it is dependent on the mass flow rate of propellant and its properties, pressure drop across injector, discharge coefficient, ease of it manufacturing, and so on. For best quality spray, and maximum effectiveness, it will be prudent to ideally use more number of orifices and thus to employ small-diameter orifices as possible from the fabrication and operational point of view. However, it is quite difficult and costly to fabricate tiny-holed orifices, which are equally difficult to maintain as they are more prone to blockages due to plugging of contaminations like soot particles, during its operation. Besides this, fine spray produced using tiny-holed orifices are more prone to combustion instabilities due to higher level of premixing mixture in the combustion chamber. Due to this, large-size-holed orifices are being employed routinely for larger rocket engines with higher level of combustion oscillations. In contrast, smaller diameter holes for injectors are being used in smaller rocket engine, which requires compact combustion chamber. The size of injectors can have a wider range from 0.1 mm for a 4.4 N thrust experimental engine to 19 mm for pressure-fed booster applications. Recall, the discharge coefficient is dependent on the type of orifices chosen for a particular design. However, the practical range of injector diameter spans 0.34 mm for F-1 booster rocket engine in the Saturn engine. In practical rocket engine injectors, orifices with the following characteristics are preferred [3–6]:

1. Orifice hole diameter 0.5–2.5 mm, depending on the size of rocket engine. The geometric ratio of orifice l/d lies between 2 and 4.
2. A chamfer is used on the upstream side to avoid flow separation.
3. Pressure drop across the injector ΔP, varies between 0.4 and 1 MPa (usually 20% of chamber pressure).
4. Injection velocity varies between 30 and 45 m/s for storable propellant and 100–120 m/s for cryogenic propellants.

Let us consider a fuel and an oxidizer jet impinging on each other, as shown in Figure 10.7, subsequently resulting in a single stream with average velocity while making an angle of θ with respect to the chamber axis. Note that fuel and oxidizer streams with the velocity of V_F and V_{Ox}, respectively, make an angle α_F and α_{Ox}, with respect to the chamber axis. By carrying out momentum balance before impingement and after impingement along both x-axis and y-axis, we can have

$$\text{x}-\text{momentum: } \dot{m}_{Ox} V_{Ox} \cos\alpha_{Ox} + \dot{m}_F V_F \cos\alpha_F = \left(\dot{m}_{Ox} + \dot{m}_F\right) V_m \cos\theta \tag{10.7}$$

$$\text{y}-\text{momentum: } \dot{m}_{Ox} V_{Ox} \sin\alpha_{Ox} + \dot{m}_F V_F \sin\alpha_F = \left(\dot{m}_{Ox} + \dot{m}_F\right) V_m \sin\theta \tag{10.8}$$

By using Equations 10.7 and 10.8, we can get,

$$\tan\theta = \frac{\dot{m}_{Ox} V_{Ox} \sin\alpha_{Ox} - \dot{m}_F V_F \sin\alpha_F}{\dot{m}_{Ox} V_{Ox} \cos\alpha_{Ox} + \dot{m}_F V_F \cos\alpha_F} \tag{10.9}$$

It can be noted that better performance of injector can be achieved only when the resultant momentum of the impinging stream occurs almost along axial direction. For this condition, tan θ happens to be almost zero, and Equation 10.9 becomes

$$\dot{m}_{Ox} V_{Ox} \sin\alpha_{Ox} = \dot{m}_F V_F \sin\alpha_F \tag{10.10}$$

In most cases for better atomization, the angle between resultant impinged jets with the chamber axis, θ can be made equal to zero. By using Equations 10.10 and 10.5 we can have,

$$\frac{\dot{m}_F V_F}{\dot{m}_{Ox} V_{Ox}} \sin\alpha_F = \frac{\dot{m}_F C_{d,F}}{\dot{m}_{Ox} C_{d,Ox}} \sqrt{\frac{\rho_{Ox}\Delta P_F}{\rho_F \Delta P_{Ox}}} \sin\alpha_F = \sin\alpha_{Ox} == \sin(\beta - \alpha_F) \tag{10.11}$$

By simplifying Equation 10.11, we can get,

$$\alpha_F = \tan^{-1}\left(\frac{\sin\beta}{\frac{1}{MR}\sqrt{\frac{\rho_{Ox}\Delta P_F}{\rho_F\Delta P_{Ox}}} + \cos\beta}\right); \quad \text{where } \beta = \alpha_F + \alpha_{Ox} \tag{10.12}$$

where

ρ_F and ρ_{Ox}, are the density of fuel and oxidizer streams, respectively
ΔP_F and ΔP_{Ox} are the pressure drop across fuel and oxidizer orifices, respectively
$C_{d,F}$ and $C_{d,Ox}$ are the discharge coefficient of fuel and oxidizer orifices, respectively
MR is the mixture ratio

For better atomization, included angle of two impinging jets, β is chosen as 90°, the fuel stream angle becomes

$$\alpha_F = \tan^{-1}\left(\frac{1}{\frac{1}{MR}\sqrt{\frac{\rho_{Ox}\Delta P_F}{\rho_F\Delta P_{Ox}}}}\right) \tag{10.13}$$

Note that this is an approximate relationship because generally the cross-sectional area of the oxidizer stream is greater than that of the fuel stream. However, it can be used for preliminary design calculation. For using this expression, we need to choose pressure drop across each injector and discharge coefficients for both fuel and oxidizer. Recall that we will have to use 20% of the chamber pressure P_c, which is considered to be a thumb rule for injector design for rocket engine.

Example 10.1

A liquid-propellant rocket engine is used to develop a thrust of 1.2 kN with characteristic velocity of 2050 m/s at chamber pressure of

6.8 MPa and MR of 1.6. If its thrust coefficient C_F happens to be 1.65, determine (1) throat area of nozzle, (2) mass flow rate of oxidizer and fuel. A doublet impinging injection system with 10 injector elements is used with injection pressure drop of 1.5 MPa and discharge coefficient of 0.76. By considering ρ_F and ρ_{Ox} to be 860 and 1100 kg/m^3, determine diameter of injection holes.

Solution

The throat area can be determined as

$$A_t = \frac{F}{C_F P_c} = \frac{1.2\times 10^3}{1.65\times 5.5\times 10^6} = 0.13\times 10^{-3}\ \text{m}^2$$

The specific impulse of this engine is estimated as

$$I_{sp} = \frac{C_F C^*}{g} = \frac{2050\times 1.65}{9.81} = 344.8\ \text{s}$$

The mass flow rate of propellant is estimated as

$$\dot{m}_p = \frac{F}{I_{sp}} = \frac{1.2\times 10^3}{344.8} = 3.48\ \text{kg/s}$$

By using Equation 10.4, we can determine mass flow rate oxidizer and fuel, as follows:

$$\dot{m}_{Ox} = \frac{MR}{1+MR}\dot{m} = \frac{1.6}{1+1.6}3.48 = 2.14\,\text{kg/s};\quad \dot{m}_F = \frac{\dot{m}}{1+MR} = 1.34\ \text{kg/s}$$

The diameter of oxidizer and fuel orifices can be estimated by using Equation 10.6, as follows:

$$d_{Ox} = \sqrt{\frac{4}{\pi}\frac{\dot{m}_{Ox}}{NC_d\sqrt{2\rho_{Ox}\Delta P}}} = \sqrt{\frac{4}{3.14}\frac{2.14}{10\times 0.76\sqrt{2\times 1100\times 1.5\times 10^6}}}$$
$$= 0.0025\ \text{m}$$

$$d_f = \sqrt{\frac{4}{\pi}\frac{\dot{m}_f}{NC_d\sqrt{2\rho_f\Delta P}}} = \sqrt{\frac{4}{3.14}\frac{1.34}{10\times 0.76\sqrt{2\times 860\times 1.5\times 10^6}}}$$
$$= 0.0021\ \text{m}$$

10.5 PERFORMANCE OF INJECTOR

10.5.1 Droplet Size Distribution

Most of the injectors used in rocket engine applications can generate droplets in size from 5 to 300 μm. In order to describe drop size distribution, one may resort to the statistical means as the spray contains large number of individual droplets [2]. Unfortunately, it has not been possible to derive such kind of function on theoretical basis. However, in engineering practice, it is essential to define the average droplet diameter such that it can represent the overall characteristics of the entire spray comprising several droplet diameters. Such kind of droplet diameter can be used in design calculation of combustor. For example, while estimating evaporation rate, ignition time, minimum ignition energy, blowoff limits, and so on, such kind of droplet diameter representing the overall characterization of the spray will be very useful. This also helps in judging the quality of spray. Let us now learn how to estimate average size of droplet in a typical spray by considering an example as given below:

Example 10.2

In a spray experiment, the droplet number and size distribution are obtained as given in the following table:

Sl. No.	Droplet Size (μm)	Number
1.	0–20	52
2.	21–30	175
3.	31–40	275
4.	41–60	185
5.	61–80	95
6.	81–110	45
7.	111–140	15
8.	141–210	5

Estimate mean diameter (MD), area mean diameter (AMD), and Sauter mean diameter (SMD) of this spray.

Solution

The various quantities from the spray data required for determination of average droplet diameter are given in tabular form as follows:

Sl. No.	d_i (μm)	ΔN_i	$d_i \Delta N_i$	$d_i^2 \Delta N_i$	$d_i^3 \Delta N_i$	CVF (%)
1.	10	0.061	0.613932	6.139315	61.39315	0.030908
2.	25.5	0.207	5.268595	134.3492	3425.904	1.755648
3.	35.5	0.325	11.52597	409.1721	14525.61	9.068431
4.	55.5	0.218	11.03011	557.0204	28129.53	23.22998
5.	70.5	0.112	7.90732	557.4661	39301.36	43.01589
6.	95.5	0.053	5.07379	484.5469	46274.23	66.31222
7.	125.5	0.018	2.22255	2710.93	35005.72	83.93553
8.	175.5	.0059	1.036009	181.8197	31909.35	100
Total =		1.0	44.67828	2609.444	198633.1	

where ΔN_i is the number fraction, which is defined as the ratio of number of droplets divided by total number of droplets. In the table, d_i represents average droplet size in the basket.

The mean diameter (*MD*) is determined as follows:

$$\text{Mean diameter} = MD = \sum d_i \Delta N_i = 44.7 \ \mu\text{m}$$

The area mean diameter (*AMD*) of this spray is determined as

$$\text{Area mean diameter} = AMD = \sqrt{\sum d_i^2 \Delta N_i} = 51.1 \ \mu\text{m}$$

The Sauter mean diameter (*SMD*) of this spray is determined as

$$\text{Sauter mean diameter} = SMD = \frac{\sum d_i^3 \Delta N_i}{\sum d_i^2 \Delta N_i} = 76.1 \ \mu\text{m}$$

From this example, we have learnt that several average droplet sizes can be defined and used to characterize the overall feature of a spray in a simple way. However, SMD is being used extensively in combustion problem as it represents the ratio of volume to surface area of the entire spray, because, the requisite heat for evaporation of the droplet of spray is provided through the surface of droplet, whereas its volume determines the liquid mass to be evaporated.

We know that drop size distribution is dependent on the type of liquid, pressure drop across the injector, chamber pressure, and diameter of the orifice. These parameters affect the drop size and its distribution nonlinearly. As mentioned earlier, the average drop size of spray, SMD, is generally related to two nondimensional numbers, namely, $Re_i = \rho V_i d_i/\mu$ and Weber number, $We_i = \rho V_i^2 d_i/\sigma$, where d_i and V_i are diameter of and velocity at exit of injector's orifice, respectively, and ρ, μ, and σ are density, viscosity, and surface tension of the liquid. The generic relationship for SMD can be expressed in terms of Re_i and We_i, as follows:

$$\frac{SMD}{d_i} = C Re_i^m We_i^n \tag{10.14}$$

where

- C can be an empirical constant
- m and n are indices that can be obtained mostly from experimental data

The Reynolds and Weber numbers for typical spray in rocket engine injectors vary in the range of 10^5–10^6. A decrease in orifice diameter results in decrease in SMD for same condition. The average drop size of spray from a particular injector varies along the distance from its orifice. For example, from an orifice of diameter of 0.8 mm, at an injection pressure of 5 MPa and chamber pressure of 0.4 MPa the SMD happens to be 70 μm at a distance of 100 mm from its orifices, which decreases further with increase in distance from its orifice due to collision with adjacent droplets. Of course, with increase in chamber pressure, lower SMD, as evident from Equation 10.14, can be obtained due to higher aerodynamic resistance imparted by the ambience.

10.5.2 Mass Distribution

It is quite important and essential to characterize the spray in terms of its mass distribution as the fuel and oxidizer mass distribution in the combustion chamber dictate its performance of the rocket engine. A simple apparatus known as rake, consisting of radially distributed tube, can be used to collect the intercepted liquid mass from the spray at certain axial distance from the exit of orifice, which can be measured using precision weight balance or any other means. By using these data, we can determine easily (1) the mass flux distribution of liquid and (2) the mixture ratio MR across a particular section. But it affects the accuracy of results as this instrument is intrusive in nature. In recent times, nonintrusive-type instrument based on optical method using LASER and CCD camera have been devised to have mass flux distribution. But it has to be calibrated by the mechanical-type mass distribution instrument.

10.5.3 Quality Factor

The mentioned performance parameters of the spray can be obtained during cold flow conditions but cannot be directly applied during firing of rocket engine. In order to overcome this problem and compare different types of injectors, a quality factor QF_b during burning, in terms of the experimental characteristic velocity and theoretical characteristic velocity of a rocket engine, has been used in practice, which can be defined as [5]

$$QF_b = \frac{\left(C^*\right)_{Exp}}{\left(C^*\right)_{Theo}} \tag{10.15}$$

This quality factor, QF_b, is dependent on the geometry of injector, combustion chamber, and spacing distance of orifices. The quality factor QF_b decreases with orifice diameter for a particular chamber length. With decrease in orifice diameter, the characteristic velocity gets enhanced, although with increase in difficulties in manufacturing as smaller diameter calls for a sophisticated manufacturing technique with enhanced cost. Besides this, smaller diameter orifices are more prone to blockage problem during operation. Note that the quality factor QF_b increases with chamber length for particular orifice diameter. Furthermore, it also decreases with orifice spacing distance δ, which, of course, can lead to difficulties in manufacturing and enhance cost. Besides these aspects, the precision of manufacturing of orifices influence of the quality factor, QF_b.

Note that this quality factor, QF_b, can represent characteristics of spray during combustion in a rocket engine but cannot take care of heat transfer, particularly through the wall, which is prevalent during the combustion process. Of course, the combustion instability is dependent to some extent on the geometry of injection system.

10.6 INJECTOR DISTRIBUTOR

Generally, multiple injectors are employed in rocket engines, particularly from medium to higher thrust levels. Of course, in very small thrust level rocket engines, single injector is preferred. A series of individual injector elements for fuel and oxidizer streams are housed in an injector head, as shown in Figure 10.6, also known as injector distributor that is located in the front portion of the combustion chamber. The shape of this injector distributor and the arrangement of injector elements on it are quite important as they influence significantly the performance of injection system. The injector distributor/head must fulfill the following requirements: (1) uniform combustion process, (2) to protect the chamber wall from excessive heating due to the presence of intense heat release during combustion, (3) to guarantee better exchange of heat between burnt gases and propellant, and (4) to avoid combustion instability. These desired effects can be achieved in the following ways: (1) ensuring better mixing of fuel and oxidizer in the central portion of combustion chamber, (2) creating a fuel-rich

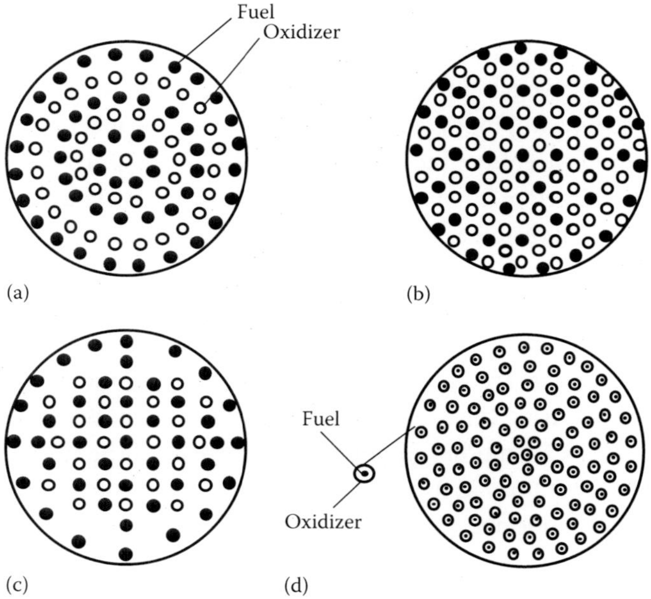

FIGURE 10.6 Four types of arrangements of injector holes in injector head: (a) concentric, (b) honeycomb, (c) alternate, and (d) co-axial.

zone near the wall region with the help of proper positions of oxidizer–fuel orifices, (3) confining the intense burning of propellants in center of combustion chamber, and (4) maintaining heterogeneous mixture at the exit of combustion such that high-frequency axial pressure oscillation along axial direction can be damped out easily [5,6].

Depending on the applications and other requirements, several types of injection distributors have been devised for rocket engines. There are several variations in the arrangements of injection holes on the injector distributor. Four of them, namely, (1) concentric, (2) honeycomb, (3) alternate, and (4) coaxial patterns, as shown in Figure 10.8, will be discussed, which are mainly used for high-thrust rocket engines [5,6]. The concentric pattern can ensure better mixing of propellant and oxidizer as compared to alternate pattern but combustion can take place even till its wall. But it is easy to manufacture. The honeycomb arrangement can provide better mixing and combustion as compared to both concentric and alternative arrangements. But the alternative arrangement can fulfill all the requirements of an injection system, although it is relatively difficult to manufacture as compared to the other three injector arrangements. It can be noted that uniform combustion can be confined to central portion of chamber, while fuel-rich zone can be maintained near the wall region. But it requires larger distance to achieve complete combustion. Figure 10.8d depicts a separate arrangement in which coaxial injectors are being used for obtaining higher thrust level, particularly for cryogenic engines. In this arrangement, hydrogen fuel in gas or two phase flow impinges radially into liquid oxygen sheet, which helps in breaking liquid sheet into fine droplets forming a nice spray. On the periphery rows, fuel injector elements are placed alone to create a fuel-rich mixture near the wall region to avoid excessive heating due to high heat release.

10.7 INJECTOR MANIFOLD

As discussed in Section 10.6, selection of injection elements and injector distribution pattern are connected to each other and have to be decided upon carefully during the designing of injection system. Besides this, injector manifold is also another important component of the injection system whose main functions are to act as a settling chamber for propellant and to supply to interconnected injector orifices that ensure formation of desired pattern of propellant spray in the combustion chamber. In order to provide uniform flow through each orifice, it is essential to design a free-flowing manifold system with minimum possible flow restriction/resistance such that there is uniform distribution of propellant mass and mixture ratio MR across the injector face. In order to achieve this uniform distribution of mass, and providing enough space for intertwined flow passages for two different propellants, we need to provide large manifold volume, which may not be feasible for the designer to provide due to design constraints, particularly for rocket engines. Even if more manifold volume can be provided, residence time of the propellant in this injector manifold could be increased, thus enhancing the priming time after the opening of inlet valve during which propellants will flow into combustion chamber in dribbling form rather than spray form. Hence, combustion will continue at a much reduced rate in the combustion chamber without raising its chamber pressure and thus it cannot achieve its desired thrust level, leading to loss in the efficiency of the rocket engine. As of now, there is no sound theoretical basis on which design of injector manifold volume can be carried out. However, in order to overcome this problem, there is a thumb rule that has been used for successful design of several rocket engines. As per this rule, the cross-sectional area of the manifold must be greater than or equal to four times the total area of all orifices placed on the injection head. In this case, the details of orifice area must be known beforehand. In order to overcome this problem, another alternate rule known as "1% rule" is being devised. It states that velocity head of the flow through the orifices must not exceed 1% of the local system pressure. In other words, the velocity head across the manifold must not exceed the 1% of the total pressure of the feed line. For example, if feed line operates at 20 MPa, then pressure drop across the manifold must not exceed 0.2 MPa. In certain cases, where higher response of feed system is called for, manifold area to total area ratio can be even reduced to 2, by which short pulse efficiency can be enhanced, although, with enhanced pressure losses.

10.8 LIQUID-PROPELLANT FEED SYSTEM

The main function of the propellant feed system is to supply requisite amount of propellant by transferring it from propellant tank to the thrust chamber at higher desired pressure by which spray of liquid propellant can be formed. For this purpose, the pressure of propellant in feed system must be raised, which can be accomplished by supplying energy. Based on this, the propellant feed systems that have been devised can be classified broadly into two categories: (1) gas pressure feed system, (2) turbo-pump system, as shown in Figure 10.7. The gas feed pressure system can be further classified based on the nature of the gases introduced into the propellant tanks: (1) cold gas pressure system, (2) hot gas pressure system, and (3) chemically generated gas feed system [3,5,6]. The chemically generated gas feed system

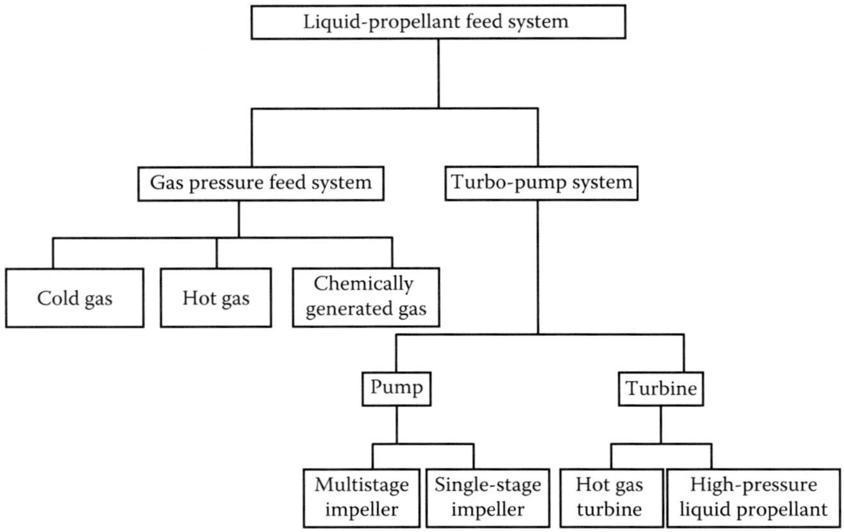

FIGURE 10.7 Types of liquid-propellant feed system.

can be further divided into two categories: (1) solid-propellant gas generator and (2) liquid-propellant gas generator. Besides this, based on the method of pressurizing gas in the tank, the pressure feed system can be divided into three independent categories: (1) direct gas pressure system, (2) flexible bag pressure system, and (3) piston pressure system. Based on the type of gas power supply, the pump feed system can be classified into three categories: (1) gas generator, (2) expander, and (3) staged combustion cycle. The separate gas generator system can be further divided into three categories: (1) solid-propellant gas generator, (2) liquid monopropellant gas generator, and (3) liquid bipropellant gas generator. We will be discussing some of these feed systems in detail. Besides these, there might be several feed systems that might have been designed and developed by several engineers across the globe. It is quite cumbersome and challenging to choose an appropriate feed system for a particular rocket engine, because the proper selection of feed system is dependent on several competitive factors, namely, vehicle's acceleration, maneuvers, weight, the thrust level and its duration, vibration levels, the requisite envelope, type of propellant, level of reliability, and cost.

10.8.1 Gas Pressure Feed System

This gas pressure feed system is one of the simplest methods of pressurizing the propellant in rocket engine in which high-pressure gas is being used to force the liquid propellants in a very controlled manner from their respective tanks. Generally, gas pressure feed system is preferred over other systems due to its simplicity and reliability for lower thrust level rocket engines with smaller quantities of propellants, as weight penalty for high-pressure tank should not outweigh the complexities of turbo-pump feed system. In recent times, with the advent of new lighter materials with higher strength for high-pressure propellant tanks, upper limit of gas pressure feed system has been enhanced considerably. Besides this, on certain occasions, low level of pressurizing the propellant tank is being used even in turbo-pump feed system to minimize pump requirements.

Let us consider a simple pressure feed system, as shown in Figure 10.8, which consists of a high-pressure gas tank, an on–off valve, a pressure regulator, propellant tanks, feed lines, and so on. Besides these components, provisions for filling and draining propellant tanks are made in this feed system with inclusion of check valves and vent valves, filters, restrictors, propellant control valves, pressure sensors, flexible elastic separators, and so on. Generally, propellant tanks are filled in the beginning followed by high-pressure gas tank. Subsequently, high-pressure gas valve is actuated to allow high-pressure gas to enter into the propellant tank in regulated manner at constant pressure through check valves as shown in Figure 10.8. It can be noted that the check valves are used to prevent mixing of fuel with oxidizer due to back flow of propellants. Once desired pressure is established in the propellant tanks, the propellants can be fed through injectors into the combustion chamber by actuating the propellant valves. Commonly, the pressurized gas is allowed to pass through, even after complete consumption of propellant, to scavenge and clean the feed lines particularly for reusable rocket engines, namely, space-maneuver rockets. For the reusable rocket systems for manned missions, certain other features, namely, thrust regulating device, tank level gauge, sniff devices to detect various hazardous vapor, isolation valves, command signal overrides system, fault detection and control system are being incorporated routinely.

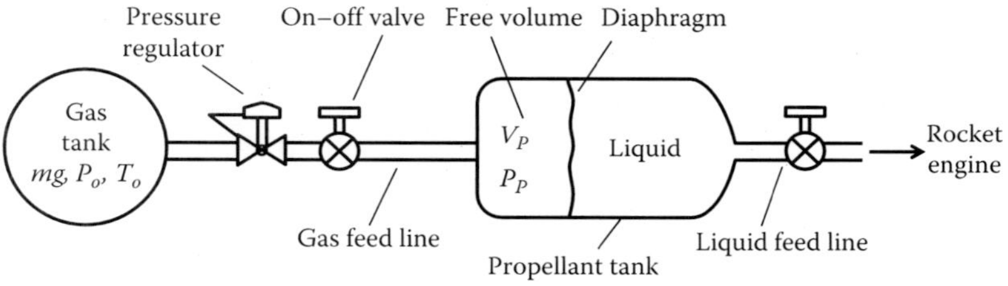

FIGURE 10.8 Schematic of a simple pressure feed system.

We can easily derive pressure balance equation for this simple pressure feed system, as shown in Figure 10.8. For this purpose, let us consider the pressure balance for fuel feed line only. The gas pressure in the pressurized gas tank P_g can be expressed in the chamber pressure P_c, pressure drop across gas line due to piping, gas pressure regulator ΔP_g, pressure drop in fuel feed line due to piping, valves, pressure regulator, injector, cooling jacket, and so on, ΔP_c, as

$$P_g = P_c + \Delta P_g + \Delta P_F \tag{10.16}$$

Similarly, we can have expression for gas pressure in the pressurized gas tank required for oxidizer feed system. It can be noted that this pressure balance relationship is not a linear equation in nature, as pressure drop across the feed lines is proportional to square of flow velocity, while chamber pressure is proportional to the total flow rate. The pressure drop across the fuel and oxidizer feed lines will be such that they can supply requisite flow rate to have intended mixture ratio in the thrust chamber. This is generally achieved by using special orifice for regulating the pressure drop across the feed line.

As mentioned earlier, the stored gas pressurizing system is used extensively due to its higher level of reliability, in which gas is generally stored in tank at initial high pressure as high as 60 MPa, although, this gas is supplied at regulated pressure to the propellant tank through pressure regulator, valves, and so on. In early systems like the V2 rocket engines, nitrogen/air was used for logistic reason. But in recent times, helium gas is being routinely used as pressurizing gas due to its ready availability, and lower molecular weight, leading to reduced total weight of pressurized gas. Generally, some of the important design requirements for gas pressure feed system are low molecular weight of the gas, high gas density at storage condition, minimum residual gas weight, and high allowable stress-to-density ratio of the propellant tank materials.

Recall that based on the nature of the gases introduced into the propellant tanks, the gas feed pressure system can be divided into three types: (1) cold gas pressure, (2) hot gas pressure, and (3) chemically generated gas feed systems [3,6], which are discussed in detail in the following.

10.8.1.1 Cold Gas Pressure Feed System

The schematic of a typical cold helium gas pressure feed system is shown in Figure 10.9a, which consists of high-pressure storage tank for helium

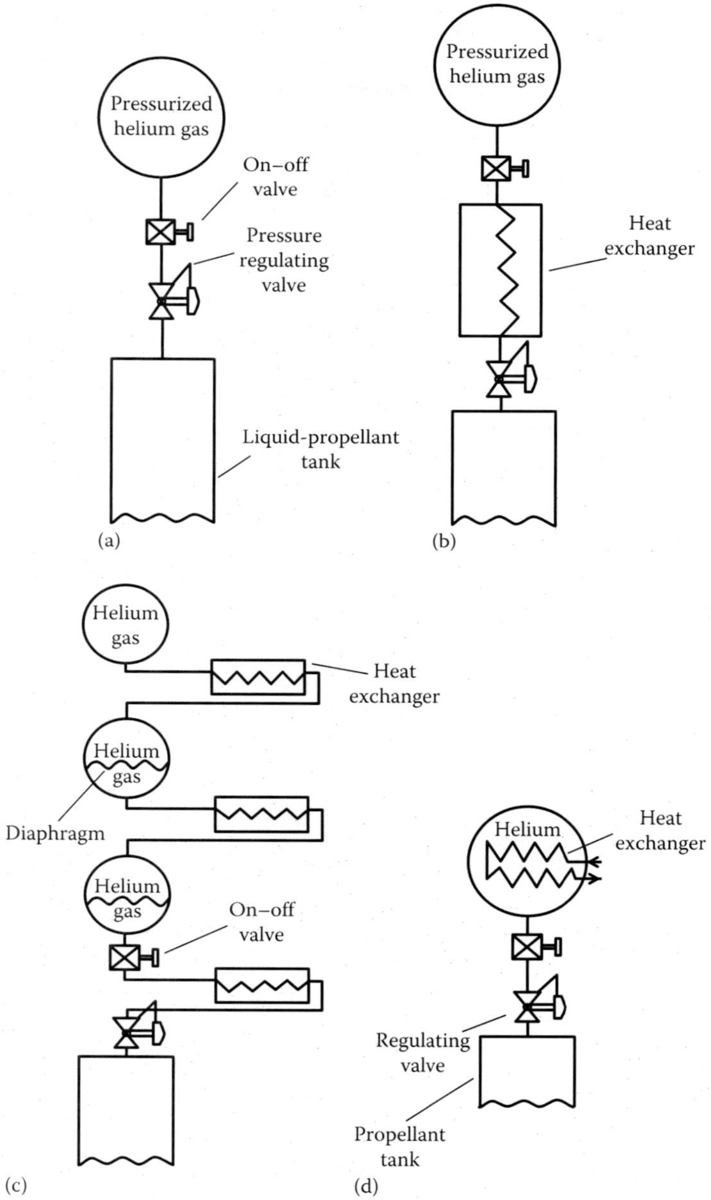

FIGURE 10.9 Schematic of gas pressure feed system: (a) cold helium, (b) hot helium with heat exchanger, (c) helium cascade system, and (d) helium hot gas using internal heater.

gas, on–off valve, and a pressure regulator. This is being preferred in lower thrust level of rocket engine in deep space due to its great simplicity. But its weight seems to be higher than other hot gas pressure feed systems due to its lower temperature and specific volume of the gas. In order to overcome this problem, several hot gas pressure feed systems have been devised, which are discussed in the following.

Determination of Mass of Pressuring Gas: Recall that pressurizing gas is used to eject the propellant through the injector from its tank for high-intensity combustion to take place in combustion chamber of engine. We need to determine the mass of pressuring gas in the gas bottle by invoking the energy conservation principle. Let us consider a simplified gas pressure feed system, as shown in Figure 10.8, in which a single propellant at pressure P_p and V_p is to be expelled completely by high-pressure gas at initial pressure P_I and temperature T_I. Keep in mind that pressurizing tank pressure and temperature will be dropping down finally to P_R and T_R with residual mass at m_R. Note that pressure regulator is meant to maintain propellant tank pressure at P_p. For this simplified analysis, the gas is assumed to follow ideal gas law. Amount of mass in the pipe line initially is assumed to be negligible. By considering expansion process to be adiabatic and invoking first law of thermodynamics, we can have

$$-dU = dW \tag{10.17}$$

We know that initial internal energy of pressurizing gas is equal to $m_I C_V T_I$. At the end of propellant expulsion, final internal energy is the sum of internal energy of gas in the pressurizing gas tank $m_R C_V T_R$ and internal energy of gas in the propellant tank $m_p C_V T_p$. The work done by the gas in the pressurizing gas tank in expelling the propellant is found to be $P_p V_p$. Then Equation 10.17 becomes

$$-\left[\left(m_R C_V T_R + m_p C_V T_p\right) - m_I C_V T_I\right] = P_p V_p \tag{10.18}$$

By invoking ideal gas law, we can express the mass of gas in propellant tank and pressurizing tank as

$$m_p = \frac{P_p V_p}{R T_p}; \quad m_R = \frac{P_R V_I}{R T_R} \tag{10.19}$$

By using Equations 10.18 and 10.19, we can have

$$m_I = \frac{P_p V_p}{C_V T_I} + \frac{P_p V_p}{RT_I} + \frac{P_R V_I}{RT_I} = (\gamma - 1)\frac{P_p V_p}{RT_I} + \frac{P_p V_p}{RT_I} + \frac{P_R V_I}{RT_I} \tag{10.20}$$

By the ideal gas law, initial volume becomes $V_I = m_I \frac{RT_I}{P_I}$. By using this in Equation 10.20, we can have

$$m_I = \frac{P_p V_p}{RT_I}\left[\frac{\gamma}{1-(P_R / P_I)}\right] \tag{10.21}$$

Note that the first terms, $P_p V_p / RT_I$, represents the mass of gas required to expel propellant using isothermal expansion of gas in pressurizing tank. The second term in the bracket represents pressure ratio through which gas undergoes adiabatic expansion during propellant expulsion. However, the actual process is nonadiabatic in nature as finite heat transfer does take place. It can be noted from Equation 10.21 that mass of high-pressure gas can be reduced by using low-molecular-weight gas. Hence, helium at high pressure of around 25 MPa is being preferred as a presiding gas. Although, hydrogen can be preferred but it is not being used in practice as it is reactive in nature. Besides this, high-temperature T_I is used to reduce mass of high-pressure gas. Hence, hot gas pressurization system is being preferred, as discussed in the following.

10.8.1.2 Hot Gas Pressure Feed System

Some of the hot gas pressure feed systems are discussed briefly in this section. The schematic of hot gas pressure feed system is shown in Figure 10.9b through d. It consists of a high-pressure helium tank, an on–off valve, a pressure regulator, and thrust chamber heat exchangers. These heat exchangers are used to extract heat from the divergent portion of the exhaust nozzle. As a result, there will be an increase in volume of the gas, leading to reduction in mass of gas required for tank pressurization. But at the end of expulsion, a considerable amount of cold and high-density helium gas remains, which incurs considerable the weight penalty due to higher amount of helium gas mass for particular mission requirement. In order to reduce this helium gas requirement further, a helium gas cascade system is being devised whose schematic is shown in Figure 10.9c. This typical system consists of helium storage tanks of equal pressure but different sizes, three heat exchangers, an on–off valve, and a pressure regulator. The second and third helium cylinders are separated internally into two compartments with the help of a flexible diaphragm. In this cascade system, the cold helium gas from the first and smallest tank is allowed to flow through a heat exchanger, which is used to push the helium gas in the middle tank. In similar manner, helium from this tank flows through the second heat exchanger and pushes the helium gas into the last pressurized tank, which eventually passes through the third heat exchanger and pressurizes the propellant into the thrust chamber. It is interesting to note that at the end of this operation, smallest storage tank contains small amount of low-density helium gas. Although mass of helium gas can be reduced considerably over the previous hot gas feed system, it is disadvantageous to adopt this system due to its higher weight and complexities. In order to overcome this problem, helium gas storage tank with internal heat exchanger can be used whose schematic is shown in Figure 10.9d. It includes a high-pressure helium gas tank containing internal heat exchanger, an on–off valve, and a pressure regulator. As the heat exchangers are placed inside the tank itself, the efficiency of the heat exchange becomes higher. Thus, it can provide easily high-temperature helium gas to propellant tank, while ensuring warm residual gas at the end of its operation. However, its size gets enhanced considerably with an increase in complexities and poses control problems during its operation. But the overall weight of the entire system may be lower than both cascade and external heating pressure feed system, depending on the type of design being adopted for any system.

10.8.1.3 Chemically Generated Gas Feed System

In this system, high-pressure hot gas is generated chemically to pressurize the storable liquid propellant by burning either solid or liquid propellants or catalytically reacted gases. There is another method particularly for noncryogenic liquid propellant, in which hypergolic propellant is directly injected into the propellant tank leading to liquid-phase exothermic chemical reactions, whose products pressurize the propellant into the thrust chamber. Note that these methods cannot be used for cryogenic propellant system as water of the combustion product would solidify into ice due to lower temperature prevailing in propellant tank. To adopt this method for feeding propellant, one has to take care of gas temperature and propellant compatibility with product gas of chemically generated gas feed system. As mentioned earlier, based on type of physical state of propellant, this chemically generated gas feed system can be divided into two categories: (1) solid-propellant gas generator and (2) liquid-propellant gas generator, which are discussed briefly.

10.8.1.3.1 Solid-Propellant Gas Generator Several types of solid-propellant gas generators have been devised and employed in practical liquid-propellant rocket engine systems due to their inherent simplicity, relatively light weight, compactness, low production cost, long-term storability, and so on. The schematic of a simple solid-propellant gas generator is shown in Figure 10.10, which consists of solid-propellant grain, igniter pellet, electrically fired initiators (squibs), safety and arming devices, and pressure relief valve. The propellant grain is housed in an aluminum alloy tank, while entire solid-propellant gas generator assembly is enclosed in an insulated steel housing. Generally, this unit has been designed and developed as a single compact package that can be made operational with minimum effort and maximum safety for its installation into the propulsion system. In order to maintain higher system reliability even after long storage, hermetically sealed outlets for gas, albeit with burst diaphragms, are being used for this design. On the initiation of ignition system, main solid-propellant grain starts producing pressurizing gas, whose rate is dependent on the burning rate of grain, grain bulk temperature, and chamber pressure, as discussed

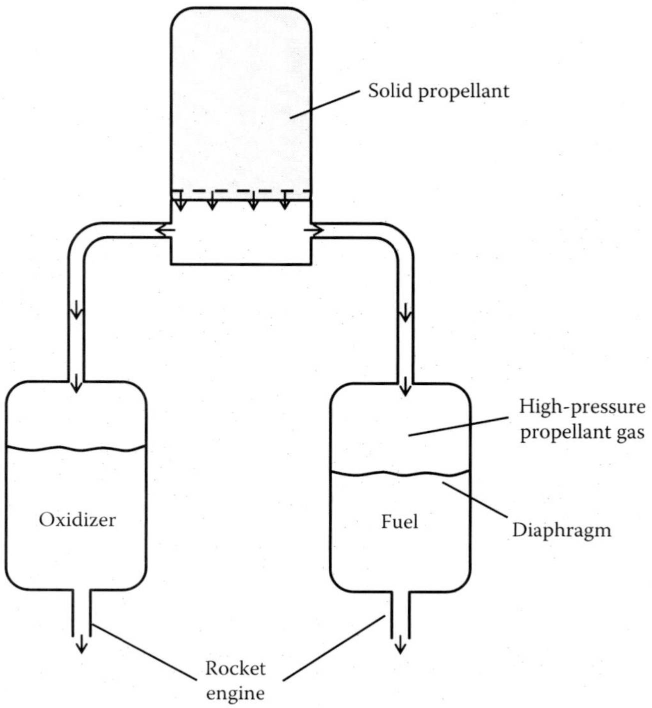

FIGURE 10.10 Schematic of a typical solid-propellant gas generator.

earlier in Chapter 9. Generally, this kind of grain is designed to produce requisite rate of gas at certain pressure, although at lower initial temperature limit. In case it is operated at higher initial temperature, a pressure regulator is used to maintain requisite chamber pressure using venting system. This kind of feed system with gas temperature even upto 1600°C and tank pressure even upto 10 MPa has been designed and developed successfully, but for short-duration applications. Designers of the solid-propellant gas generator during its design and development must take care of the following issues:

1. Chemical compatibility of hot gas with the liquid propellant. Impingement of the gases on the liquid propellant must be avoided. However, slight interactions between hot gas and liquid propellant may be permitted that may take additional gas generation during this interaction, provided it takes place in a controlled and reproducible manner.
2. Hot gases must not be soluble in liquid propellant, which may lead to uncontrolled reaction. In order to avoid such problems, certain devices, expulsion/flexible bags, or layers of insulating fluid floating between liquid propellant and hot gas or pressure release valve can be used.
3. The difficulties of regulating propellant tank pressure that may vary with variation in pressure and temperature of hot gases must be avoided.
4. The feed pipe line and valves must not be clogged with the solid particles accompanied by the hot gas.
5. The solid-propellant gas generator must take care of problems during lack of restart capability of the main thruster.
6. It is essential to provide a rapid venting outlet during premature shutting of rocket engine.

As mentioned earlier, due their simplicity and reliability, solid propellant gas feed systems are being used extensively to pressurize liquid-propellant tank. As a result, several types of solid-propellant gas feed systems are being designed and have developed over the years. Four of these commonly used solid-propellant gas generating systems, namely, (1) solid-propellant gas generator without cooling, (2) solid-propellant gas generator with solid cooling, (3) solid-propellant gas generator with azide cooling pack, and (4) helium system with solid-propellant gas generator heating [6], as shown in Figure 10.11, are discussed briefly. The solid-propellant gas generator without cooling, as shown in Figure 10.11a, includes solid-propellant grain, igniter, filter, and

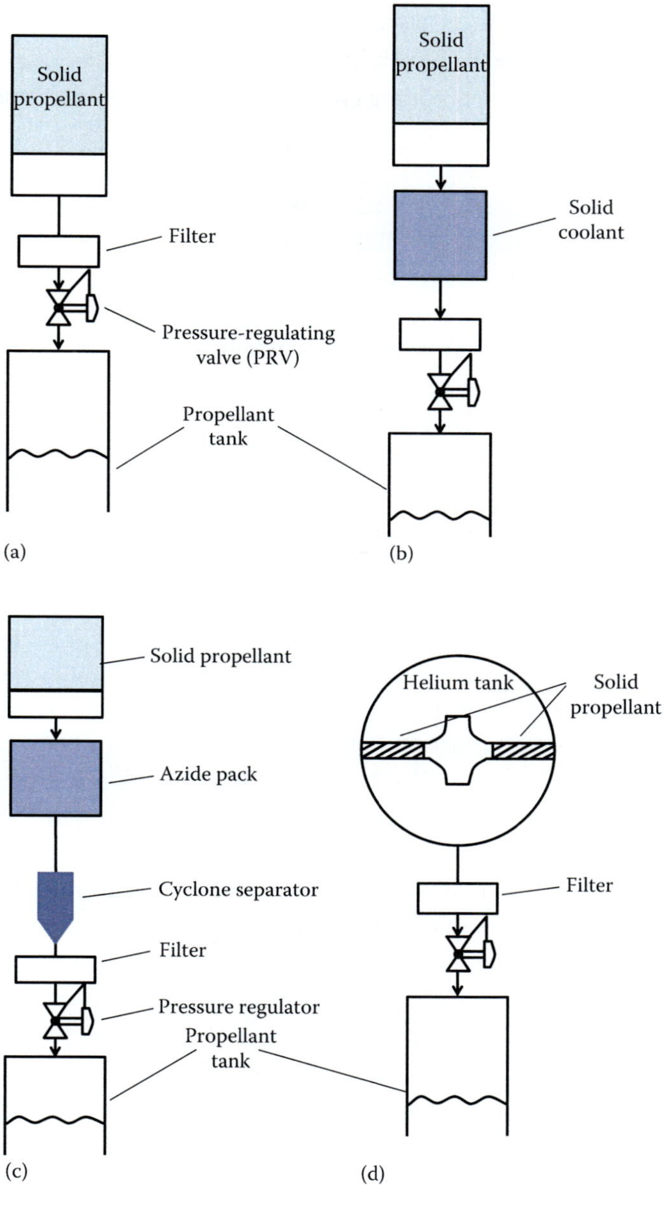

FIGURE 10.11 Schematic of four solid-propellant gas generating systems: (a) solid-propellant gas generator, (b) solid-propellant gas generator with solid coolant, (c) solid-propellant gas generator with azide cooling pack, and (d) helium system with solid-propellant gas generator heating.

pressure regulator. On ignition of the solid propellant, hot gases are formed due to burning of propellant. These high-temperature and high-pressure hot gases are passed through the filter to the propellant tank at a constant pressure with the help of pressure-regulating valve. The pressure regulator controls the pressure by diverting excess amount of gases overboard for which venting line must be provided. This kind of simple system is generally used for short-duration application as it may lead to problem of excess heating of liquid propellant. In order to overcome this problem, solid-propellant gas generator with solid cooling, as shown in Figure 10.11b, has been designed and devised. It has similar components as that of previous gas-generating system (Figure 10.11a), in addition to sublimating solid coolant, as shown in Figure 10.11b. When the hot gas from solid-propellant chamber passes through the solid coolant chamber in which solid materials undergoes decomposition with reduction in its temperature accompanied with additional rise in its pressure. In a typical ammonium nitrate base propellant system, pellets of oxalic acid is being employed as solid coolant, which decomposes endothermically at temperature above 120°C into a mixture of CO, CO_2, and H_2O. Of course, the desired temperature of the gas can be achieved easily by controlling the ratio of propellant–coolant pellet. By this kind of system, gas temperature as low as 200°C can be achieved easily. Note that subsequently, the hot gas at a constant pressure, with the help of pressure-regulating valve, from this gas generator can enter into propellant tank through filter, as in previous solid-propellant feed system. The decomposed gas may be reactive enough to react with liquid propellant. In order to overcome this problem, another gas generator known as solid-propellant gas generator with azide cooling pack as shown in Figure 10.11c has been devised and developed, which contains azide pack cooler in place of solid coolant. It is interesting to note that whenever hot gas from solid-propellant chamber passes through azide, it decomposes to produce pure nitrogen gas, but under ideal conditions. However, in practice, the decomposition of azide produces certain metal particles as the azide pack is generally contaminated with metal particles. Apart from filters, cyclone separator is used to eliminate these metal particles and others that can manage to deliver relatively pure nitrogen gas at gas temperature as low as 320°C to the propellant tank. A more compact solid-propellant gas generator with helium gas, as shown in Figure 10.11d, has been developed, which consists of high-pressure helium tank with solid-propellant gas generator being mounted internally, a filter, and a pressure regulator. In this case, helium gas is heated up with help of hot gas produced from burning of solid-propellant gas. Interestingly, the burning of the solid-propellant grain not only enhances the temperature but also the pressure of helium gas. Although this system needs relatively large high-pressure storage tank, it is preferred over others due to its enhanced compactness and reduced overall weight (as high 30%) as compared to the cold helium system.

Analysis of solid-propellant feed system: Recall that the mass of gas produced due to burning of solid propellant enhances the pressure in the propellant tank, which helps in expelling propellant through injector in liquid-propellant rocket engine. This mass of gas produced by burning of solid propellant must be equal to increase in mass in the free volume of propellant tank, as shown in Figure 10.12. Hence, by continuity, we can have

$$\frac{d\left(\rho_{p,t}V_{p,t}\right)}{dt} = \eta\rho_p \dot{r} A_b \tag{10.22}$$

where

$\rho_{p,t}$ is the density of gas produced by burning of solid propellant in the free volume

$V_{p,t}$ of liquid-propellant tank

η is the factor taking care of heat losses and interaction between liquid propellant and hot gases, whose value varies from 0.25 to 1

ρ_p is the density of solid propellant

$\dot{r}$ is the linear burning rate

A_b is the burning surface area of solid propellant

Note that temperature of product gases is assumed to be equal to adiabatic temperature T_c, which may not vary much with pressure. By using ideal gas law, the first term in Equation 10.22 becomes

$$\frac{d\left(\rho_{p,t}V_{p,t}\right)}{dt} = \frac{1}{RT_c}\frac{d\left(P_{p,t}V_{p,t}\right)}{dt} = \frac{1}{RT_c}\left[V_{p,t}\frac{dP_{p,t}}{dt} + P_{p,t}\frac{dV_{p,t}}{dt}\right] = \eta\rho_{p,t}\dot{r}A_b \tag{10.23}$$

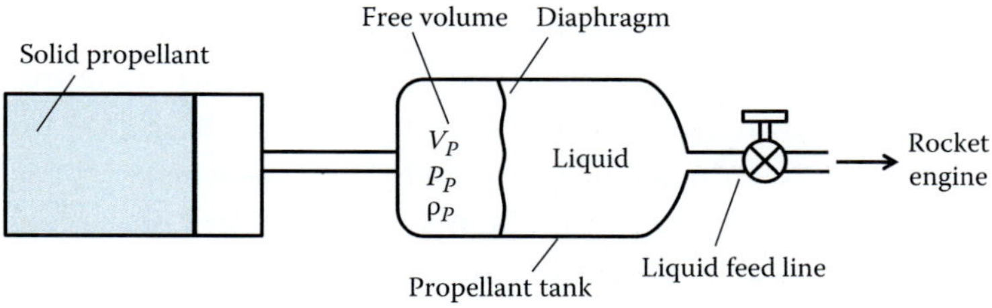

FIGURE 10.12 Schematic of a simple solid-propellant gas pressure feed system.

Generally, constant pressure is maintained in the propellant. Then, Equation 10.23 becomes

$$\frac{d\left(\rho_{p,t}V_{p,t}\right)}{dt}=\frac{1}{RT_c}\left[P_{p,t}\frac{dV_{p,t}}{dt}\right] \tag{10.24}$$

By combining Equations 10.22 and 10.24, we can have

$$\frac{dV_{p,t}}{dt}=\dot{Q}_L=\frac{\eta\rho_{p,t}\dot{r}A_bRT_c}{P_{p,t}} \tag{10.25}$$

where $\dot{Q}_L$ is the volumetric flow rate of liquid.

By using Bernoulli's equation and correction factor, we can derive an expression for the volumetric flow rate of liquid propellant, as given in the following:

$$\dot{Q}_L=C_dA_L\sqrt{\frac{2\left(P_{p,t}-P_L\right)}{\rho_L}} \tag{10.26}$$

where

P_L is the liquid pressure
A_L is the injector area
ρ_L is the liquid propellant density
C_d is the discharge coefficient whose value varies from 0.65 to 0.9.

The variation of the volumetric flow rate $\dot{Q}_L$ of liquid propellant from Equations 10.25 and 10.26 is plotted as solid line in Figure 10.13, with same tank pressure. It can be noted that $\dot{Q}_L$ from Equation 10.25 decreases gradually with tank pressure, while $\dot{Q}_L$ from Equation 10.26 increases steeply with tank pressure for certain $(P_{p,t}-P_L)$. The operating point would be when the volumetric flow rate, $\dot{Q}_L$ of liquid propellant determined from both equations would match, as shown in Figure 10.13, which is also known as the stability point. In case the liquid pressure is increased due to enhanced combustion chamber pressure, propellant flow rate will get reduced. If there is mismatch between these two, then it may lead to unstable conditions. In order to avoid this, higher combustion pressure index n is preferred. But higher combustion pressure index n is favored in actual system. In certain design, an auxiliary nozzle

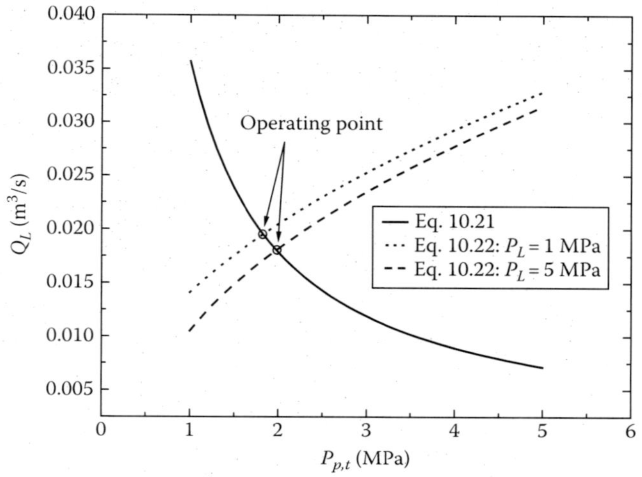

FIGURE 10.13 Variation of liquid propellant flow rate with tank pressure during operation of solid-propellant gas pressure feed system.

is used to allow excess gas produced by solid propellant to avoid this unwanted unstable operation.

Example 10.3

A solid-propellant pressure feed system with chamber pressure of 6.75 MPa is used in liquid rocket engine to supply liquid propellant with density of 850 kg/m³ at pressure of 2.5 MPa through 12 orifices with diameter of 2.5 mm. The discharge coefficient of the orifice is equal to 0.91. The regression rate of solid propellant with density of 1950 kg/m³, adiabatic temperature of 3200 K, and molecular weight of 25 g/mol is 2.5 mm/s with interaction loss of 0.91. Determine the liquid propellant flow rate and total burning surface area of solid propellant.

Solution

The mass flow rate of liquid propellant can be determined by using Equation 10.26, as follows:

$$\dot{m}_L = C_d A_L N \sqrt{2\rho_L\left(P_{p,t} - P_L\right)}$$
$$= 0.91\frac{3.14}{4}12\left(2.5\times10^{-3}\right)^2\sqrt{2\times850\left(6.75-2.5\right)10^6} = 4.55 \text{ kg/s}$$

By using Equation 10.25, we can determine burning surface area of solid propellant as follows:

$$A_b = \frac{MWP_{p,t}\dot{m}_L}{\eta\rho_s\rho_L\dot{r}R_uT_c} = \frac{25\times6.75\times10^6\times5.06}{0.91\times1450\times850\times2.5\times10^{-3}\times8314\times3200}$$
$$= 0.01\ \text{m}^2$$

10.9 TURBO-PUMP FEED SYSTEM

The gas pressure feed system would not be suitable for high-thrust, long-duration rocket engine systems with high total specific impulse. For this purpose, the turbo-pump feed system is usually being preferred due to its lower system weight and higher performance level. This kind of feed system is preferred for booster, and sustainers of space vehicles, as also for long-range missile systems and aircraft performance augmentation. This turbo-pump feed system has several advantages over the gas pressure feed system, as follows [5–7]:

1. Better flexibility in its operation due to easier pump speed control.
2. Stable pressure as high as 6–8 MPa can be obtained easily.
3. Smaller volume requirements even for higher-thrust rocket engines.
4. Higher power-to-weight ratio from 15 to 50 kW/kg.

Let us consider a simplified turbo-pump system, as shown in Figure 8.2b, which consists of propellant tanks, propellant inlet and discharge ducts, pumps, turbine, speed reduction gear box, gas generator, and heat exchanger followed by a nozzle. In this case, the propellants are pressurized with the help of pumps. These pumps are run by the turbines whose powers are obtained due to expansion of hot gases. In the present case, the hot gases are supplied from the gas generator, which produces hot gases due to burning of liquid propellants (1%–5% of total flow rate). The gear transmission system consisting of gear box, rotational, coupling, and so on is used to transfer the torque from the turbine system to the pump at reduced angular speed (rpm).

10.9.1 Types of Turbo-Pump Feed System

Based on the configuration of turbine and pump drive and modes of discharge of exhaust gases from turbine, several rocket engine cycles have been designed and developed over the years. These can be primarily classified into four categories: (1) monopropellant, (2) bipropellant, (3) expander,

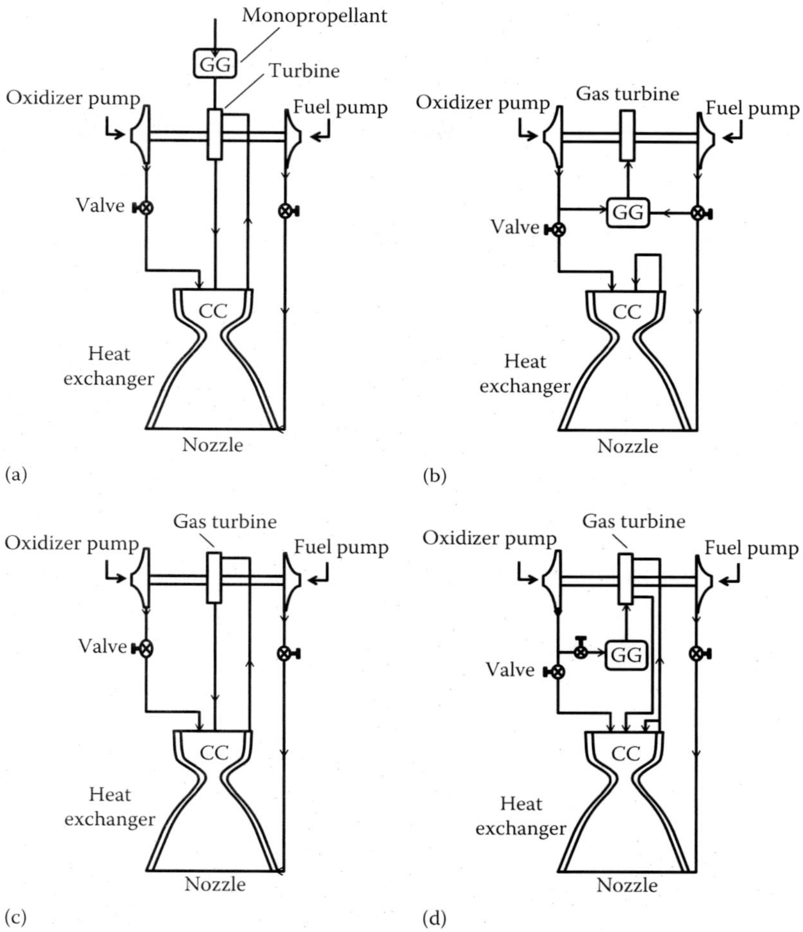

FIGURE 10.14 Schematic of four types of turbo-pump feed systems: (a) monopropellant, (b) bipropellant, (c) expander, and (d) staged combustion cycles.

and (4) staged combustion cycles [6–7], as shown in Figure 10.14, which are described briefly as follows:

1. *Monopropellant cycle*: This system being the simplest one is being used extensively in pressurizing the liquid propellants in several rocket engines. This monopropellant system, as shown in Figure 10.14a, consists of monopropellant tank, pressurized gas cylinder, gas generator, solid propellant cartridge starter, turbine and pumps, liquid propellant tanks, and so on. The monopropellant is entered into

gas generator as spray, which can be ignited with the help of solid-propellant cartridge starter. On the initiation of ignition, monopropellant gets decomposed exothermically and produces high-pressure hot gases that are expanded in a turbine to run pumps. The exhaust gas from the turbine is expanded further to ambient pressure using a separate nozzle. Hence, this is termed as an open cycle. This kind of turbo-pump feed system is independent of altitude, as turbine operating condition would not be affected by change in altitude during its operation.

2. *Bipropellant cycle*: This is one of the most widely used turbo-pump feed systems because engine main propellant is being used and thus is resulted in simpler and reliable system as compared to other systems. In other words, there is no need to have separate monopropellant tank and its feed line system along with associated arrangements, as in monopropellant system. The schematic of bipropellant system is shown in Figure 10.14b, which consists of gas generator, turbine and pumps, liquid-propellant tanks, and so on. As mentioned earlier, in this case, certain amount of fuel and oxidizer from main liquid propellant feed lines at certain flow rate is diverted into gas generator, which is burnt to produce high-pressure and high-temperature gas to drive turbine. Subsequently, the hot gases from the turbine are expanded further to certain low-pressure point. As a result, this system can operate at high pressure ratio with low flow rate to reduce its performance parameter, namely, specific impulse. Sometimes, this system is also termed as bootstrap method as the same propellants are being used to run gas generator.

3. *Expander cycle*: The schematic of expander cycle is shown in Figure 10.14c, which is devoid of gas generator, compared to the other two systems. Note that this kind of cycle can be used for hydrogen-fuelled rocket engine system in which heated hydrogen gas emanated from the thrust chamber cooling jacket is expanded directly in the turbine to drive propellant pumps. Subsequently, the expanded hydrogen gas is fed into thrust chamber to be burnt in the presence of liquid oxygen. It must be emphasized that the pressure at the exit of turbine must be higher than the requisite thrust chamber pressure for smooth operation rocket engine. Obviously, either outlet pressure from hydrogen pump must be quite high or thrust chamber must be of lower pressure for successful operation of this kind of system.

Note that this kind of expander feed system has been used in RL 10 engine, whose chamber pressure is around 2.73 MPa.

4. *Staged combustion cycle*: This cycle is basically a combination of both bipropellant and expander cycle methods. The schematic of staged combustion is shown in Figure 10.14d, which consists of gas generator, pump and turbine, and propellant tanks. In this case, the entire fuel is diverted into the gas generator, in which it is reacted with oxidizer fuel at rich fuel–oxidizer ratio to produce high-temperature and high-pressure gas. Subsequently, the hot gases from the turbine are expanded further to certain pressure, which is fed directly into thrust chamber. As combustion takes place in a staged manner, both in gas generator and thrust chamber, hence it is termed as staged combustion cycle. A variation of this cycle is used in SSME engine, in which an oxidizer-rich gas turbine for oxidizer pump and a fuel-rich turbine for fuel pump from two separate gas generators are employed to achieve high-pressure system. A peak system pressure of 55 MPa and chamber pressure of 22.2 MPa can be achieved in this system.

10.9.1.1 Propellant Turbo-Pumps

Generally, light weight, high-performance, high-reliability, and small-volume (compact) turbines and pumps for pressuring liquid propellants are preferred for design of rocket engine systems. We all know that a turbomachine consists of rotating elements, stationary elements, and hub, which are housed in casing. Based on the types of predominant flow in machine (geometry of rotating elements), turbomachines can be classified into three categories: (1) radial (centrifugal), (2) mixed (helico-centrifugal), and (3) axial (helicoidal). Among these types, radial pumps are preferred for almost all operational liquid-propellant rocket engines and hence is discussed in detail in the following.

10.9.2 Propellant Pumps

The schematic of a typical centrifugal pump is shown in Figure 10.15a, which consists of a rotating element (inducer and impeller), diffuser, volute, and casing. The impeller mounted on the rotational with the help of hub has several vanes that whirl the fluid around due to its rotation. Generally, liquid enters axially into the inducer, which guides the fluid to enter into the rotating impeller with minimum pressure losses. The fluid gets accelerated in the impeller, while moving against the centrifugal force

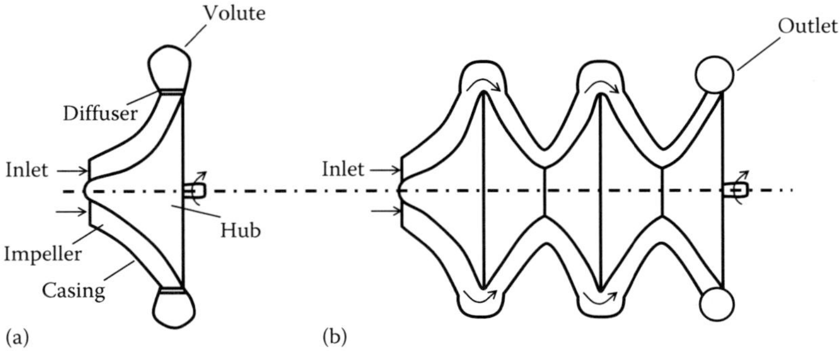

FIGURE 10.15 Schematic of a centrifugal pump: (a) single stage and (b) multistage.

field due to imparting of kinetic energy and leaves the impeller in radial direction. Subsequently, the fluid enters into stator known as diffuser, in which fluid flow is decelerated to enhance its static pressure. In order to analyze the pumping processes in this centrifugal pump, we can make the following assumptions:

1. Incompressible steady flow.
2. Friction and viscous forces are neglected.
3. No swirl in the incoming flow.

Let us consider a control volume containing the impeller, as shown in Figure 10.15a. We can apply the Euler's equation for this centrifugal pump between inlet station (1) and outlet station (2) of the impeller due to whirling of fluid. By assuming axial flow at inlet station (1), we can have an expression for power input $\dot{W}$ to the rotor in terms of a change in angular momentum, as given in the following:

$$\dot{W} = \dot{m}\left[U_2 V_{\theta_2} - U_1 V_{\theta_1}\right] = \dot{m} U_2 V_{\theta_2} \quad (10.27)$$

where

$\dot{m}$ is the mass flow rate of fluid
U is the impeller tip velocity
V_θ is the tangential velocity of fluid

The velocity triangles at the exit of three types of impellers: (1) radial, (2) forward and (3) backward-leaning vanes are shown in Figure 10.16.

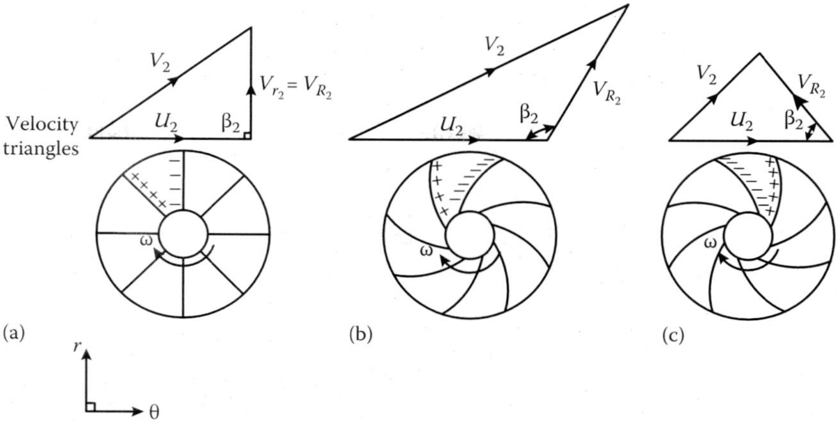

FIGURE 10.16 Velocity triangles for three types of centrifugal impellers: (a) radial, (b) forward, and (c) backward-leaning vanes.

For incompressible flow, the tangential velocity of fluid, V_{θ_2} leaving the impeller can be expressed easily in terms of relative velocity V_{R_2}, impeller tip velocity U_2, and fluid angle between them β_2, as follows:

$$V_{\theta_2} = U_2 + V_{R_2} = U_2 + V_{r_2} \cot\beta_2 \tag{10.28}$$

where V_{r_2} is the radial velocity. By substituting Equation 10.28 in Equation 10.27, we can have

$$\dot{W} = \dot{m}U_2V_{\theta_2} = \dot{m}U_2^2\left(1 + \frac{\dot{m}}{2\pi R_2 b\rho U_2}\cot\beta_2\right) \tag{10.29}$$

where

b is the radial width of vane at its exit

r_2 is the radius at impeller's exit

Note that negligible thickness of vanes at its exit is assumed. By using energy equation (Equation 2.2) for steady flow, Equation 10.29 can be rewritten as

$$\frac{\dot{W}}{\dot{m}} = U_2^2\left(1 + \frac{\dot{m}}{2\pi R_2 b\rho U_2}\cot\beta_2\right) = \Delta h_t \tag{10.30}$$

where Δh_t is the total enthalpy change of the fluid during pumping process. By invoking the definition of enthalpy in terms of internal energy u, pressure P, and specific volume v as $h = u + Pv$, we can have for stagnation/total state

$$T_t ds_t = dh_t - \frac{dP_t}{\rho_t} \tag{10.31}$$

As the pumping process is assumed in the present analysis for reasons of simplicity to be isentropic in nature, we can derive an expression for ideal change in total pressure across the impeller during pumping process by using Equations 10.30 and 10.31, as follows:

$$\Delta h_t = \frac{\Delta P_t}{\rho_t} = U_2^2\left(1 + \frac{\dot{m}}{2\pi R_2 b\rho U_2}\cot\beta_2\right) \tag{10.32}$$

For an ideal situation, for same pressure rise across a pump, less amount of work input must be supplied. But in actual case, more amount of work output has to be supplied for getting same amount of pressure rise due to occurrences of losses during pumping process. In order to determine the performance of the pump, a quantity known as pump efficiency η_p is defined as the ratio of ideal (isentropic) work input to real work input for same pressure rise, expressed as

$$\eta_p = \frac{\Delta h_{ts}}{\Delta h_t} = \frac{\Delta P_{t,actual}}{\Delta P_{t,ideal}} \tag{10.33}$$

Note that the pump efficiency η_p is usually determined from experimental data as it is quite difficult to predict it. In literature, two nondimensional performance parameters, namely, pressure coefficient ψ, and flow coefficient φ, are quite often used, which are defined as follows:

$$\psi = \frac{\Delta P_t}{\rho U_2^2} \quad \text{and} \quad \varphi = \frac{\dot{m}}{2\pi R_2 b\rho U_2}, \tag{10.34}$$

By using these coefficients, we can rewrite Equation 10.32 as

$$\psi = \frac{\Delta P_t}{\rho_t U_2^2} = \left(1 + \varphi\cot\beta_2\right) \tag{10.35}$$

Note that losses during pumping of fluid are entirely ignored and this process is assumed to be adiabatic in nature. Hence, the pumping efficiency under this condition will be 100%. As mentioned earlier, flow is assumed to leave the blade with the blade angle itself under an ideal condition. The variations of pressure coefficient ψ with the flow coefficient φ for all three types of impellers under an ideal condition are plotted in Figure 10.17. It can be noted that the pressure coefficient ψ remains constant with the flow coefficient φ for an ideal condition. But for backward-swept impeller vanes, the pressure coefficient ψ decreases with the flow coefficient φ, due to reduced transfer of work input caused by the vane angle. In contrast, for forward-swept impeller vanes, the pressure coefficient ψ increases with the flow coefficient φ, due to reduced transfer of work input caused by the vane angle. Note that for the same mass flow rate, and tip speed U_2, the forward-swept vane impeller will do more work on the fluid. Then, you may be tempted to conclude that it produces higher pressure rise in comparison to other two impellers. From the velocity diagram, it can be observed that higher value of kinetic energy is seldom desired as its reduction to static pressure by diffusion is difficult to carry out efficiently in a reasonable size of casing. Thus, it may result in lowering the overall compressor efficiency. Besides this, a positive slope as in the curve in Figure 10.17 may lead to flow instability, which is not preferred for rocket engine. Now let us compare the radial blade and backward-swept vaned impellers by considering the velocity diagram shown in Figure 10.17. It can be noted that the radial blade impeller has greater pressure rise capability at given tip speed but

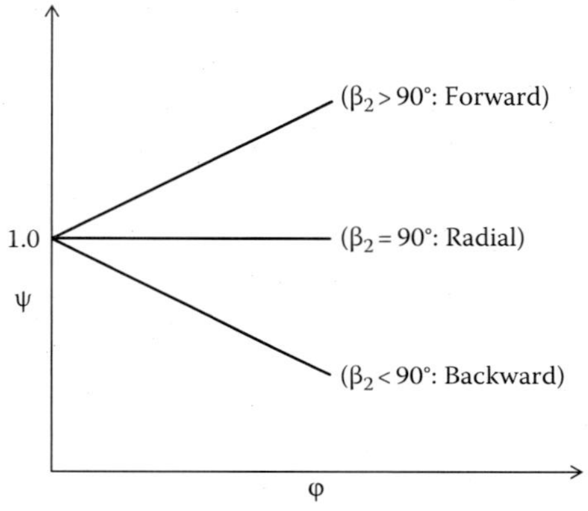

FIGURE 10.17 Variation of pressure coefficient ψ with the flow coefficient φ.

results in higher velocities with which the fluid will be entering into the diffuser. Thus, the diffusion process can be less efficient as compared to backward swept impeller. However, the straight radial vaned impellers were used extensively for many years in rocket engine, due to their simpler construction. But with an improved material strength, backward-swept vane particularly for higher stage pressure are being preferred in recent times.

By carrying out nondimensional analysis, it can be shown that the performance parameters of the pump, namely, the pressure coefficient ψ and pump efficiency η_p, is dependent on the nondimensional quantities, namely, Reynolds number Re and the flow coefficient φ, as given in the following:

$$\psi = F(Re, \varphi, \text{design}) \tag{10.36}$$

$$\eta_p = G(Re, \varphi, \text{design}) \tag{10.37}$$

It has been observed experimentally that beyond certain Re, both the pressure coefficient ψ and pump efficiency η_p would vary with Re as the flow in the pump is turbulent in nature. Hence, the turbo-pumps in rocket engine are operated at higher Re much beyond $Re = \rho V_{r_2} b/\mu = 10^5$. But the actual pressure coefficient ψ and pump efficiency η_p in a pump will exhibit lower values than ideal pump due to two reasons: (1) Frictional losses in the impeller and diffuser vanes casing, and so on; these losses get magnified considerably when flow gets separated, leading to stalling of flow due to energy losses during mixing process between slow and fast moving streams; (2) the fluid flow would not leave at the exit of impeller at same angle as that of the vane, resulting in a slip between vane and fluid flow, which reduces the work transfer that results in lower pressure rise across the pump for same work input.

In order to select a pump for a particular rocket engine, we need to compare geometrically similar pumps in terms of volumetric flow (discharge) rate Q, the pressure rise ΔP, rotational speed N, and specific mass ρ. Let us consider two geometrically similar pumps A and B, for which b/R_2 will be same and $V_r \cos\beta_2$ is proportional to the $U_2(= 2\pi N R_2)$. As the volumetric flow (discharge) rate Q is equal to $2\pi R_2 b V_{R_2} \cot\beta_2$, we can have

$$\frac{Q_A}{Q_B} = \frac{N_A R_A^3}{N_B R_B^3} \tag{10.38}$$

The ratio of stagnation pressure rise between two geometrically similar pumps A and B can be related to

$$\frac{(\Delta P_t/\rho)_A}{(\Delta P_t/\rho)_B} = \frac{N_A^2 R_A^2}{N_B^2 R_B^2} \tag{10.39}$$

By eliminating R_A/R_B in Equations 10.38 and 10.39, we can get an expression for rotational speed as follows:

$$\frac{N_B}{N_A} = \frac{\sqrt{Q_A/Q_B}}{\left[\frac{(\Delta P_t/\rho)_A}{(\Delta P_t/\rho)_B}\right]^{3/4}} \tag{10.40}$$

Let us consider the pump B as our reference pump with $Q_B = 1\ \text{m}^3/\text{s}$ and $(\Delta P_t/\rho)_B = 1\ \text{J/kg}$. Then the rotational speed of pump B is termed as the pump-specific speed N_s, which can be expressed from Equation 10.40, by omitting the indices as follows:

$$N_s = \frac{N\sqrt{Q}}{(\Delta P_t/\rho)^{3/4}} \tag{10.41}$$

Note that the pump-specific speed N_s is a dimensional performance parameter that does not vary significantly for geometrically similar impellers. For typical impellers, the general range of the pump-specific speed N_s is shown in Figure 10.18 along with their respective schematic ranging from axial to centrifugal through mixed flow pumps, which indicates the characteristics

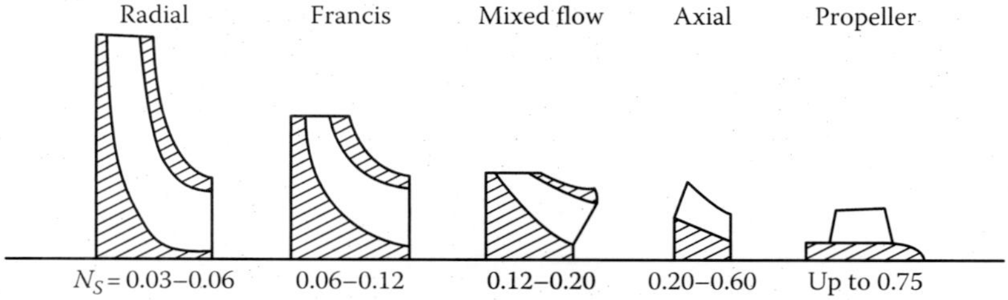

FIGURE 10.18 Schematic of certain impellers along their respective specific speed ranges.

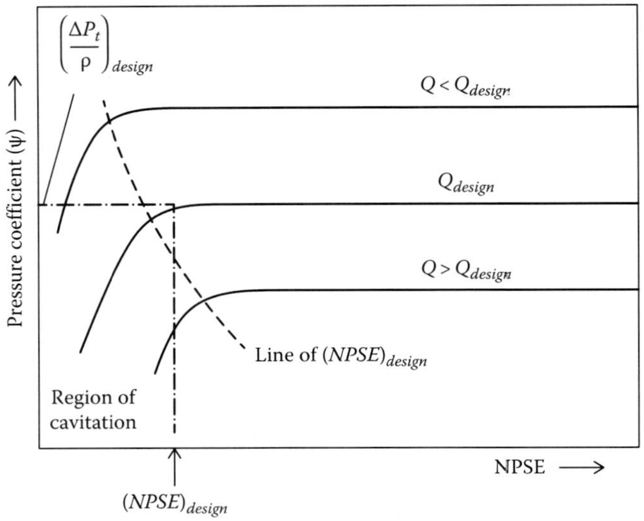

FIGURE 10.19 The effect of cavitation on the pump performance.

of pump shape. It can be observed from Equation 10.41 that the pump-specific speed N_s is dependent on the volumetric flow (discharge) rate Q and $(\Delta P_t/\rho)$. More specifically, the pump-specific speed N_s increases with volumetric flow rate, while it decreases with work input rise for the same pressure ratio. Although it is required to enhance the rotational speed N for reducing the size of pump particularly for rocket engine, it is limited by allowable material stresses or by cavitation (boiling of liquid). Note that the shape of the impeller is influenced by the maximum rotational speed N, the volumetric flow (discharge) rate Q, and total enthalpy rise Δh_t. When designer chooses the rotational speed N, the volumetric flow (discharge) rate Q, and $(\Delta P_t/\rho)$ of a pump for particular application, he or she can select easily the type of impeller by determining the pump-specific speed N_s from Figure 10.19. Now we will further discuss the phenomena of cavitation in the next subsection.

10.9.2.1 Cavitation

We have ignored the occurrence of cavitation during the discussion on the operation of pump for rocket engines, although its performance is limited by the phenomenon of cavitation. Cavitation occurs due to spontaneous bubble formation from the vapor of the liquid as soon as the local static pressure p in any portion of pump falls below the vapor pressure p_v of the liquid [7]. Generally, these gas bubbles grow in liquid and subsequently collapse quickly, sometimes within a few milliseconds. This phenomenon is accompanied with a high rise in temperature, as high as 10,000 K and local pressure as high as 400 MPa. The occurrence of such high local temperature and pressure can damage the various parts of the pump wherever it occurs. This sudden formation and collapse of bubbles in the pump can lead to pressure fluctuation that can be transmitted across the rocket engine, leading to the combustion instabilities that must be avoided at any cost. Furthermore, the phenomenon of cavitation also affects the overall performance of pump in two ways: (1) the pressure in the pump may be reduced to very low or even zero value, particularly when formation of bubbles occupy the large portion of impeller as the pressure rise is proportional to the density of fluid as per Equation 10.32; (2) in case of cavitation being confined to smaller region of the impeller, the sudden rise and fall of local pressure and temperature can also erode the impeller vanes, if the occurrence of cavitation persists for a longer period of time. The damage caused due to occurrence of cavitation in pump is not of significant concern in rocket engine as its duration of operation is limited to few minutes only. But the decrease in delivery pressure of the pump due to formation of bubbles is the most critical concern for the designer as it affects the overall performance of the rocket engine. Hence, we need to learn how to evaluate the extent of cavitation that occurs in a pump.

We know that the cavitation can be avoided only when the local pressure in the pump is greater than or at least equal to the saturated vapor pressure of the liquid p_v. Generally, the minimum pressure in a pump occurs near the leading edge of the impeller. Hence, the extent of cavitation that occurs in a pump is evaluated in terms of net positive suction energy (*NSPE*), which is defined as the ratio of difference between the stagnation pressure at inlet P_{t_1} and the saturated vapor pressure of the liquid p_v, and density of liquid, given by

$$NPSE = \frac{P_{t_1} - p_v}{\rho} \tag{10.42}$$

The value of *NPSE* at which cavitation occurs in a pump for the first time is known as critical net positive suction energy $(NSPE)_c$. In order to avoid occurrence of cavitation in a pump, *NPSE* must be greater than $(NSPE)_c$. The effect of cavitation on the performance of pump in terms of pressure coefficient ψ and *NPSE* is shown in Figure 10.20 for three different volumetric flow rates around the design condition. The $(NPSE)_c$ indicates the points below which cavitation is likely to occur, which can be avoided even

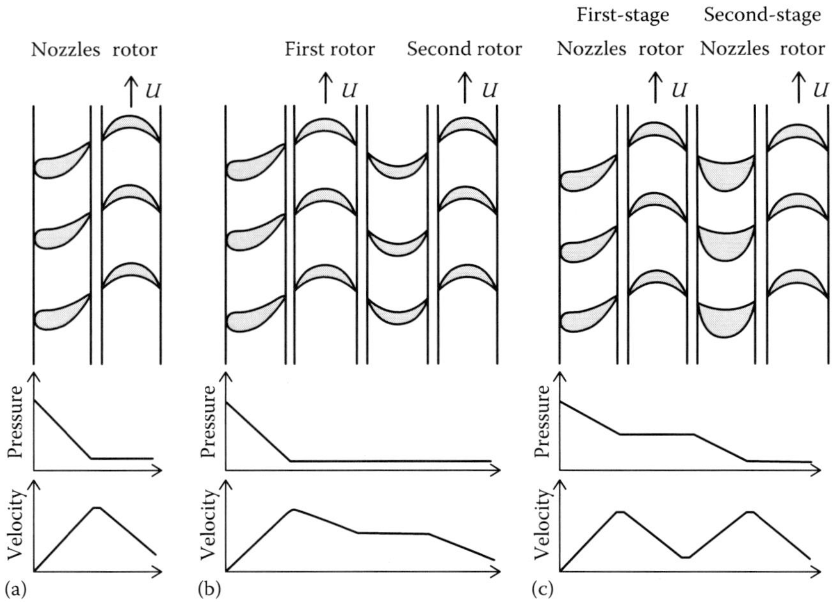

FIGURE 10.20 Schematic representation of three types of impulse turbines along with their respective pressure and velocity profiles (a) single-stage, single-rotor impulse turbine, (b) single-stage, two-rotor velocity compound turbine, and (c) two-stage, two-rotor pressure compound turbine.

during design condition as there is certain loss of certain portion of total pressure across the pump. As the cavitation is dependent on the rational speed of the pump, similar to the pump-specific speed N_s, another term known as the suction-specific speed N_{ss} is defined as

$$N_{ss} = \frac{N\sqrt{Q}}{\left(\Delta P_t/\rho\right)^{3/4}} = \frac{N\sqrt{Q}}{\left(NPSE\right)^{3/4}} \tag{10.43}$$

By using, Equations 10.41 and 10.43, we can have

$$\frac{N_s}{N_{ss}} = \left[\frac{NPSE}{\left(\Delta P_t/\rho\right)}\right]^{3/4} \Rightarrow \tau = \left[\frac{NPSE}{\left(\Delta P_t/\rho\right)}\right] = \left(\frac{N_s}{N_{ss}}\right)^{4/3} \tag{10.44}$$

where τ is the Thomas parameter, which is often used during the design of pump. The maximum rotational speed of a pump can be determined from the pump suction characteristics for given suction pressure and requisite flow rate. The specific speed N_s for inferior pumps occurs around 5000 rpm. But in recent times, high-performance pumps can operate between 10,000 and 15,000 rpm even without cavitation [7]. However, we generally choose a higher value of the suction-specific speed N_{ss} for rocket engine pumps, as it permits higher angular velocity, which has resulted in the design of smaller, light-weight, and compact pump. For example, the pump for space shuttle vehicle operates at 60,000 rpm. For this kind of reusable vehicle, cavitation during pumping operation must be avoided at any cost, although damage due to the occurrence of cavitation may not be of serious consequence for short-duration rocket engine applications.

As mentioned earlier, the cavitation in the pump can be avoided by ensuring the occurrence of higher local pressure above the vapor pressure of the liquid. This can be accomplished easily by avoiding high flow velocities or by using high fluid pressure or by combination of both. Higher inlet fluid pressure in turbo-pumps of rocket engines can be obtained using high tank pressure possibly in combination with booster pumps. Besides this, dissolved gas from the liquid propellant can be removed before being fed into the pump. It is advisable to avoid impurities/dust particles in the liquid propellant as they act as precursors for formation of bubbles. It has also been observed that minute cracks on the blade surface can trigger the nucleation of vapor bubbles.

Example 10.4

A liquid rocket engine is designed to supply 105 kN of thrust with $I_{sp} = 250$ s, during which oxidizer–fuel ratio is maintained at 2.75. A single turbine with mechanical efficiency of 0.92 is used to drive oxidizer ($\eta_{Ox} = 0.65$) and fuel ($\eta_F = 0.75$) pumps. The combustion chamber pressure and temperature are 7.5 MPa and 3150 K, respectively. The injection velocity for oxidizer and fuel is maintained at 50 and 75 m/s, respectively. Determine the total power of turbine if propellant tank pressure for both fuel and oxidizer is maintained at 0.25 MPa. Consider $\rho_F = 850\ \text{kg/m}^3$, $\rho_{Ox} = 1200\ \text{kg/m}^3$.

Solution

The power consumed by oxidizer and fuel pumps can be estimated as

$$\dot{W}_{Ox} = \frac{\dot{m}_{Ox}}{\eta_{Ox}} \frac{\Delta P_{Ox}}{\rho_{Ox}}; \quad \dot{W}_F = \frac{\dot{m}_F \Delta P_F}{\eta_F \rho_F} \tag{A}$$

But the pressure drop across oxidizer pump can be determined as

$$\Delta P_{Ox} = P_c + \Delta P_{inj} - P_T = P_c + \frac{\rho_{Ox} V_{Ox}^2}{2} - P_T$$

$$= 7.5 + \frac{1200 \times 50^2}{2 \times 10^6} - 0.25 = 8.50 \text{ MPa}$$

Similarly, the pressure drop across fuel pump can be determined as

$$P_F = P_c + \frac{P_F V_F^2}{2} - P_T = 7.5 + \frac{850 \times 65^2}{2 \times 10^6} - 0.25 = 8.07 \text{ MPa}$$

Total mass flow rate of propellant $\dot{M}_P$ can be determined as

$$\dot{m}_P = \frac{F}{gI_{sp}} = \frac{105 \times 10^3}{9.81 \times 250} = 42.8 \text{ kg/s}$$

The oxidizer mass flow rate can be determined by using Equation 10.4 as

$$\dot{m}_{Ox} = \dot{m}_P \frac{Ox/F}{1 + Ox/F} = 42.8 \frac{2.75}{3.75} = 31.4 \text{ kg/s}$$

The fuel mass flow rate can be determined as

$$\dot{m}_F = 42.8 - 31.4 = 11.4 \text{ kg/s}$$

By substituting the values in Equation A, we can determine P_{ox} and P_f as follows:

$$\dot{W}_{Ox} = \frac{31.4}{0.65} \frac{8.75 \times 10^3}{1200} = 352.2 \text{ kW}$$

$$\dot{W}_F = \frac{11.4}{0.75} \frac{8.05 \times 10^3}{850} = 144.3 \text{ kW}$$

The turbine power can be estimated as

$$\dot{W}_t = \frac{\left(\dot{W}_{Ox} + \dot{W}_F\right)}{\eta_m} = \frac{352.2 + 144.3}{0.92} = 539.67 \text{ kW}$$

10.9.2.2 Propellant Turbines

The turbo-pumps are to be driven by the power delivered from one or more turbines. The high-temperature and high-pressure gas from the gas generator is expanded in turbine to produce power that is directly coupled to the shafts of the pumps for most of the cases. In order to avoid the cavitation during the pumping actions, the pumps are allowed to rotate at lower angular speed than that of the turbine, which results in decrease in overall efficiency of the turbine. For rocket engine applications, the number of stages for requisite power level is restricted to one or two and, therefore, turbine with high power per stage is to be designed. The turbines employed in rocket engines are of axial type, in which direction of flow is predominantly in axial direction. These can be further divided into: (1) impulse and (2) reaction turbines. In the case of reaction turbine, the expansion of gas takes place only in the rotor. The reaction force on the rotor due to expansion of gas drives the rotor. The gas velocity prevailing in the rotor is relatively low as compared to the impulse turbine, and thus it results in higher turbine efficiency due to lower losses. However, pure reaction type of turbine calls for more number of stages for same power level as compared to impulse-type turbine. Hence, most turbines used to drive turbo-pumps in rocket engine applications are of the impulse type. In an ideal impulse-type turbine, gas is expanded entirely in the nozzle (stationary blades) and thus potential energy is converted into kinetic energy. Subsequently, the gas is passed through the rotor blades in which kinetic energy of the gas is imparted to the rotating blades. As a result, there will be decrease in gas velocity, while static pressure remains constant, as depicted in Figure 10.20. Note that impulse turbine can be either single or multistage. Generally impulse turbine is being preferred in rocket engine applications due to its higher power density.

Three types of impulse turbines are shown in Figure 10.20 along with their respective profiles of velocity and pressure of gas. It can be noted that the gas is expanded in the stationary nozzle of the single stage of single rotor impulse turbine shown in Figure 10.20a. The kinetic energy of the gas is expended to drive the rotor, during which gas velocity decreases, while static pressure remains almost constant. The schematic of a single-stage two-rotor velocity compound impulse turbine is shown in Figure 10.20b. This is sometimes called single-stage turbine as all pressure drops occur only in the first rows of nozzle. Of course the gas leaves first rotor with swirl in the direction opposite to the rotation, and pass through a row of stator such that it can enter smoothly into the second rotor in which no expansion takes place. Note that both pressure and velocity of the gas remain almost constant while passing through the second rows of stator blades. If the gas happens to leave the last rotor without any swirl component, it can be easily demonstrated that the power output from two rotors is four times that of a single-stage impulse turbine. For this reason alone, single-stage two-rotor velocity compound impulse turbine (see Figure 10.20b) is commonly employed in turbo-pumps for higher power rocket engines, although it is less efficient as compared to the single-stage impulse turbine. In order to reduce its weight, two rows of rotor blades are mounted often on a single disc, in which branches are placed around the blade roots as the centrifugal stresses are not of significant value. In the case of a two-stage pressure compound impulse turbine (Figure 10.20c), gas is expanded in both rows of stationary nozzles as the main objective is to have identical entrance velocities at both rotors such that both stages can produce same amount of power output. The problem of pressure differences between two stages calls for better sealing mechanism to reduce losses due to gas leakages. In order to overcome the exhaust problem, the designer is attempted to have the exit absolute gas velocity, which is almost parallel to the axis of rotation.

Let us consider gas flow between two rotor blades of an ideal impulse turbine in order to determine its power output. The schematic of velocity triangle is shown in Figure 10.21, along with the magnitude and direction of gas respective velocities at both inlet and outlet of an ideal impulse turbine. The velocity is assumed to be uniform along entire circumstance as the width of the blades is considered to be small as compared to the blade diameter. By applying the Euler's Equation for this impulse turbine between inlet station (1) and outlet station (2) of the rotor due to whirling

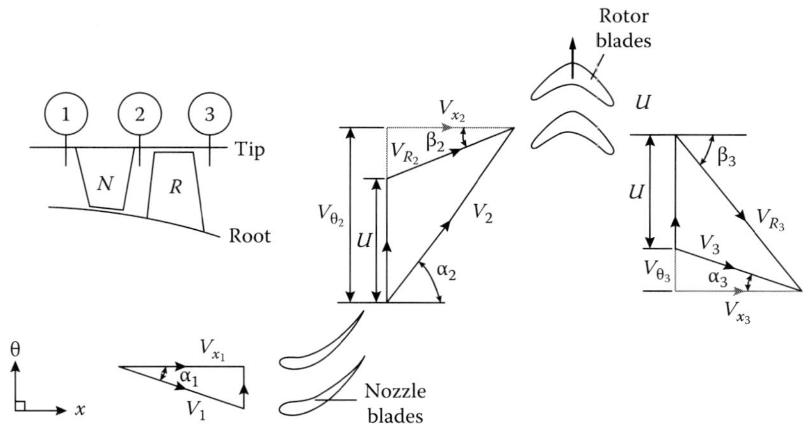

FIGURE 10.21 Schematic of velocity triangles for a single turbine.

of fluid, we can have an expression for power input, $\dot{W}_t$ to the rotor in terms of a change in angular momentum, as given in the following:

$$\dot{W}_t = \dot{m}\left[h_{t_1} - h_{t3}\right] = \dot{m}\left[U_2 V_{\theta_2} - U_3 V_{\theta_3}\right] = \dot{m} U V_x \left[\tan\alpha_2 + \tan\alpha_3\right] \quad (10.45)$$

where

$\dot{m}$ is the mass flow rate of fluid
U is the blade velocity
V_θ is the tangential velocity of fluid
V_x is the axial velocity
α and β are the absolute velocity (V) and relative velocity (V_R) air angles with reference to axis, respectively

From the velocity triangle, we know that

$$\frac{U}{V_x} = \tan\alpha_2 - \tan\beta_2 = \tan\beta_3 - \tan\alpha_3 \quad (10.46)$$

By using Equation 10.45, Equation 10.46 can be expressed in terms of relative velocity angles (β), as follows:

$$\dot{W}_t = \dot{m}\left[h_{t_1} - h_{t3}\right] = \dot{m} U V_x \left[\tan\beta_2 + \tan\beta_3\right] \quad (10.47)$$

Note that the above the expression is obtained for ideal conditions. But there will be losses in the actual turbine, which can be considered by using the stage efficiency $\eta_{t,st}$ as defined in the following:

$$\eta_{t,st} = \frac{T_{t1} - T_{t3}}{T_{t1} - T_{t3s}} \quad (10.48)$$

We can now derive the expression for pressure ratio across a stage, as given in the following:

$$\frac{P_{t3}}{P_{t1}} = \left(\frac{T_{t3s}}{T_{t1}}\right)^{\frac{\gamma}{\gamma-1}} = \left(1 - \frac{T_{t3s} - T_{t1}}{T_{t1}}\right)^{\frac{\gamma}{\gamma-1}} = \left(1 - \frac{U V_x \left[\tan\beta_2 + \tan\beta_3\right]}{\eta_{t,st} C_P T_{t1}}\right)^{\frac{\gamma}{\gamma-1}} \quad (10.49)$$

The pressure ratio (P_{t1}/P_{t3}) across an axial turbine can be enhanced by increasing either blade velocity or axial velocity or both. It can also be enhanced by increasing the blade angles. However, larger turning angles can lead to losses in the turbine. Let us define three important nondimensional variables: (1) stage loading coefficient (ψ), (2) flow coefficient (φ), and (3) reaction ratio (RR), which are being used routinely for preliminary design.

$$\psi = \frac{C_P \Delta T_t}{U^2} = \frac{V_x}{U}\left[\tan\beta_2 + \tan\beta_3\right] = \varphi\left[\tan\beta_2 + \tan\beta_3\right] \quad (10.50)$$

For turbine stage design, stage loading coefficient in the range of 1.5–2.5 is being preferred, while flow coefficient in the range of 0.5–1.2 is being used. The reaction ratio RR is defined as

$$RR = \frac{h_2 - h_3}{h_1 - T_3} \approx \frac{T_2 - T_3}{T_1 - T_3} \quad (10.51)$$

Assuming the velocity V_3 is equal to V_2, we can get

$$C_p\left[T_{t_1} - T_{t3}\right] = C_p\left[T_1 - T_3\right] = UV_x\left[\tan\beta_2 + \tan\beta_3\right] \quad (10.52)$$

As no work is obtained from flowing fluid relative to rotor blade, we can have

$$C_p\left[T_2 - T_3\right] = \frac{1}{2}\left(V_{3R}^2 - V_{2R}^2\right) = \frac{1}{2}V_x^2\left[\tan^2\beta_3 - \tan^2\beta_2\right] \quad (10.53)$$

By using Equations 10.52 and 10.53 in Equation 10.51, we can have an expression for RR:

$$RR = \frac{T_2 - T_3}{T_1 - T_3} = \frac{\frac{V_x^2}{2}\left[\tan\beta_3^2 - \tan\beta_2^2\right]}{\frac{V_x}{U}\left[\tan\beta_2 + \tan\beta_3\right]} = \frac{\varphi}{2}\left[\tan\beta_3 - \tan\beta_2\right] \quad (10.54)$$

For 50% RR axial turbine, by using Equations 10.54 and 10.46, turbine blade is found to be symmetric as given below;

$$\alpha_2 = \beta_3; \quad \beta_2 = \alpha_3 \quad (10.55)$$

The stage loading coefficient for 50% RR axial turbine can be expressed as

$$\psi = \varphi\left[\tan\beta_2 + \tan\beta_3\right] = 2\varphi\tan\alpha_2 - 1 \tag{10.56}$$

For impulse blade axial turbine, RR will be zero and angle β_2 is equal to β_3. The stage loading coefficient for impulse blade can be expressed as

$$\psi = \varphi\left[\tan\beta_2 + \tan\beta_3\right] = 2\varphi\tan\beta_2 - 1 = 2\left(\varphi\tan\alpha_2 - 1\right) \tag{10.57}$$

For larger blade loading coefficient, we need to provide large nozzle angle, which may incur higher losses, leading to lower stage efficiency. Hence, one has to use optimum value of blade angle for higher stage efficiency.

REVIEW QUESTIONS

1. What are the stages of liquid fuel disintegration in a simple orifice during atomization process? Explain them in detail.
2. What do you mean by Ohensorge number? What is its physical meaning?
3. What do you mean by Weber number? What is its physical meaning?
4. What are the classifications of injectors used in LPRE?
5. What are the functions of injectors in LPRE? Describe four non-impinging injectors with neat sketches.
6. What are the impinging types of injectors? Describe them with their respective advantages and disadvantages?
7. What do you mean by SMD? Why is it used in rocket engine?
8. What is MMD? When are they preferred over SMD?
9. How is AMD different from SMD?
10. What do you mean by quality factor for a spray?
11. What are the types of arrangements for injector holes used in injector head of LPRE?
12. Which type of injector hole arrangement is preferred for HPRE?

13. Why is injector distributor used in modern rocket engines?

14. What are the functions of injector manifold?

15. What are the types of feed systems that can be used for LPRE and HPRE, respectively?

16. Describe a simple gas pressure feed system and state when it is preferred over pump feed system?

17. What are the types of gas pressure feed system devised for LPRE? Describe each of them with neat sketch.

18. Describe four types of solid-propellant gas pressure feed system with neat sketches.

19. What are the types of turbo-pump systems used in LPRE? Describe each of them with neat sketch.

20. What do you mean by pump-specific speed?

21. What do you mean by cavitation? Why is it to be avoided? How is it to be avoided?

22. What do you mean by NSPE? How is it related to suction-specific speed?

PROBLEMS

10.1 A small monopropellant liquid hydrazine rocket engine is designed to produce thrust of 25 N and I_{sp} of 198, which is to be operated for 5 h. Two layers of alumina granules are used for the catalyst bed with blade loading B_L of 3.75 g/cm$^2 \cdot$ s. Determine diameter of catalyst bed and amount of hydrazine required for this operation.

10.2 A liquid-propellant rocket engine is used to develop a thrust of 0.95 kN with characteristic velocity of 1750 m/s at chamber pressure of 4.5 MPa and MR of 1.1. If its thrust coefficient C_F happens to be 1.25, determine (1) throat of nozzle, (2) mass flow rate of oxidizer and fuel. A doublet impinging injection system with 12 injector elements is used with injection pressure drop of 1.2 MPa and discharge coefficient of 0.66. By considering ρ_F and ρ_{Ox} to be 1020 and 1150 kg/m^3, determine the diameter of injection holes.

10.3 In a spray experiment, the droplet number and size distribution are obtained as given in the following table:

Sl. No.	Droplet Size (μm)	Number
1.	0–20	25
2.	21–30	155
3.	31–40	375
4.	41–60	275
5.	61–80	124
6.	81–110	75
7.	111–140	25
8.	141–200	5

Estimate mean diameter (MD), area mean diameter (AMD), Sauter mean diameter (SMD) and Volume mean diameter (VMD) of this spray.

10.4 A liquid-propellant (MMH + N_2O_4) rocket engine is used to develop a thrust of 1.2 kN with characteristic velocity of 2050 m/s at chamber pressure of 0.85 MPa and MR of 1.6. If its thrust coefficient C_F happens to be 1.65, determine (1) throat of nozzle, (2) $\dot{m}_{\text{MMH}}$ and $\dot{m}_{N_2O_4}$. For this, 10 elements doublet impinging injection system is used with injection pressure of 1.5 MPa and discharge coefficient of 0.96. By considering ρ_{MMH} and $\rho_{N_2O_4}$ to be 868 and 1442 kg/m^3, determine diameter of fuel and oxidizer orifices.

10.5 A liquid-propellant (UDMH + N_2O_4) rocket engine is used to develop a thrust of 4.5 kN with characteristic velocity of 2250 m/s at chamber pressure of 3.85 MPa and *OFR* of 1.6. If its thrust coefficient C_F and characteristic length happen to be 1.45 and 1.5 m, respectively, determine combustion chamber volume and volumetric flow rate of each propellant by considering ρ_{UDMH} and $\rho_{N_2O_4}$ to be 791 and 1442 kg/m^3.

10.6 A semicryogenic (liquid kerosene + LO_2) liquid-propellant rocket engine based on gas generator feed cycle is designed to produce I_{sp} of 350 s and thrust of 750 kN. The MR for rocket engine is 2.5. The temperature of gas generator is restricted to 750 K. The pressure change across the kerosene (ρ = 950 kg/m^3) and liquid oxygen (ρ = 1150 kg/m^3) pumps can be 4 and 5.5 MPa, respectively. The isentropic efficiencies of kerosene and LOX pumps are 0.85 and 0.9, respectively. The product gas from the gas generator is expanded in turbine with pressure ratio of 15 and isentropic efficiency of 0.9. If the specific heat and specific heat ratio of the combustion products from the gas generator happen to be 1.2 kJ/kg K and 1.25, respectively, determine the power of fuel and oxidizer pump and percentage of propellants to be supplied to gas generator. Assume mechanical efficiency for both fuel and oxidizer pumps to be 0.98.

10.7 A solid-propellant pressure feed system with its chamber pressure of 1.75 MPa is used in liquid rocket engine to supply liquid propellant with density of 950 kg/m^3 at pressure of 1.1 MPa through 10 orifices with diameter of 2.1 mm. The discharge coefficient of the orifice is equal to 0.85. The regression rate of solid propellant with density of 1750 kg/m^3, adiabatic temperature of 3000 K, and molecular weight of 24 g/mol is 2.8 mm/s with interaction loss of 0.88. Determine liquid propellant flow rate and total burning surface area of solid propellant.

10.8 Determine the volume of N_2 gas tank at 13 MPa and 298 K required to pressurize H_2O_2 propellant tank at 2.5 MPa in rocket engine that produces 10 kN with I_{sp} of 245 s at chamber pressure of 1.8 MPa for 40 s. Assume that 1.5% extra propellant is to be provided for safe operation and density of propellant is 1150 kg/m^3.

10.9 A radial centrifugal pump is used to pressurize 105 L/s of liquid N_2H_4 with relative density of 1.02 g/cm^3 from 2.5 to 21 MPa, in which impeller tip speed should not exceed 100 m/s. Assume the width of the impeller is 10 mm and impeller diameter 150 mm. Determine number of stages and rotational speed.

10.10 A centrifugal pump is used to pressurize 1050 L/s of liquid UDMH with density of 791 kg/m^3 from 2.1 to 7.5 MPa. Determine the impeller speed if its specific speed is restricted to 0.5. If the vapor pressure of UDMH at 20°C in its storage tank is 13.7 kPa, determine net positive suction head and its suction-specific speed.

10.11 A liquid rocket engine is designed to supply 135 kN of thrust with I_{sp} = 275 s, during which oxidizer–fuel ratio is maintained at 2.75.

A single turbine with mechanical efficiency of 0.95 is used to drive oxidizer ($\eta_{Ox}=0.75$) and fuel ($\eta_F=0.85$) pumps. The combustion chamber pressure and temperature are 7.5 MPa and 3150 K, respectively. The injection velocity for oxidizer and fuel is maintained at 65 and 85 m/s, respectively. Determine the total power of turbine if propellant tank pressure for both fuel and oxidizer is maintained at 0.25 MPa. Consider $\rho_{\dot{F}} = 950\ \text{kg/m}^3$, $\rho_{Ox} = 1100\ \text{kg/m}^3$.

REFERENCES

1. Lefebvr, A.H., *Atomization and Spray*, Taylor & Francis, New York, 1989.
2. Mishra, D.P., *Gas Turbine Propulsion*, MV Learning, London, U.K., 2015.
3. Sutton, G.P. and Biblarz, O., *Rocket Propulsion Elements*, 7th edn., John Wiley & Sons Inc., New York, 2001.
4. Ramamurthi, K., *Rocket Propulsion*, MacMillan Published India Ltd, New Delhi, India, 2010.
5. Barrere, M., Jaumotte, A., de Veubeke, B.F., and Vandenkerckhove, J., *Rocket Propulsion*, Elsevier Publishing Company, New York, 1960.
6. Huzel, D.K. and Huang, D.H., *Modern Engineering for Design of Liquid Propellant Rocket Engines*, Vol. 147, AIAA Publication, Washington, DC, 1992.
7. Hill, P.G. and Peterson, C., *Mechanics and Thermodyanamics of Propulsion*, 2nd edn., Addison-Wesley Publishing Company, Boston, MA, 1999.

CHAPTER 11

Nonchemical Rocket Engine

11.1 INTRODUCTION

We have discussed extensively about chemical rockets in which energy is stored in chemical propellants to produce high-temperature and high-pressure gas that can be expanded in the nozzle to produce requisite thrust for space vehicles. Although it is simple to design chemical rocket engines for space applications, there is a fundamental limitation to enhance the energy level for deep-space applications. In other words, it would not be possible to enhance the thrust level and exhaust velocity beyond a certain value, as the power level of rocket is limited by the chemical energy level per unit mass due to the requirement of product gas with high combustion temperature and low molecular mass for obtaining higher exhaust velocity. We know that the maximum specific impulse is limited to 465 s for cryogenic liquid-propellant engine. Of course, the concept of multistaging has been used to overcome this problem but with higher effective mass ratio. However, nonchemical rocket engines can be used to achieve higher exhaust and specific impulse for deep-space applications. Let us look at typical data pertaining to certain characteristics of both chemical and nonchemical rockets, as shown in Table 11.1. It can be noted that nonchemical rockets have higher specific impulse and lower thrust/weight ratios as compared to chemical rocket engines.

Based on the source of energy, the nonchemical rocket engines can be classified into three categories: (1) electrical rocket engine, (2) nuclear

TABLE 11.1 Comparison of Rocket Engines

Sl. No.	Types	Specific Impulse(s)	Thrust/ Weight Ratio	Thrust Duration
1.	Chemical Rocket Engine	170–465	1–10	Minutes
2.	Electrothermal	300–1,500	$<10^{-3}$	Months (steady) Years (intermittent)
3.	Electromagnetic	1,000–10,000	$<10^{-4}$	Months (steady) Years (intermittent)
4.	Electrostatic	2,000–100,000	$<10^{-4}$–10^{-6}	Months/years (steady)
5.	Nuclear (thermal)	750–1,500	1–5	Hours

Source: Humble, R.W. et al., *Space Propulsion Analysis and Design*, McGraw Hill, New York, 1995 [2].

rocket engine, and (3) solar rocket engine. In this chapter, the basic principles of electrical rocket along with their respective design aspects are discussed extensively, highlighting their limitations. Besides this, both solar and nuclear rocket engines are covered briefly as they are in the infancy state of development.

11.2 BASIC PRINCIPLES OF ELECTRICAL ROCKET ENGINE

We know that in a chemical rocket engine, energy contained in the chemical bonds is utilized to enhance the exhaust velocity of product gas with minimum possible low molecular weight for producing higher specific impulse. But in case of electrical rockets, electrical energy can be utilized to impart higher flight velocity of space vehicle.

11.2.1 Classifications of Electrical Rockets

The simplest way of utilizing electrical energy for rocket propulsion is to heat the propellant flow by electrical heated surface. This kind of engine is known as resistance heating rocket/resistojet. Besides this, thermal energy can be added to propellant flow using electric arcs for rocket propulsion, which is known as arc-heating rocket/arcjet. Both resistojet and arcjet categories are known as electrothermal rockets. But the specific impulse achieved by these engines is in the range of 300–1500. In order to improve the specific impulse further, charge particles are accelerated by applying electrostatic forces with the help of electrical energy. This kind of technique for propelling a space vehicle is termed as electrostatic propulsion. If the charge particles happen to be ionized atoms or molecules, then such rocket is known as ion propulsion engine. Besides this, magnetic fields are utilized to accelerate the particles to a very high speed for obtaining higher impulse. In some cases, ionized gas known as plasma is employed to produce thrust, which is termed as plasma propulsion. Generally, it can be categorized as electromagnetic propulsion. All these electrical rocket engines can be classified broadly into three categories: (1) electrothermal, (2) electrostatic, and (3) electromagnetic. It can be noted that usual expansion of high-pressure and high-temperature gas through a nozzle is not used for producing thrust in case of electrostatic and electromagnetic rockets.

11.2.2 Background Physics of Electrical Rockets

From the preceding discussion it is clear that we need to get acquainted with the fundamentals of electrostatic and electromagnetic forces that are essential for electrical rockets. Besides this, the physics of ionized gases, electrical discharge of the particles encountered during the operation of electrical rockets are considered in this section briefly. It is important to have fundamental understanding of these basic concepts while dealing with design and development of new electric rockets. Interested readers may refer to the classic books on electric propulsion [1]. Of course, some of these aspects of electrostatic and electromagnetic forces might have been covered in your earlier courses. However, we will revisit briefly the four important topics relevant to electric rockets: (1) electrostatic force and its field, (2) electromagnetic force and its field, (3) ionization, and (4) electric discharge behavior.

11.2.2.1 Electrostatic and Electromagnetic Forces

We can recall from our high school physics that forces do exist between electrically charged particles governed by the famous Coulomb's law. As per this law, the electrostatic force F_e between two point charges is directly proportional to the product of two charges q_1 and q_2 and inversely proportional to the square of distance R between them, which is expressed as

$$F_e = K\frac{q_1 q_2}{|R|^3}R; \quad \text{where } K = \frac{1}{4\pi\varepsilon_0} \tag{11.1}$$

where

K is the constant that is equal to 9×10^9 Nm²/C²

ε_0 is the electrical permittivity of free space

This electrical permittivity represents the ability to store a charge in free space and is expressed by C/Vm. The permittivity of any other medium can be determined easily for the electrical permittivity of free space ε_0 and the relative permittivity ε_R as given in the following:

$$\varepsilon = \varepsilon_0 \varepsilon_R \tag{11.2}$$

The electric field E around a charge particle is formed when a unit charge experiences force, which is expressed using Equation 11.1:

$$E = K \frac{q_1}{|R|^3} R \tag{11.3}$$

When a charge is placed in this electrostatic field (V/m), the electrostatic force experienced by it would be expressed as

$$F_e = E q_2 \tag{11.4}$$

But when a charge q moves with a velocity V_q, it experiences a force F_m perpendicular to the direction of V_q and magnetic field B. This magnetic force F_m is expressed in mathematical form as

$$F_m = q V_q \times B; \tag{11.5}$$

where B is the magnetic field, which is expressed as Tesla (Weber/m^2). The magnetic field B around a charge particle is formed when a unit charge experiences a magnetic force. It is defined as the vector field that helps in describing the motion of a charged particle that satisfies the Lorentz force. The magmatic force becomes zero when a charge moves along the direction of magnetic force.

This magnetic force is governed by the ability of material to magnetize in presence of magnetic field, which is known as the permeability μ. Note that the magnetic permeability, of any other medium can be determined easily for permeability of free space μ_0 and the relative permittivity μ_R, as given in the following:

$$\mu = \mu_0 \mu_R \tag{11.6}$$

Interestingly, the permeability of free space μ_0 is related to the permittivity of free space ε_0 and velocity of light in vacuum (C_0) as

$$\mu_0 = \frac{1}{C_0^2 \varepsilon_0} \tag{11.7}$$

The permeability of free space μ_0 is equal to 1.257×10^{-6} H (Henry)/m. The induced magnetic field B in a medium of permeability μ can be related to the auxiliary magnetic field H as

$$B = \mu H \tag{11.8}$$

Note that H can be conceived to be formed by a certain amount of charge per unit area traveling at a given velocity. However, a charge particle will experience both electric and magnetic forces simultaneously, which can be expressed mathematically by combining Equations 11.4 and 11.5, as follows:

$$F = q\left(E + V_q \times B\right); \tag{11.9}$$

But in a real situation, several particles will be interacting with each other in the presence of, both, the electric and magnetic fields. Hence, the total electromagnetic force per unit volume vector acting on the ith charge particle with mass m that causes a motion can be expressed in scalar form as

$$m\frac{dV_i}{dt} = q_i\left(E + V_i \times B\right) + \sum_k F_{ik} \tag{11.10}$$

where

- ρ_q is the net charge density
- I_q is the electric current vector density
- F_{qk} is the collision force per charge particle

The summation term on the right-hand side represents all momentum changes due to collision of charge particles including same type. The charge particle energy does not get affected by the magnetic field as Lorenz force is perpendicular to particle's velocity as shown in Figure 11.1. In this case, propellant flows through the electric field, which is caused due to current

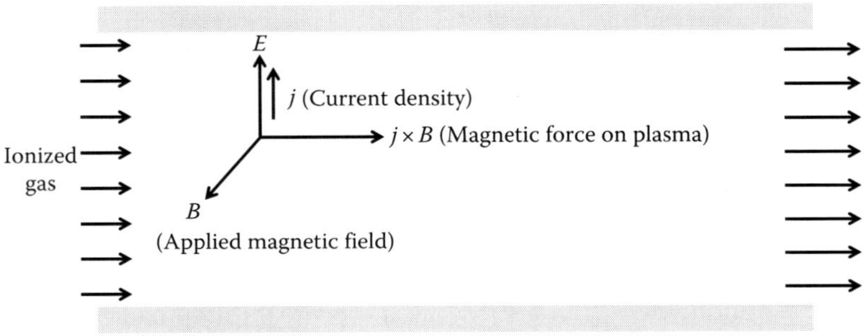

FIGURE 11.1 Electromagnetic force on a charge particle.

flows from the anode to the cathode. The magnetic field produced perpendicular to the current emanates from the poles of permanent magnets not shown in Figure 11.1. The current flowing through the gas also induces a magnetic field. The magnetic fields caused due to external sources add to the magnetic field induced by the current itself. In other words, electromagnetic forces are generated due to the combined magnetic fields that accelerate the charge particle. In case of electric field alone, the charge particles accelerate being lighter in nature, as per the following expression:

$$\frac{dV_i}{dt} = \frac{q_i}{m} E \tag{11.11}$$

But in the case of magnetic field alone, without any collision, the particle undergoes gyration due to balance between the centrifugal and Lorenz force, as given in the following:

$$\frac{V_i^2}{R_g} = \frac{q_i}{m} |V_i \times B| \tag{11.12}$$

where R_g is the radius of gyration. The frequency of gyration ω_g, is given by

$$\omega_g = \frac{q_i B}{m} \tag{11.13}$$

When an electric field is applied to a gyrating particle, the center of gyration drifts perpendicular to the magnetic field with the following drift velocity V_d:

$$V_d = \frac{E \times B}{B^2} \tag{11.14}$$

Note that the electrostatic force accelerates the charge particle upward, while the magnetic field accelerates the particle perpendicular to the electric field that results in the drift velocity perpendicular to both electric and magnetic fields.

For electrostatic acceleration, the net charge density ρ_q is finite and is often known as space charge density. In the case of ion rocket, space charge density is instrumental in imparting electrostatic force for the acceleration of charge particle. In contrast, in plasma rockets, electromagnetic accelerator is used for acceleration of propellants. We need to understand the processes involved during ionization of particles in electric rocket engine, as discussed in the following.

11.2.2.2 Ionization

We know that ionization is the process of producing charge particle from an atom by heat, electrical discharge, radiation, or chemical reaction. During this process, electron from the atom will be lost due to collision with subatomic particle or with other atoms, molecules, or ions, or interactions of photons with adequate high frequency or by strong electric field. Ionization of molecules in a gas can take place not only through the mentioned process but with additional possibilities such as internal rotation vibrational modes related to molecular structure, and so on. For ionization to be successful, the electron is to be removed from its ground state by supplying requisite energy, which is known as the ionization potential E_i. Note that the ionization potential expressed in terms of electron volt varies from one gas to another, as shown in Table 11.1. An electron volt is the energy one electronic charge acquires while falling through a potential difference 1 V, (1eV = 1.6×10^{-19} J). Of course, ionization of gas can be possible at lower relative energies through multiple collisions by exciting the bound electron first to higher state and then to provide sufficient energy to escape. The ionization energy is also dependent on the ionization cross section, as shown in Table 11.2, for each gas. Besides this, the ionization of gas is dependent on the high-energy tails of the Maxwellian distributions of electron energies. It also depends on the electron temperature rather than atom/ion temperature.

11.2.2.3 Electric Discharge Behavior

We know that electric discharge in gases occurs when electric current flows through a gaseous medium due to ionization of the gas. When the neutral gas is exposed to electrical field, the resistance will be quite high. But if the

TABLE 11.2 Ionization Potentials of Various Gases

Gas	Ionization Potential E_i (eV)	Maximum Cross Section (10–20 m²)/ Max Electro Energy (eV)
Argon (A)	15.75	2.75/75
Hydrogen (H)	13.59	0.7/75
Hydrogen (H_2)	15.42	1.0/70
Helium (He)	24.58	0.38/90
Nitrogen (N)	14.54	
Nitrogen (N_2)	15.58	3.0/100
Neon (Ne)	21.56	0.8/200
Oxygen (O)	13.61	
Oxygen (O_2)	12.2	
Krypton (Kr)	13.9	4.1/80
Xenon (Xe)	12.1	5.3/100
Cesium vapor (Cs)	3.89	9.4/20

potential difference across anode and cathode, as shown in Figure 11.2, is enhanced, the gas becomes partially ionized and thus allows conducting electricity with drop in its resistance. The processes involved during arc formation are shown in Figure 11.2. The electrons created by initial ionization of gas molecules acquire sufficient energy from the electric field between collisions and thus ionize other atoms/molecules. Subsequently, positive ions acquire from the electric field during collisions to emit electrons from the cathode by bombardment. This process may be augmented by internal radiation due to photoemission from the cathode, which can

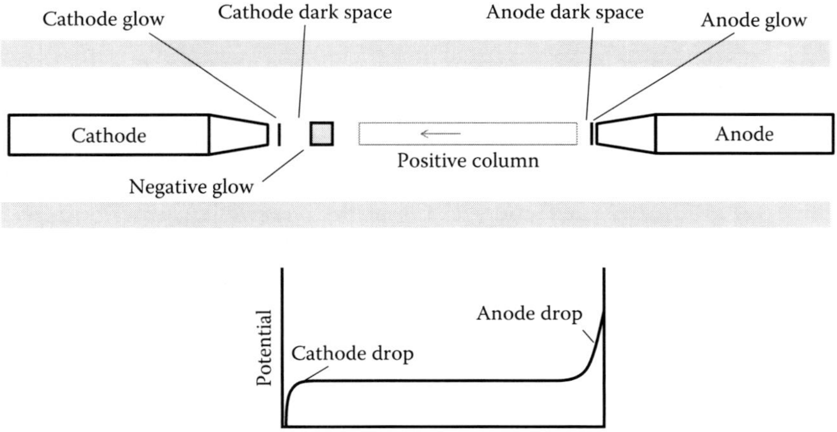

FIGURE 11.2 Processes during arc formation.

form a spark/arc. Note that spark/arc can occur only when potential difference goes beyond the certain breakdown voltage, which is dependent on discharge gap and gas pressure. As per Paschen's law, the breakdown voltage is dependent on the product of the discharge gap and gas pressure. In vacuum, the breakdown voltage is the lowest, as the product of the discharge gap and gas pressure is zero. Beyond this breakdown voltage across the cathode and anode, the discharge behavior depends on the voltage, shape of electrodes, and gas pressure. If the gas pressure is high and electrodes have sharp points, then Corona discharge will occur. But if the electrodes are blunt and low gas pressure prevails, glow discharge will occur depending on the current supplied by the power source. This glow discharge is mainly caused due to ion-bombardment emission from the cathode. Electrons and positive ions move in opposite directions and transfer their charge to their anode and cathode, respectively. Note that neutral gas atoms/molecules get heated due to their collisions with ions and electrons. This quasi-neutral plasma is often known as the positive column, which occupies larger portion of the discharge space, as shown in Figure 11.2. The voltage drop across this column is quite small. However, bulk of voltage drops occur in regions near to the anode and cathode, as shown in Figure 11.2. Although the voltage fall near the cathode is quite low as compared with that at the anode, its surface emits electrons profusely due to combination of thermionic, photoelectric, and field emission processes.

11.3 ELECTROTHERMAL THRUSTERS

In case of electrothermal rocket/thrusters, electrical energy is utilized to heat the propellant to high-temperature and high-pressure exhaust gas, which is expanded in a convergent–divergent nozzle to produce thrust. The propellant can be heated by using either electrical resistive power or by electrical arc discharge. Hence, electrothermal thrusters are broadly divided into two categories: (1) resistojet and (2) arcjet, which are discussed in the following.

11.3.1 Resistojets

This is the simplest of all electrical thrusters that consists of electrically heated pressure chamber and a convergent–divergent nozzle similar to the chemical rocket engine. In this case, the propellant gas is heated by making it come in contact with electrically heated solid surface. Hence, several types of resistojets can be devised to augment heat transfer from electrically heated solid surface to the incoming propellant. Some of the heating elements for resistojets are shown in Figure 11.3. In the case of straight coiled

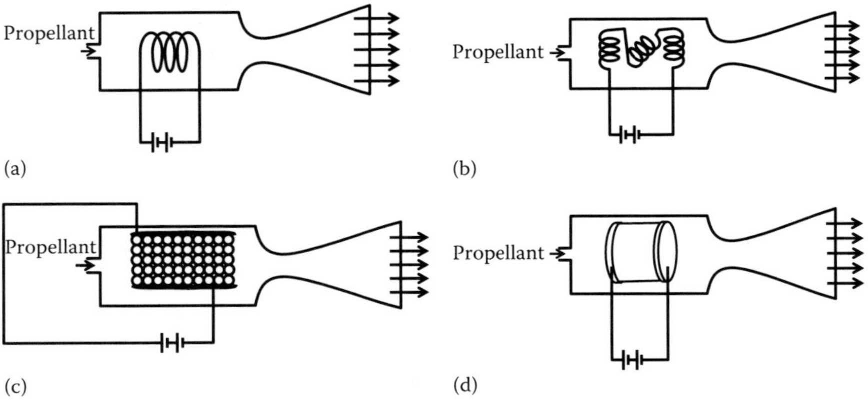

FIGURE 11.3 Four types of electrical heating elements for resistojets: (a) straight coiled wire, (b) transverse coiled wire, (c) sphere bed, and (d) chamber wall.

wire resistojet, propellant passes along the heating element, which restricts the heating rate of the propellant. In order to improve the heating rate of the propellant, heating coils can be placed in transverse direction of flow, as shown in Figure 11.3b. A bed of tungsten spheres (Figure 11.3c) is used to heat the propellant by passing current through the contact resistances of spheres. However, it enhances the bulkiness while incurring significant pressure losses. The heating chamber can be used as resistance to heat the incoming propellant while passing over it. Both DC and AC power supplies at a power level of 1 W to 100 kW are being used over a wide range of supply voltage. The performance of such kind of resistojets depends on the extent of heat transfer from resistance element to the propellant gas stream and radiative heat transfer from the complete assembly as heat loss through the walls is basically a loss of power in the thruster. In order to reduce radiative heat transfer, radiative shields are provided around the active tungsten heating element, as shown in Figure 11.4, which not only reduces the heat losses but also transfers heat to the incoming propellant in the reentrant gas flow passages. Thus, it can be called regenerative resistojet, as multichamber heat exchanger is used to recover heat loss through the chamber wall. Another way of reducing heat transfer is to place an array of resistive heating ducts in a honeycomb manner, as shown in Figure 11.3b. Of course, such types of methods may increase its weight, which may be quite small as compared to the power supply system particularly for higher power resistojets. Besides this, heat losses from the nozzle throat and heater cavity are to be minimized if not eliminated altogether.

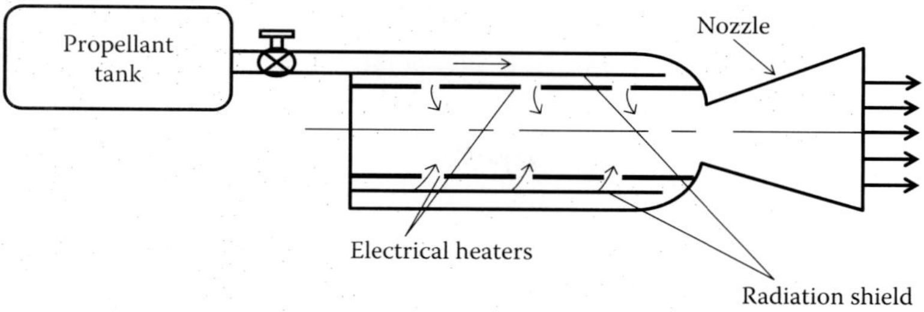

FIGURE 11.4 Schematic of a typical resistojet.

The performance of electrothermal thrusters can be evaluated by using the following parameters: thrust coefficient C_F, characteristic velocity C^*, specific impulse, various efficiencies, as used in chemical rocket engines (see Chapter 3). The specific impulse I_{sp} can be related to thrust coefficient C_F and characteristic velocity C^*, as given in the following:

$$I_{sp} = \frac{C^* C_F}{g} \tag{11.15}$$

Note that the characteristic velocity C^* does not depend on the method of heating of the propellant gas, which can be observed from the following expression (see Chapter 4):

$$C^* = \frac{\sqrt{R_u T_{t2}/MW}}{\Gamma} \quad \text{where } \Gamma = \sqrt{\gamma\left(\frac{2}{(\gamma+1)}\right)^{\frac{\gamma+1}{\gamma-1}}} \tag{11.16}$$

In electrothermal rockets, the characteristic velocity C^* depends only on the molecular weight specific heat ratio and temperature of propellant gas as in the case of chemical rocket. In the case of a chemical rocket, temperature of propellant gas is dependent on the propellant composition and not on its mass flow rate. In contrast, in the electrochemical rocket, temperature is dependent inversely on the mass flow rate. If the mass flow rate is too small for a particular heat input, then the temperature may rise to spoil the heating element itself. Hence, an electrothermal thruster has a maximum flow rate/power ratio that is limited by the materials of filaments and chamber wall. Generally, high melting point temperature materials such as tungsten and platinum are used as filament materials. The power output P_J

of the electrothermal thruster is basically the kinetic energy of jet per unit time, generated due to expansion of resistive heated propellant through the exhaust nozzle. The kinetic power/thrust ratio for the electrothermal thruster can be expressed as

$$\frac{P_J}{F} = \frac{0.5\dot{m}V_e^2}{\dot{m}V_e} = 0.5V_e = 0.5I_{sp}g \tag{11.17}$$

Note that for the electrothermal thruster with higher specific impulse, the electrical power requirement will be quite high per unit thrust. All electrical resistive power transferred to the propellant cannot be converted into the thrust power. The electrical energy can be lost due to (1) stay currents and ohmic resistance, (2) heat losses, (3) nonaxial component of exhaust, and (4) nonuniform heating of propellant. Hence, thruster efficiency η_T can be defined as the ratio of thrust producing kinetic energy rate from the exhaust nozzle to the total electrical power, given as follows:

$$\eta_T = \frac{\dot{m}V_e^2}{2P_J} = \frac{\dot{m}V_e^2}{2Vi} = \frac{FI_{sp}g}{2Vi} \tag{11.18}$$

where

V is the voltage

i is the current

This efficiency indicates how effectively the electrical energy is converted into the thrust power. Generally, the thruster efficiency of resistojets varies from 65% to 90%, which depends on the nature of propellant and its exhaust temperature.

The major engineering problems of designing and developing resistojets is the fabricating and maintaining of the conducts and insulators that can keep electrical integrity and vacuum seals at high temperature in the range of 2500–3000 K. Of course, tungsten and boron nitride insulators can be used as conductor and insulator at high temperature. Besides this, the refractory metals such as tungsten, tantalum, molybdenum, rhenium platinum, and their alloys are being used for resistance element. Hence, it can be concluded that resistojet is a temperature-limited device, which property can be overcome with the development of high-temperature materials in future. The chamber pressure of electrothermal thruster can be chosen properly by considering the propellant type, and its dissociation and recombination characteristics, heat exchanging performance, and rate of nozzle throat erosion. As high chamber pressure improves the heat exchange between the electrical element and propellant with reduction in gas dissociation losses, it incurs a higher rate of heat transfer rate, nozzle throat erosion, and higher stress on chamber wall. Hence, the high chamber pressure for electrothermal thruster is restricted to 1.5 MPa. In spite of lower values of thrust and mass flow rate, these thrusters are preferred for space applications. They are generally used for space mission with moderate levels of velocity increment that are limited by power level, thrusting time, and plume effects. They are deployed for auxiliary but important applications during satellite operation, station keeping, drag neutralization, altitude control, and so on, where compactness, high reliability, and low thrust are called for. Several resistojets such as Satcom-1R, Gstar-3, and Meteor 3-1 are being used on several occasions for space applications due to higher performance level.

Example 11.1

A 1.75 kW electrochemical rocket with helium as propellant is to be designed and developed using tungsten filament resistive heating element for obtaining the thrust coefficient of 1.8. Assuming the mass flow rate of propellant to be 0.12 g/s, determine the exit velocity, characteristics velocity, temperature of propellant gas, thrust, and specific impulse for electrochemical rockets with thruster efficiency of 0.85.

Solution

The exit jet velocity can be determined by using Equation 11.18 as follows:

$$V_e = \sqrt{\frac{2P_J\eta_T}{\dot{m}}} = \sqrt{\frac{2\times 1750\times 0.85}{0.00012}} = 4979 \text{ m/s}$$

The characteristics velocity can be estimated as

$$C^* = \frac{V_e}{C_F} = \frac{4979}{1.8} = 2766.1 \text{ m/s}$$

The temperature of the propellant gas can be obtained using Equation 11.16 as follows:

$$T_{t2} = \frac{\left(C^*\Gamma\right)^2 MW}{R_u} = \frac{4\left(2766.1\times 0.73\right)^2}{8314} = 1961.7 \text{ K}$$

$$\Gamma = \sqrt{\gamma\left(\frac{2}{\left(\gamma+1\right)}\right)^{\frac{\gamma+1}{\gamma-1}}} = \sqrt{1.66\left(\frac{2}{2.66}\right)^4} = 0.73$$

The thrust produced by this engine is determined as

$$F = \dot{m}V_e = 0.0012\times 4979 = 5.97 \text{ N}$$

The specific impulse of this engine is determined as

$$I_{sp} = \frac{V_e}{g} = \frac{4979}{9.81} = 507.5 \text{ s}$$

It can be noted that there is higher specific impulse for this electrothermal thruster as compared to the chemical rocket engine, of course, with very low thrust value. Generally, the electrothermal thruster can provide much higher exhaust velocity as compared to chemical rockets but with low mass flow rate and lower thrust. But its advantage is that it can operate with lower molecular weight gas without much constraint, as in chemical rocket engines.

In order to improve the specific impulse further, monopropellant like hydrazine is dissociated with the help of resistive heating elements. During this process, heat from the surface of resistive heating surfaces is transferred to the propellant and its dissociated products due to conduction, convection, and radiation mode of heat transfer. As a result, the temperature of propellant gas is enhanced further due to heat release of chemical reactions during hydrazine dissociation apart from resistive heating. For this purpose, adequate residence time must be provided by utilizing proper design to effect proper heat transfer to the product gas. The augmentation of product gas temperature can enhance the specific impulse further. Such kind of thruster is known as augmented electrothermal hydrazine thruster (AEHT).

11.3.2 Arcjets

The limitation of the electrothermal rocket to have higher specific impulse can be overcome by releasing the heat to the propellant gas at higher temperature using arc discharge. Of course, the high-temperature and high-pressure gases are accelerated to supersonic speed by expanding in a nozzle as in a chemical rocket. Such kind of thruster is known as arc-heating rocket/arcjet, in which electrical energy is converted into thermal energy by producing an arc. A typical arcjet is shown in Figure 11.5, which consists of anode and cathode and CD nozzle. It must also contain both power supply and propellant feed systems. Both, anode and cathode are generally made of tungsten material as it can withstand high temperature. The anode is made in the shape of nozzle that can serve as nozzle for expansion of propellant gas. A steady DC voltage is to be maintained across the anode and cathode using a suitable power supply system such that current can be changed depending on the propellant flow rate. In order to initiate the arc, higher voltage as compared to steady operation voltage is to be supplied to ensure breakdown of cold gas. The current passes through the ionized gas adjacent to the cathode. The arc discharges from the tip of cathode forming an arc column, as shown in Figure 11.5 in the throat region of nozzle. The maximum temperature occurs in the region of arc core in an arcjet thruster, because energy input to the propellant occurs in the narrow region of smaller diameter laminar flow within the throat of nozzle. Hence, the hot gas at the core of arc must mix quickly with cold propellant gas either with help of creating turbulence or vortex in the propellant flow.

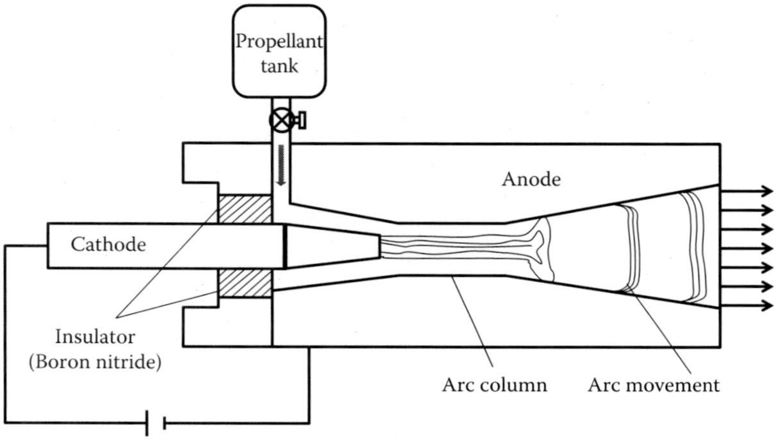

FIGURE 11.5 Schematic of a typical arcjet.

Generally, the propellant is introduced to the annular plenum chamber from propellant tank with a control valve, as shown in Figure 11.5, closer to the cathode so that it can keep the anode at lower temperature. When the working fluid is passed through the arc, it attains high temperature due to mixing of hot and cold propellant gas. It is desirable for the arc to attach itself in the divergent portion of nozzle just downstream of its throat. However, sometimes the arc may move downstream of divergent portion of nozzle getting detached to the arc column which is dependent on the magnitude of arc voltage and propellant flow rate. Generally, high-current arcs are inherently unstable as the ions are affected not by electric field but also the magnetic field caused by motions of ions themselves. As a result, arc may break down into a number of columns, forming pinches and wiggles known as *pinching effect* that may reduce the energy transfer to the propellant gas. This problem can be avoided by using either imposing external magnetic field or by imparting swirling flow to the propellant gas. This swirling propellant gas also helps in improving heat transfer from walls and electrodes and this keeps electrode cooler while enhancing mixing and heat transfer between arc and propellant gas. This pinching effect may cause hot spots at arc attachment points, resulting in local erosions of electrodes leading to reduction in their life. Such problems can be avoided by using high-temperature materials, external electric field, or by imparting swirling to the propellant gas flow. Besides this, the rate of erosion of electrodes is also influenced by the nature of propellants. For example, hydrogen gas has lower erosion rate as compared to nitrogen and argon gas. Of course, several propellants such as H_2, He, Ar, Ne, N_2, NH_3, and N_2H_4 (catalytic decompositions products) can be used for this thruster. However, H_2 gas is the most preferred among these gases due to its better heat transfer and other desirable properties.

In the case of an arcjet thruster, almost half of the electrical energy is lost to the jet, while 10%–20% electrical energy input is also lost in the form of heat losses. As a result, the typical efficiency of arcjet thruster ranges from 30% to 40%, which is much lower than an electrothermal thruster. However, it is preferred over the electrothermal one as the propellant gas can attain higher temperature in the range of 4000–5000 K because higher electrical energy can be transferred during the formation of arc due to direct interaction of electric current and propellant. As a result, exhaust velocities in the range of 7,500–20,000 m/s can be achieved with enhanced specific impulse. Generally, the thrust-to-power ratio of arcjet thruster is around six times that of the resistojets. High power arcjets in the range of

2–3 kW are being used for station keeping thrusters. But the power supply unit for arcjets is quite complex and bulky as compared to that of the electrothermal thrusters. Several arcjets are being used routinely as station keeping thrusters. A lifetime as high as 800 h for an arcjet has been established; this makes it viable for space applications.

As in the case of resistojet, the catalytic decompositions of hydrazine can be used in an arcjet to further augment specific impulse. This is known as chemical arcjet thruster (CAJT). Liquid hydrazine is decomposed catalytically into gaseous products which enters the arc at around 1000 K. Subsequently, the heat transfer to this hydrazine gas can enhance product gas temperature and results in higher specific impulse. Note that hydrazine being a liquid is easy to store in smaller volume space and hence is preferred over other gaseous propellants. The specific impulse in the range of 400–600 s can be easily achieved for a chemically arcjet thruster.

11.4 ELECTROSTATIC THRUSTERS

The specific impulse of electrical thrusters can be increased further beyond that of the electrothermal thrusters by the expansion of high-temperature exhaust gas in a CD nozzle. Rather, the specific impulse of electrical thrusters can be enhanced considerably by accelerating the nonneutral charged particle in the exhaust gas with the help of electric field. This electric force on the charged particle of propellants depends on the strength of electric field, charge particles, and their direction. Generally, positive ions are preferred over the electrons due to their heavy molecular mass. For this purpose, the propellant has to be ionized first and these ions are accelerated further using electromagnetic fields to very high bulk velocity. Of course, the number of ions in the exhaust should be small such that they would collide or recombine with electrons very often to have lower bulk velocity. Positive ions can be produced by bombarding monatomic gas with electrons from a heated cathode. This kind of thruster is known as the *electron bombardment thruster*. In some cases, positive ions are produced when propellant comes in contact with hot porous contact ionizer, which is labeled as *ion contact thruster*. Besides this, charged colloidals also produced by charging of propellant droplets are being used in electrostatic thrust. This kind of thruster is known as *colloid thruster*.

Let us consider a typical ion thruster, as shown in Figure 11.6, which consists of three main components, namely, ionization chamber, accelerating chamber, and neutralizing electron gun. The ionization chamber

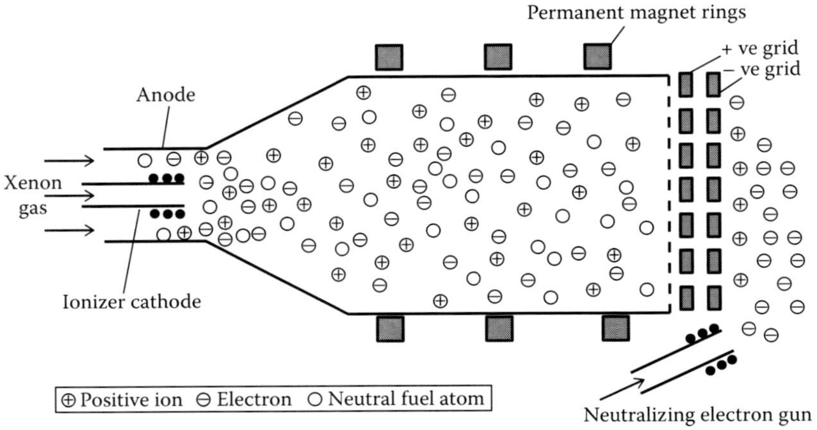

FIGURE 11.6 Schematic of accelerating positive charge ion across the grid.

consists of cathode and anode, which ionizes the propellant gas, as shown in Figure 11.6. This is also known as thermionic emitter as high temperature prevails on the cathode face, which augments electron generation. Subsequently, these electrons are accelerated due to the presence of the radial magnetic field created by strong permanent magnets, as shown in Figure 11.6, and attain higher energy levels in the order of tens of electron volts, which is sufficient to ionize the neutral propellant gas. Generally, ionization of propellant gas is achieved during its collisions with high-energy electrons.

In some cases, axial magnetic field is provided in conjunction with radial field, making the electrons move in a spiral direction. As a result, the residence time in a finite chamber increases, which enhances the collisions of electrons with neutral propellant, and this efficiency of ionization gets increased considerably. There is an increase in ions production, which enhances electron current.

Subsequently, the positive ions get accelerated further to V_e while passing through two grids, as shown in Figure 11.6. The first grid extracts the ion particle, which is known as extraction grid, shown in Figure 11.7. The second grid is known as the accelerating grid as it accelerates the ions to very high exhaust velocity V_e as opposite polarity is maintained with high potential difference with respect to the extraction grid. The grids get easily eroded over time due to bombardment of charge particles on their surface during the operation of ion thruster. This kind of erosion of grid is known as sputter, which must be avoided to have a better life. Generally,

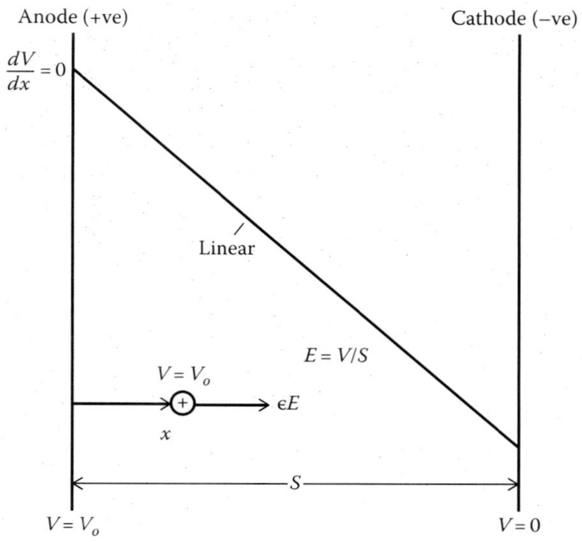

FIGURE 11.7 Potential distribution in a planar electrical field.

high erosive resistant material, namely, molybdenum is preferred due to its lower sputter. Generally, the gap between two grids is in the range of 2–3 mm. The ions become highly energized and leave the thruster with a very high velocity in the order of 3,000–60,000 m/s, depending on the design and operational conditions. The expelled ion beam provides the thrust on the accelerating grid due to momentum change. As a result, there is no need to have a nozzle for producing thrust as in thermal rockets. But these ions are to be neutralized in the downstream of the thruster, which can be accomplished using neutralizing electron gun, as shown in Figure 11.6. The electrons released from this gun combine with positive ions to make them neutral so that the accumulation of ions around spacecraft can be avoided.

11.4.1 Basic Principles of Electrostatic Thrusters

In order to analyze the operation of electrostatic thruster, let us consider a planar electrical field of strength E created by potential difference of V across a distance S, as shown in Figure 11.7. Let us consider a single charge particle with mass m_q with unit charge q moving in this homogeneous electric field experiencing a force F_q, which is equal to qE. The work done qES by this force on it is equal to the change in kinetic energy from anode to cathode, expressed as

$$\frac{m_q V_q^2}{2} = qES = qV \tag{11.19}$$

Note that in this expression, it has been assumed that charge leaves the anode with negligible velocity. Then the first term represents the kinetic energy gained by the charge. Using Equation 11.19, the velocity of charge can be expressed as

$$V_q = \sqrt{2\left(q/m_q\right)V} \tag{11.20}$$

Note that the velocity of charge particles passing through the same electric field will be the same for all. Then, the thrust experienced by the thruster would be equal to the product of total mass flow rate of ions $\dot{m}_p$:

$$F = \dot{m}_p V_q \tag{11.21}$$

The total mass flow rate of ions $\dot{m}_p$ can be expressed in terms of current i, charge q, and mass of charge m_q, as given in the following:

$$\dot{m}_p = \frac{m_q i}{q}; \quad \text{as } i = \frac{\text{Charge}}{\text{Time}} = \frac{\text{Charge}}{\text{Particle}} \cdot \frac{\text{Particle}}{\text{Mass}} \cdot \frac{\text{Mass}}{\text{Time}} = \frac{q\dot{m}_p}{m_q} \tag{11.22}$$

By using Equations 11.20 and 11.22, we can rewrite the thrust relation (Equation 11.21) as

$$F = \frac{m_q i}{q}\sqrt{2\left(q/m_q\right)V} = i\sqrt{2\left(m_q/q\right)V} \tag{11.23}$$

The thrust can be expressed in terms of beam power P_b as it is equal to kinetic energy rate $\dot{m}_p V_q^2/2$:

$$F = \frac{2P_b}{V_q}; \quad \text{as } P_b = \dot{m}_p V_q^2/2 = Vi \tag{11.24}$$

It can be noted that the thrust is dependent on the beam power P_b and charge velocity but not on the nature of propellant, unlike in the thermal rocket engine. By using Equation 11.23, the expression for thrust can be written in terms of the beam power P_b and voltage V and mass-to-charge ratio as follows:

$$F = \sqrt{2P_b i\left(m_q/q\right)} = \sqrt{2P_b^2 \frac{\left(m_q/q\right)}{V}} \tag{11.25}$$

It can be noted that in a propellant with lighter ions, the higher current is to be used for achieving higher thrust. However, current per unit area, known as current density j, is limited for electrostatic thruster as higher current means a larger cross-sectional area and, thus, heavier thruster.

We need to understand the limitation of current density j further. We know that when positive ion density increases, the electrostatic field created by it counteracts the accelerating field. We know that the constant electric field created by the accelerating grid in the absence of ions depends on the potential difference and separation distance between grids. When ions are passed through the grids, the electric field gets changed due to its presence. In order to increase the thrust for a given ion thruster, current flowing between grids is to be increased, which creates an opposing electrostatic field due to the ion field. If this electrostatic field due to ion flow reaches certain critical value, it neutralizes the applied potential across the grids and thus ions would move from the anode ($dV/dx = 0$ at $x = 0$). This condition is known as the space charge limit, indicating the maximum current that can be used for design of electrostatic thruster. We can derive an expression for this maximum current density by invoking the Gauss Law, which relates the field strength E and charge density ρ_q, considering the field to be one dimension, as given in the following:

$$\frac{dE}{dx} = \frac{\rho_q}{\varepsilon_0} \tag{11.26}$$

By using the definition of field E, the Poisson's equation can be derived from Equation 11.26, as follows:

$$\frac{d^2V}{dx^2} = -\frac{\rho_q}{\varepsilon_0}; \quad \text{as } E = -\frac{dV}{dx} \tag{11.27}$$

Equation 11.27 can be expressed in terms of current density j, charge mass mq, charge q, permittivity ε_0, and voltage difference $(V_0 - V)$ across the field as

$$\frac{d^2V}{dx^2} = -\frac{j}{\varepsilon_0}\sqrt{\frac{m_q}{2q(V_0 - V)}}; \quad \text{as } \frac{m_q V_q^2}{2} = q(V_0 - V); \; j = \rho_q V_q \tag{11.28}$$

By integrating Equation 11.28 at limiting condition, we get the limiting current density as follows:

$$\frac{d}{dx}\left(\frac{dV}{dx}\right)^2 = -2C\sqrt{\frac{1}{(V_0 - V)}}\frac{dV}{dx}; \quad \text{as } C = \frac{j}{\varepsilon_0}\sqrt{\left(\frac{m_q}{2q}\right)} \tag{11.29}$$

By applying limiting boundary condition $dV/dx = 0$ at $x = 0$ and $V = V_0$, Equation 11.29 can be integrated to get

$$\frac{dV}{dx} = 2\sqrt{C}(V_0 - V)^{1/4}; \quad \text{as } C_1 = \text{Integration constant} = 0 \tag{11.30}$$

By integrating Equation 11.30 applying boundary condition ($V = V_0$, at $x = 0$), we can get the following expression for V:

$$V = V_0 - \left(\frac{3}{2}x\right)^{4/3} \varepsilon_0 \left(\frac{m_q}{2q}\right)^{1/2} \left(\frac{j}{\varepsilon_0}\right)^{2/3} \tag{11.31}$$

The maximum current density corresponding to potential difference of V_0 and electrode spacing of s can be obtained by using Equation 11.31 as follows:

$$j_{\max} = \frac{4\varepsilon_0}{9}\left(\frac{2q}{m_q}\right)^{1/2} \frac{V_0^{3/2}}{S^2} \tag{11.32}$$

Note that this maximum current density is caused due to space charge effect and is used as an important design parameter. The maximum current density limited by the space charge effect can be enhanced by placing decelerated grid downstream of the acceleration grid. The maximum thrust per unit area F'' can be determined as

$$F''_{max} = \frac{F}{A} = \frac{\dot{m}_p V_q}{A} = j_{max}\left(\frac{m_q}{q}\right)V_q; \quad \text{as} \quad \frac{\dot{m}_p}{A} = \rho V_q = \rho_q \frac{m_q}{q} V_q$$
$$\text{and} \quad j = \rho_q V_q \tag{11.33}$$

The maximum thrust per unit area F'' can be expressed in terms of electric field by using Equations 11.32 and 11.20, as follows:

$$F''_{max} = \frac{8\varepsilon_0}{9}\left(\frac{V_0}{S}\right)^2 = \frac{8\varepsilon_0}{9}(E)^2 \tag{11.34}$$

It can be noted that the maximum thrust per unit area is dependent strongly on voltage and inversely on electrode spacing. Hence, the maximum thrust can be enhanced by decreasing the electrode spacing S for same potential difference. In order words, field strength must be increased either by increasing voltage or by decreasing the electrode spacing S, or by combining both parameters. Note that spacing between electrodes cannot be decreased beyond 0.5 mm as electrical breakdown may occur due to distortion in the electric field. The potential difference between electrodes for getting maximum current density can vary between 1.5 and 2.5 kV.

Example 11.2

An ion thruster with charge-to-mass ratio of 475 C/kg is designed to produce specific impulse of 2200 s. If the distance between the grids is 2 mm and maximum allowable voltage is 8550 V, determine space charge current density limit and the diameter of the round beam for producing a thrust of 6.5 N.

Solution

Considering the one-dimensional model and using Equation 11.32, we can determine maximum space charge current density as follows:

$$j_{\max} = \frac{4\varepsilon_0}{9}\left(\frac{2q}{m_q}\right)^{1/2}\frac{V_0^{3/2}}{S^2} = \frac{4\times 8.75\times 10^{-12}}{9}(2\times 475)^{1/2}\frac{(8550)^{3/2}}{(2\times 10^{-3})^2}$$

$$= 23.69 \ \mathrm{A/m^2}$$

By using Equation 11.33, we can determine the cross-sectional area of the beam as follows:

$$A = \frac{F_{\max}}{\dot{m}_p V} = \frac{F_{\max}}{j_{\max} I_{sp} g}\left(\frac{q}{m_q}\right) = \frac{6.5}{23.69\times 2200\times 9.81}(475) = 0.006 \ \mathrm{m^2}$$

The diameter of round beam for producing a thrust of 0.65 N is estimated as

$$d = \sqrt{\frac{4A}{\pi}} = \sqrt{\frac{4\times 0.006}{3.14}} = 0.087 \ \mathrm{m}$$

11.4.2 Propellant Choice

Several propellants such as nitrogen (N_2), hydrogen (H_2), mercury (Hg), cesium (Cs), krypton (Kr), and xenon (Xe) can be used for this thruster. The atomic mass and number of these gases are given in Table 11.2. For getting higher exhaust velocity as per Equation 11.20, propellant with higher charge-to-mass ratio is required along with larger potential difference across the grids. As this kind of thruster produces very high exhaust velocity as compared to other thermal thrusters, there is no need to use hydrogen gas. The nitrogen gas is not used as it is quite difficult to ionize. Among all other propellants, Hg, and Cs, and Xe, are toxic in nature, which can contaminate the optical surfaces of spacecraft due to their vapors. Besides this, extra energy is required to evaporate these metallic propellants as they are quite difficult to ionize in liquid form. Hence, in modern times, xenon is preferred as propellant as it has reasonable charge-to-mass ratio and, being nontoxic by nature, does not pose any problem of contamination of spacecraft.

11.4.3 Performance of Ion Thruster

The performance of ion thruster can be evaluated in terms of thrust per unit area, specific impulse, and power efficiency. We can use the expression Equation 11.33 for determining the thrust per unit area. The specific impulse of this kind of thruster can be expressed in terms of charge-to-mass ratio and voltage using Equation 11.20, as follows:

$$I_{sp} = \frac{V_q}{g} = \frac{\sqrt{2\left(q/m_q\right)V}}{g} \tag{11.35}$$

The specific impulse I_{sp} is dependent on charge-to-mass ratio and voltage, which can be enhanced for obtaining higher performance. The charge-to-mass ratio is dependent on the type of propellant. Generally, cesium is preferred, as already discussed. The practical limit of the voltage that can be used for this kind of thruster is 50,000 V, with field strength of 10^7 V/m. The specific impulse in the range of 20–10,000 can be achieved for ion thruster.

The ratio of power to thrust is an important performance parameter that can be expressed using Equation 11.2:

$$\frac{P_b}{F} = \frac{V_q}{2} = \frac{I_{sp}g}{2} \tag{11.36}$$

where V_q is the exit velocity of charges leaving the ion thruster. This parameter is quite handy to judge the relative performances of electrical propulsion devices.

The power efficiency of an ion thruster is defined as the ratio of beam power P_b, to the sum of beam power and power losses P_l, expressed as

$$\eta_{P_b} = \frac{P_b}{P_b + P_l} \tag{11.37}$$

In order to find out the performance of ion thruster, we need to understand the mode of losses occurred during its operation. During ionization, neutral gas molecules get excited by the electron collision but do not get ionized. As a result, energy is radiated from these inert molecules. Besides this, power is lost due to radiation from the ions. Although the ionization potential of propellant gases is quite small, in the range of 10–20 eV, but much higher level energy per ion in the range of 400–700 eV is to be expended for getting requisite high temperature. As a result, the heat loss due to radiation is quite large. Besides this, the neutral atoms absorb energy from the ions and defocus the ion beam, leading to lowering the efficiency of ion thrusters. Generally, the thermal losses due to radiation heat losses are predominant, which can be estimated as

$$P_l = \varepsilon_R \sigma A \left(T^4 - T_0^4\right) \tag{11.38}$$

where

ε_R is the emissivity
σ is the Stefan–Boltzmann constant
A is the effective area of the ionizing sources
T is the absolute temperature of gas
T_0 is the temperature of free space, which is almost zero or negligible as compared to T

Under this assumption, by using Equation 11.38, we can rewrite Equation 11.37 in terms of kinetic energy as

$$\eta_{P_b} = \frac{\dot{m}_p V_q^2/2}{\dot{m}_p V_q^2/2 + \varepsilon_R \sigma A T^4} = \frac{1}{1 + 2\varepsilon_R \sigma T^4 \Big/ \left[\left(m_q/q\right) j I_{sp}^2\right]} \tag{11.39}$$

It can be noted from this expression that the power efficiency of an ion thruster is dependent on the mass of ion. In order to have higher power

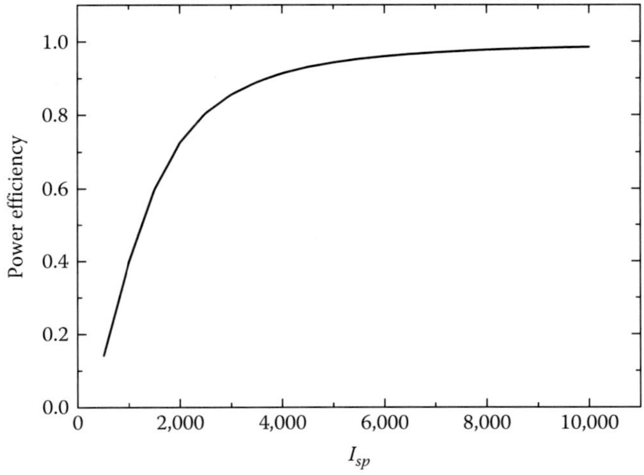

FIGURE 11.8 Potential distribution in a planar electrical field for j = 75 A/m², T = 1600 K, m_q/q = 200 kg/C.

efficiency, we need to use heavier ion. Besides this, power efficiency increases with specific impulse and current density for a given propellant and source temperature. The variation of power efficiency with specific impulse for $T = 1600$ K, $j = 125$ A/m^2, and cesium ion is shown in Figure 11.8. It can be noted that higher specific impulse of 5000 s is to be achieved in order to have reasonable power efficiency of 90%.

11.5 ELECTROMAGNETIC THRUSTER

We have learnt that electrostatic thruster can produce very low thrust with a very high exhaust velocity; of course, with a higher efficiency than that of the thermal thrusters. As a result, these thrusters are not preferred for certain space applications. The thrust level cannot be increased beyond a certain limit as only positive ions can contribute to the thrust and the ion current is limited by the space charge effect. Besides these, sputtering is another problem when higher current density is being used for obtaining higher thrust level. Hence, it will be prudent to accelerate instead the plasma of only positive ions using electromagnetic field. Basically, plasma is a mixture of electrons, positive ions, and neutral atoms that conduct electricity at a higher temperature—beyond 5000 K. Generally, plasma can be formed by either heating propellant gas or applying strong electromagnetic fields or both. Then the plasma is ejected out at a higher velocity (15,000–50,000 m/s) and thus thrust is obtained. As both positive and negative ions present in equal amounts, they can be accelerated together in the engine and produce neutral exhaust beam, unlike ion rocket engine. It has also higher thrust per unit area (10–100 times higher) as compared to ion rocket engine. Several types of electromagnetic/plasma thrusters, namely, electrodynamic, pulsed-plasma, magnetoplasma dynamic (MPD), inductive plasma, pulse ablative engines, stationary plasma (Hall effect) thruster, and so on, have been designed and developed in the laboratory.

11.5.1 Basic Principles of Electromagnetic Thruster

In the electromagnetic thruster, the ionized gas is accelerated using both electric and magnetic fields arranged orthogonally, as shown in Figure 11.1. The current carried by the plasma gas from anode to cathode (not shown in Figure 11.1) along the electric field vector generates a propulsive force while interacting with applied magnetic field. This electromagnetic force per unit volume acting on the plasma is expressed as

$$F_m = j \times B \tag{11.40}$$

Note that this electromagnetic force is perpendicular to the local current density and to the local magnetic field. The current flowing through the plasma gas also induces a magnetic field. In the case of the electromagnetic thruster, the entire plasma gas is accelerated, unlike in electrostatic thruster, as this electromagnetic force acts not only on ions, but on electrons and neutral atoms.

11.5.2 Types of Plasma Thruster

The concept of electromagnetic acceleration of plasma, unlike other electrical thruster concepts, offers several possibilities of devising thrusters for space applications. On the basis of applied fields and internal currents, the electromagnetic thruster can be divided into steady-state and pulse devices. In case of steady thruster, the current density patterns in the plasma, magnetic field, the flow velocity, and thermodynamic properties remain constant with time, while in pulsed thruster, all these change with time. Based on type of magnetic field, namely, externally applied or induced by current patterns, they can be classified as self-field and applied-field thrusters, particularly in case of steady thruster. The pulsed thruster can be further divided into inductive and direct current types, not shown in Figure 11.9.

Based on possible permutations of the electromagnetic concept, several types of electromagnetic thrusters have been designed and developed, but only a few have met the performance requirement, namely, efficiency, reliability, range of performance, and system compatibility for space applications. Of these, some of the advanced thrusters, namely, (1) magnetoplasmadynamic thruster (MPDT), (2) pulsed plasma thruster (PPT), and (3) the Hall effect thruster (HET) are discussed in the following.

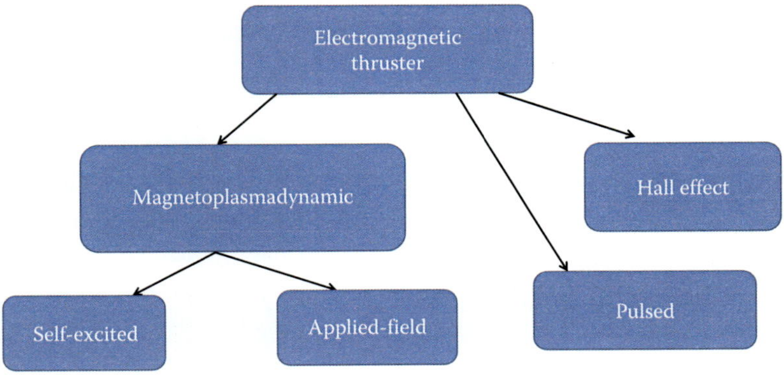

FIGURE 11.9 Types of electromagnetic thrusters.

11.5.2.1 Magnetoplasmadynamic Thrusters

The magnetoplasmadynamic thruster (MPDT), as shown in Figure 11.10a, consists of a central cathode, an annular anode, and some form of inter-electrode insulator. Gaseous propellants enter into the coaxial chamber over the cathode and get ionized by passing through an intense, azimuthally uniform electric arc standing in the gap between anode and cathode. Note that this kind of thruster is similar to the arcjet discussed in Section 11.3. This is also known as the self-excited MPDT, as the magnetic field is a self-excited one due to the presence of arc. At high arc current, azimuthal magnetic field is created, which exerts both axial and radial forces on the propellant flow. The accelerating propellant passing over the hot cathode produces plasma just beyond its tip. Subsequently, the plasma gets expanded along the axial direction to produce desired thrust. The thrust

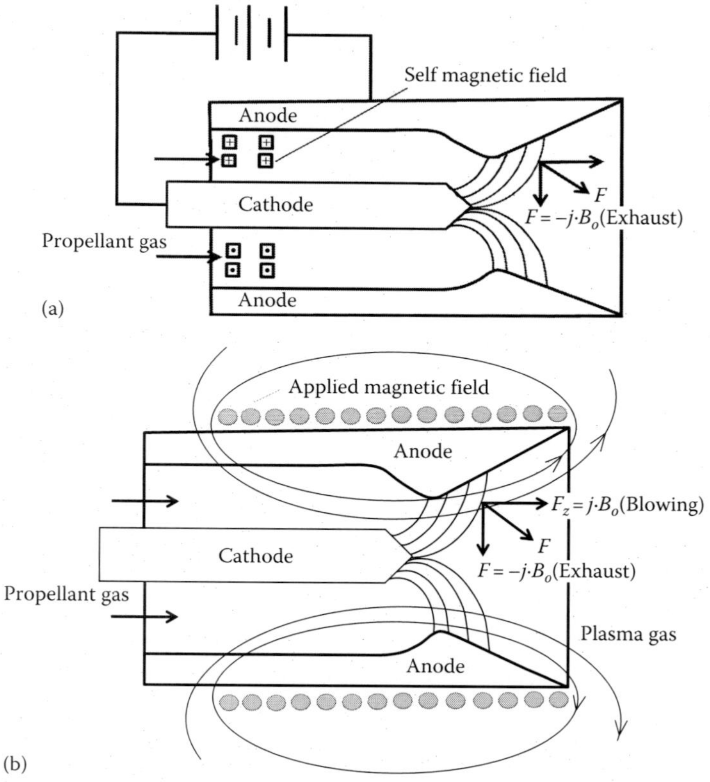

FIGURE 11.10 Schematic of (a) self-excited magnetoplasmadynamic thruster (MPDT) and (b) applied magnetic field MPDT.

is proportional to the magnetic field strength and hence to current density square (j^2). For this self-excited MPDT, a generic thrust relation can be derived by using electromagnetic tensor analysis [1], as given in the following:

$$F = \frac{\mu_0 j^2}{4\pi}\left(\ln\left(\frac{R_a}{R_c}\right) + A\right) \tag{11.41}$$

where

- F is the total thrust
- μ_0 is the vacuum magnetic permeability
- j is the total arc current
- R_a and R_c are the effective arc attachment radii on the anode and cathode
- A is a parameter that is slightly less than unity and depends on the details of the current attachment patterns on the electrodes

Note that this relation agrees reasonably with experiments. Interestingly, it is independent of the mass flow rate and properties of the propellant. Thus, the exhaust velocity can be scaled to the ratio of current density square to mass flow rate ($j^2/\dot{m}$) and the back emf (useful part of the voltage) scales as ($j^3/\dot{m}$). In other words, at low current level, ohmic and near-electrode voltage losses will be more and efficiency will be low. In practice, MPD thrusters for noble gases have maximum efficiency of only 35%, even at megawatt power levels. Of course, one can get better efficiencies for hydrogen gas as propellant at much higher specific impulse. On the other hand, at high current and low flow rate, a strong plasma pinching force associated with the Hall currents along axial direction is acted in opposite direction to the pressure gradient of propellant flow rate. As a result, poor contact of plasma with anode leads to lower specific power and, eventually, lower efficiency.

The power level of self-field MPDT is restricted to 100 kW. In order to enhance power level, external magnetic field is applied, as shown in Figure 11.10b. This is similar to that of the self-excited MPDT but with an external magnetic field, as shown in Figure 11.10b, which enhances the magnetic field in the plasma, leading to higher thrust density with higher efficiency. Both these MPDTs face the problem of cathode erosion, which can be overcome using the thruster in pulse mode.

11.5.2.2 Pulsed Plasma Thruster

In this kind of thruster, a pulse flow of plasma in the order of 10 ms is generated by using a rapid electric discharge across solid propellant (Teflon), unlike in steady MPDT. The schematic of a typical pulse plasma thruster (PPT) is shown in Figure 11.11, which consists of spring-loaded solid propellant bar (Teflon), anode, cathode, capacitance power supplier, pulse generator, and spark plug. In this case, a capacitor is charged from the power supply system that can produce 1–2 kV across electrodes of spark plug that will impinge into the Teflon. As a result Teflon gets degraded into carbon and fluorine atoms and ions in the pulse form due to the combined effects of thermal flux, particle bombardment, and surface reactions, leading to the formation of plasma. Of course, the instantaneous current is in the tens of kA, which can produce high enough self-induced magnetic field to accelerate the plasma to produce thrust. This kind of thruster can provide specific impulses in the range of 1000–1500 s, particularly at higher power level. The advantage of PPT is that it has a nontoxic propellant feed system with the thruster in a single compact unit. It can also be operated over a very wide range of mean power or thrust by simply varying the repetition rate. As a result, it has been in use for precise orbital or attitude-control applications. For example, PPTs were used in early Zond-2, LES-6 satellite, and US Navy's *TIP* and *NOVA* spacecraft. However, the overall of the PPTs is quite low as compared to other electrical thrusters. Its efficiency can be improved by pulse tailoring, minimizing the thermal energy loss, and operation at higher instantaneous power. It is still preferred for relatively light propulsion applications, due to its higher reliability and precision control capability.

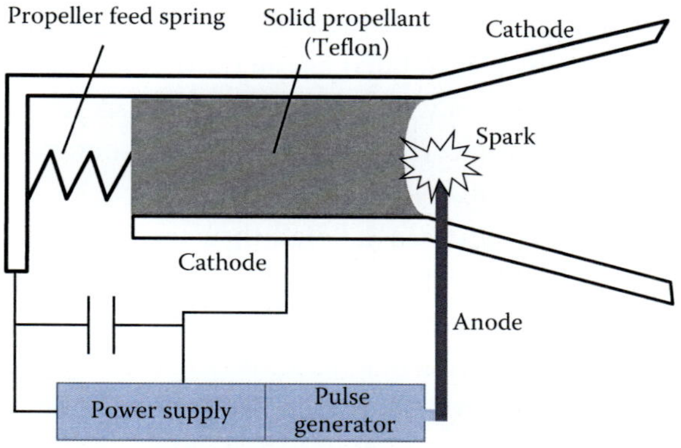

FIGURE 11.11 Schematic of a pulse plasma thruster.

In order to exploit the advantages of MPD propulsion with power-limited spacecraft, the MPDT can be designed such that it can be operated in a quasi-steady (QS) pulsed mode. For this purpose, long enough high-current pulses are used so that it can be operated under almost steady-state condition. This kind of thruster is known as quasi-steady magnetoplasmadynamic thruster (QS-MPDT), which provides higher efficiency along with the instantaneous high power, while drawing low steady-state power from the spacecraft power unit. This kind of thruster with power level of 1 kW has been designed and developed in Japan, which was used onboard the Japanese Space Flyer Unit.

11.5.2.3 Hall Effect Thruster

In this case, the Hall current produced due to Hall effect in an ionized gas can be utilized to accelerate the plasma gas by segmenting electrodes and applying a stream-wise potential gradient. Let us understand the concept of Hall effect in plasma further by considering flow of plasma, as shown in Figure 11.12a. In this, the electric field is in axial direction and magnetic field in transverse direction. The electrons and ions present in the plasma will behave differently due to large difference between their charge-to-mass ratios. When the gyro frequency of the electrons is quite large, they drift in direction perpendicular to both electric and magnetic fields and a Hall current density j is produced along vertical direction (see Figure 11.12a). The interaction of this current with the magnetic field produces an axial force and accelerates the plasma. This can be possible when the current-carrying electrons travel significant portions of their cycloidal motions in the crossed fields before transferring their momentum to the ion particles during collision. Hence, extent of the Hall Effect can be characterized by the Hall parameter, which is defined as the ratio of the gyro frequency to the particle collision frequency. When the collisions are less frequent, then Hall parameter will be large. This is possible only when plasma densities are low enough or magnetic fields are high enough.

The schematic of a typical coaxial Hall thruster developed by Russians is shown in Figure 11.11, which consists of anode, cathode, magnetic coils, and power supply. This is also known as stationary plasma thruster (SPT) as the cylindrically symmetric magnetic field is used across the annular discharge cell. When the propellant (xenon) is passed through small fine holes on the uniformly distributed plate, not shown in this figure, it gets partially ionized due to discharge between annular cathode and the cathode placed downstream of the discharge cell. The electrons generated by

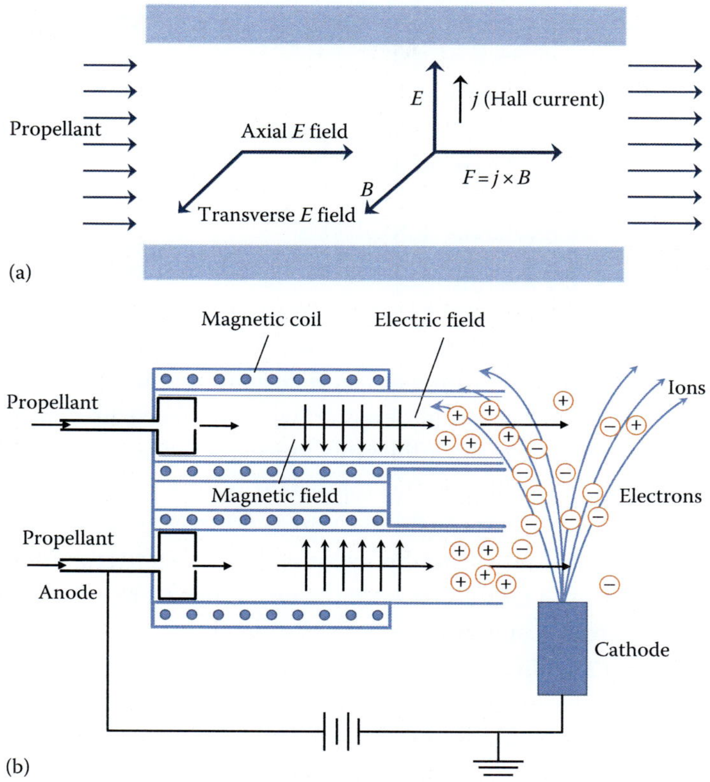

FIGURE 11.12 Schematic of (a) Hall effect and (b) a typical Hall effect thruster.

a thermionic cathode, being attracted by the anode, are passed through this annular along its length. As a result, these electrons, being lighter than ions, are acted upon by the crossed electric and magnetic fields and get deflected along the circumferential direction, forming a swirling flow generating a circular Hall current. The interaction of this Hall current with the radial magnetic field can generate a force on the plasma. As a result, the ions get accelerated in the neutral plasma by the electrostatic field along the length of annular cell. Hence, the problems of space charge limitation and sputtering in ion thruster are eliminated in this thruster. Rather it can produce relatively high thrust density as compared to conventional electrostatic thruster. Generally, the exhaust velocity of this kind of thruster varies between 15,000 and 25,000 m/s, which can be produced with voltage in the range of 200–500 V. The flow rate of propellant lies between 2 and 10 mg/s. The overall efficiency of thrusters lies between 50% and 60% as electrons do not contribute to the generation of thrust. In this case, higher thrust per unit area in the order of 10 times as compared to the electrostatic thruster is achieved due to absence of space charge limitation. For example, The Russian SPT-140 (5 kW) xenon Hall thruster produces a thrust of 250 mN with exhaust velocity of 22.5 km/s, operating with discharge potential 450 V, current of 10, and efficiency of 57%.

11.6 NUCLEAR ROCKET ENGINES

In certain space applications, we need large amount of energy density, which is the order of magnitude that can be released by chemical reactions. In such situation, we can think of nuclear energy as a viable source as its energy density is quite enormous as compared to that of chemical energy source. Note that nuclear rocket engine has never been used in practice due to risk of spreading of radioactive emission, although considerable amount of research is going on this topic [4,5].

In nuclear rocket engine, nuclear energy is utilized in heating the working fluid for producing thrust by expanding the high-temperature and high-pressure gas in a CD nozzle, as in chemical rocket engine. The nuclear sources used in rocket engine are mainly through three methods: (1) fission energy release, (2) radioactive isotope decay, and (3) fusion energy release. In all three types of engines, the propellant is heated by release of energy within the nucleus of atoms, and in all these, generally liquid hydrogen is used as working fluid. Note that in nuclear rocket engine, energy source is different from that of the propellant, unlike in chemical rocket engines.

In the nuclear fission reactor rocket, heat is generated by the fission of enriched uranium (U^{233}) or plutonium (P^{239}) in the solid form in the nuclear reactor. The excited U^{233} or P^{239} produces neutrons namely polonium (Po^{210}), plutonium (P^{238}) during nuclear reactions and these neutrons are absorbed by the fuel to generate more number of neutrons as a chain reaction. Then the reactor attains the critical mode to produce large amount of heat. The runaway of nuclear reactions must be contained by moderators that slow down the neutron production by controlling the multiplications of neutrons during nuclear chain reactions.

A typical solid-core fission thermal reactor is shown in the Figure 11.13, which is similar to that of electrothermal engine. It consists of nuclear reactor, neutron reflector, propellant feed system, and CD nozzle. The nuclear reactor core contains the nuclear fuel components where the nuclear reactions take place and the heat is generated. The enriched uranium is contained in the fuel elements, which can be in rod or pebble bed form. Generally, uranium in metal form is not used as a nuclear fuel in the fuel

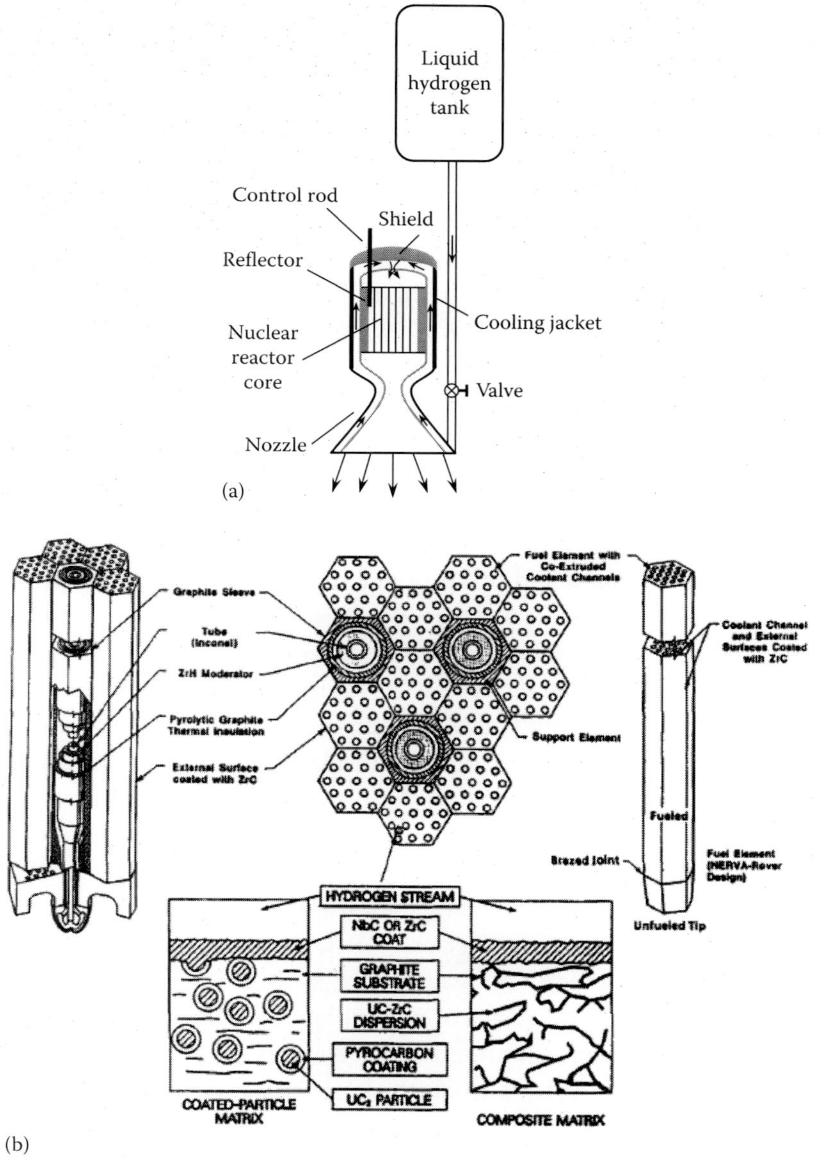

FIGURE 11.13 Schematic of (a) a typical solid-core fission reactor [3] and (b) fuel element of KIWI reactor core [6]. ([a]: From Sutton, G.P. and Ross, D.M., *Rocket Propulsion Elements*, 5th edn., John Wiley & Sons, New York, 1975; [b]: From Borowski, S.K. et al., Nuclear thermal rocket vehicle design options for future NASA missions to the Moon and Mars, NASA Technical Memorandum 107071, AIAA-93-4170, 1993.)

element, as its melting temperature is around 1400 K and dangerous for rocket engine applications. Rather, uranium compounds, namely, uranium carbide (UC_2), uranium dioxide (UO_2), and uranium nitride (UN) are being used as nuclear fuel, as interaction between neutrons and uranium nuclei is not influenced by chemical combination of uranium with other elements in its compound form. Of course, dimensions of the fuel elements for same power level will increase as compared to pure uranium, because the uranium compound contains fewer uranium nuclei per volume. The most preferred compound of uranium is uranium dioxide, as it is quite stable with melting temperature of 3075 K. Besides this, it is quite compatible with hydrogen gas from the stability point of view. As discussed earlier, the chained nuclear reactions must be contained to release heat in a controlled manner. The most common material for moderator is graphite, which can scatter away neutrons, which have low molecular mass, and arrest the multiplication process further. Besides this, graphite is being preferred as moderator material as it can maintain structural and dimensional ingenuity at high temperature and pressure due its higher sublimation temperature (3990 K) at reasonable pressure. But it can react with hydrogen at high temperature to produce hydrocarbon. Hence, the fuel element must be coated with certain refractory materials to prevent the graphite of the fuel matrix to come in contact with hydrogen at high temperature. Generally, niobium carbide (NbC) or zirconium carbide (ZrC) are being used as protective coating of fuel elements, as these two refractory materials do not react chemically even at high temperature and remain neutral in the neutron environment. As a result, the lifetime of fuel elements is increased with coating of carbides even when they are operated at high-temperature and high-pressure condition. Let us look at the cross section of fuel element matrix used in the KIWI reactor core, as shown in Figure 11.13b [6]. The uranium oxide in the forms of small spheres is dispersed in the graphite matrix. These graphite uranium oxide matrices are used to fabricate the hexagonal-shaped rods with 19 holes along its length for hydrogen flow, which can be used as fuel elements in the nuclear core. But both inner and outer surfaces of these fuel elements are coated with niobium carbide. All these six hexagonal rods are locked by a stainless steel tie rod to form a fuel assembly. Several such fuel assemblies can be mounted together to form the reactor core.

This nuclear thermal reactor has several components, namely, liquid hydrogen tank, pump, turbine, reaction chamber, and CD nozzle, as shown in Figure 11.13, which are similar to that of the liquid-propellant chemical rocket engine. The liquid hydrogen passes through the cooling passages of the nozzle to extract heat from the nozzle. Subsequently, hydrogen comes in contact with core of solid fission reaction and gets heated to high-temperature and high-pressure gas, which is expanded in a nozzle to obtain thrust. The nuclear fission engines with high thrust level in the range of 30–989 kN have been demonstrated in the past. It has been reported that high specific impulse, as high as 1000 s, can be achieved at least for shorter duration, which can be used for manned planetary exploratory mission.

But in the isotope decay engine, a radioactive material imparts radiation that is converted into thermal energy. Subsequently, this thermal energy can be used in raising the temperature of a working fluid. Generally, Pu^{238} and Po^{210} are used to heat hydrogen and ammonia as propellant. But it provides lower thrust level as compared to other nuclear engines but at a lower propellant temperature level. Of course, such kind of engines has not been used for any space applications to date.

The third type of nuclear engine is based on the principle of fusion in which the released nuclear energy is used to heat working fluid in similar way as in other kind of nuclear engines. Although several concepts have been attempted, none of them have been tested due to poor understanding of controlling nuclear fusion process in present time. It is hoped that it will emerge as contender for space power plant in future with advent of matured knowledge in future.

11.7 SOLAR ENERGY ROCKETS

Solar energy is an attractive source of energy as it is available freely in space. Interestingly, the power of solar radiation at any particular location in space remains almost constant and hence can be used for generating power for spacecraft and its propulsion. Several methods have been devised for utilizing solar energy to propel spacecraft. Solar cells can convert solar energy into electricity directly, with efficiency in the range of 15%–20%. In spite of the low efficiency, they are extensively used nowadays to generate electricity for spacecraft power generation. Generally, the photovoltaic cells that absorb the Sun's radiation can be used to propel small rocket engines using electrical propulsion system. However, it incurs huge mass penalty along with cost escalation when it is scaled for bigger solar rocket engines. Besides this, its collection efficiency is small and there is lower allowable surface temperature of collector. As a result, massive structure per unit of electrical power level has to be installed for larger power application. Hence, its application is restricted to low power devices for space applications. However, with improved solar cells like gallium arsenide and use of improved solar collectors, solar energy can be used in future not only for power generation but also for propulsion in deep space.

Another method is the solar heating of propellants like hydrogen. Let us consider a schematic of a solar rocket engine in conceptual form, as shown in Figure 11.14, which consists of hydrogen propellant tank, propellant feed system, parabolic reflector, heat receiver, rocket engines, and other accessories. There are several types of solar reflectors that can be used for concentrating solar radiation to the heat receiver. Generally, parabolic mirrors or Fresnel lenses with cooling system are used for collecting and concentrating solar radiation. The receiver must be made of high-temperature materials, namely, metals like tungsten, aluminum alloy with a provision of exchanging heat with propellants. When liquid hydrogen is passed through heat exchanger cum receiver, it is converted into gas whose temperature can go up to as high as 2200°C, which is subsequently expanded in the nozzle for obtaining the thrust. Generally, low thrust level in the range of 1–10 N can be achieved but with higher performance level, as high as three times in comparison to chemical rocket engines. But it can only be deployed in space as, being large in size and of lighter weight, it cannot experience atmospheric drag. There is also the problem of hydrogen propellant storage. In addition, it has several technological challenges

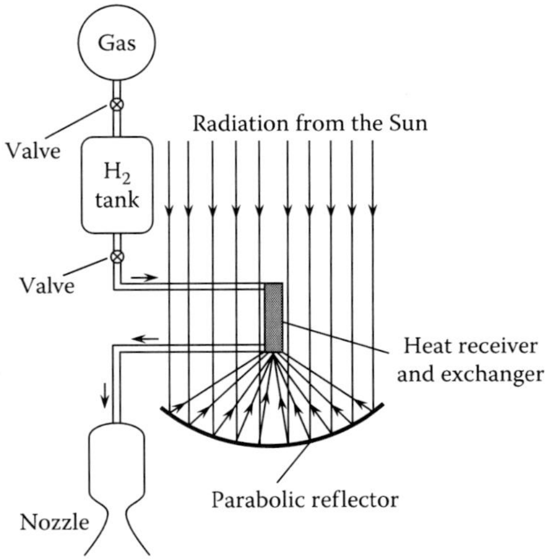

FIGURE 11.14 Schematic of solar rocket engine. (From Sutton, G.P. and Ross, D.M., *Rocket Propulsion Elements*, 5th edn., John Wiley & Sons, New York, 1975.)

that need to be addressed. As a result, solar thermal rocket engine has not been employed in actual mission to date.

There is another concept known as solar sail, in which a big photon reflector is used to collect the solar radiation pressure of light from a star by which a small amount of thrust can be produced. In 1924, Friedrich Zander, a Russian engineer, had provided the idea of using this concept of producing thrust without using any fuel for propelling a spacecraft in space. Although large solar sails can produce only tiny amount of acceleration, it can achieve considerable velocity over a longer period of time as the force on the sails from the Sun varies inversely with square of its distance from the Sun. Recently, in May 21, 2010, Japan Aerospace Exploring Agency launched the IKAROS (Interplanetary Kite-craft Accelerated by Radiation of the Sun) and was the first to demonstrate solar sail technology interplay in space. It is believed that more efforts will be made to advance this technology further in future.

REVIEW QUESTIONS

1. What are the types of electrical rocket engines you are aware of?
2. Define the Coulomb force. How is it used in electrical thruster?
3. What is the Lorentz force? Can it be used for electrical thruster? Explain it with an example.
4. What do you mean by arc heating of electrical rocket engine?
5. Describe the discharge behavior with a neat diagram.
6. What are the types of resistojets? Describe each and compare them.
7. What are the components of a typical resistojet?
8. Why is helium preferred in a resistojet over other gases?
9. What is the basic principle of arcjet? How is it different from resistojet?
10. Explain the basic principle of electrostatic thruster?
11. What are the gases that can be used as propellant for electrostatic thruster?
12. Define power efficiency of an ion thruster. How does it depend on specific mass to charge density?

13. How is ion rocket engine different from arc heating of electrical rocket engine?

14. What are the types of electromagnetic thrusters you are aware of?

15. Why is swirling electron field used in a Hall thruster? Explain it.

16. What are the types of electromagnetic thruster you are aware of?

17. What is MPDT? Describe it with a diagram.

18. How is self-excited MPDT different from applied magnetic MPDT?

19. What is PPT? How is it different from Hall thruster?

20. What do you mean by solar thermal rocket engine? Explain its working principle.

21. Explain the basic principle of solid-core fission nuclear rocket engine.

22. What is the speciality of KIWI fuel element? Describe it with a diagram.

PROBLEMS

11.1 A cesium ion rocket is designed to produce a specific impulse of 4500 s. Determine the voltage and beam power per unit thrust. If in place of Cs, Argon gas is used, determine I_{sp} and power per unit thrust with same voltage.

11.2 A 1.5 kW electrothermal rocket with hydrogen as propellant is designed and developed for obtaining the thrust coefficient of 1.8. The temperature should not exceed the temperature limit of tungsten filament (3200 K). Assuming the mass flow rate of propellant to be 0.2 g/s and thruster efficiency of 0.75, determine the exit velocity, characteristic velocity, temperature of propellant gas, thrust, and specific impulse for electrochemical rockets.

11.3 An ion thruster is designed and developed to produce 75 mN with specific impulse of 2750. The cesium gas with atomic mass of 2.21×10^{-25} and charge-to-mass ratio of 7.25×10^{5} is used as propellant. If the thruster efficiency is 0.75, determine loss per ion. If I_{sp} is increased by 50%, what would be its efficiency? Take 1 eV = 1.6×10^{-19} J.

11.4 A satellite with payload of 85 kg is to be placed in low earth orbit using an ion rocket with I_{sp} of 2200, thruster efficiency of 0.85, and thrust of 0.35 N. If the operating duration is 20 days and ratio between electrical power and mass of power plant is 150 W/kg, then determine the total propellant mass, velocity increase, and acceleration of vehicle considering the ideal vacuum condition.

11.5 A resistojet with throat diameter of 2.4 mm is designed and developed with helium as propellant gas for maneuvering of a satellite. The propellant at the chamber pressure of 0.35 MPa and temperature 1200 K is expanded in a CD nozzle with expansion ratio of 81. Determine mass flow rate, exit velocity, characteristic velocity, and specific impulse for this rocket.

11.6 In an ion thruster, 1.5 g of positively charged xenon gas is charged to average charge density of 15,750 C/kg. Determine the electrostatic force experienced by these charges while moving across electrostatic field of 1.75 MV/m. If the distance between the grid is 1.5 mm, determine the work done on the charge and final velocity.

11.7 In an electrostatic thruster, xenon gas with charge density of 2.7×10^4 C/kg is passing through the accelerator grids with voltage of 3.75 kV to produce the thrust. If the ion current happens to be 0.35 A, determine velocity of ions and thrust.

11.8 An ion thruster is designed to produce specific impulse of 3500 s. The xenon gas with charge density of 2.5×10^4 C/kg is passing through the accelerator grids with a gap of 2.1 mm. If the maximum allowable voltage is 10,550 V, determine space charge current density limit and the diameter of the round beam for producing a thrust of 7.5 N.

11.9 In an ion thruster with charge-to-mass ratio of 7145 C/kg, a grid voltage of 4500 V is applied, which consumes 125 kg xenon propellant for 8 days. Determine the exit velocity, thrust, and I_{sp} of this ion thruster.

11.10 In an ion thruster, the xenon gas with charge density of 7.4×10^5 C/kg is passing through the accelerator grids with a gap of 0.5 mm and a voltage difference of 1050 V. Determine the exit velocity, specific impulse, efficiency, and thrust-to-power ratio. If the maximum voltage difference between the grids happens to be 1900 V, determine the charge current density limit of this thruster. Determine the exit velocity and specific impulse.

REFERENCES

1. Jahn, R.G., *Physics of Electric Propulsion*, McGraw-Hill, Inc, New York, 1968.
2. Humble, R.W., Henery, G.N., and Larson, W.J., *Space PropulsionAnalysis and Design*, McGraw Hill, New York, 1995.
3. Sutton, G.P. and Ross, D.M., *Rocket Propulsion Elements*, 5th edn., John Wiley & Sons, New York, 1975.
4. Gunn, S., Nuclear propulsion—A historical perspective, *Space Policy*, 17, 291–298, 2001.
5. Bussard, R.W. and Delauer, R.D., *Nuclear Rocket Propulsion*, McGraw Hill, New York, 1965.
6. Borowski, S.K., Corban, R.R., McGuire, M.L., and Beke, E.G., Nuclear thermal rocketvehicle design options for future NASA missions to the Moon and Mars, NASA Technical Memorandum 107071, AIAA-93-4170, 1993.

Appendices

APPENDIX A

TABLE A.1 Physical Constants

Universal gas constant:

$R_u = 8.314\ \text{J/(g} \cdot \text{mol} \cdot \text{K)}$

Standard acceleration due to gravity:

$g = 1\ \text{kg m/(N s}^2\text{)} = 9.80665\ \text{m/s}^2$

Stefan–Boltzmann constant:

$\sigma = 5.6697 \times 10^{-8}\ \text{W/(m}^2 \cdot \text{K}^4\text{)}$

Boltzmann constant:

k_B = Boltzmann constant = 1.38054×10^{-16} erg/(molecule K)

Avogadro's number:

$N_A = 6.02252 \times 10^{23}$ molecules/(g · mol)

Electric charge:

$e = 1.602 \times 10^{-19}$ Coulomb

Permittivity:

$\varepsilon_0 = 8.854187 \times 10^{-12}$ F/m

Planck's constant:

$h = 6.256 \times 10^{-27}$ erg · s

Speed of Light:

$c = 2.997925 \times 10^8$ m/s (in vacuum)

Gravitational constant:

$G = 6.67 \times 10^{-11}\ \text{m}^3\text{/kg} \cdot \text{s}^2$

Standard atmosphere pressure:

P_{atm} = 101,325 Pa = 1 atm

TABLE A.2 Conversion Factors

Energy	1 J	$= 9.47817 \times 10^{-4}$ Btu
		$= 2.3885 \times 10^{-4}$ keal
Energy rate	1 W	= 3.41214 Btu/h
Force	1 N	= 0.224809
Heat flux	1 W/m^2	= 0.3171 $Btu/(h \cdot ft^2)$
Kinematic viscosity and diffusivities	1 m^2/s	$= 3.875 \times 10^4$ $ft^{2/}h$
Length	1 m	= 39.370 in.
		= 3.2808 ft
Mass	1 kg	= 2.2046 lb_m
Mass density	1 kg/m^3	= 0.062428 lb_m/ft^3
Mass flow rate	1 kg/s	= 7936.6 lb_m/h
Pressure	Pa	1 = 1 N/m^2
		= 0.020885 4 lb_f/ft^2
		$= 1.4504 \times 10^{-4}$ $lb_f/in.^2$
		$= 4.015 \times 10^{-3}$ in water
	1×10^5 N/m	= 1 bar
Specific heat	$1 \cdot J/kg \cdot K$	$= 2.3886 \times 10^{-4}$ $Btu/(lb_m \cdot °F)$
Temperature	K	= (5/9) °R
		= (5/9) (°F + 459.67)
		= °C + 273.15
Time	3600 s	1 h

APPENDIX B

TABLE B.1 Isentropic Flow Table ($\gamma = 1.4$)

M	T_t/T	P_t/P	ρ_t/ρ	A/A^*	MFP
0.1	1.002000	1.007018	1.005008	5.821829	1.998998
0.2	1.008000	1.028281	1.02012	2.963520	3.986079
0.3	1.018000	1.06443	1.045609	2.035065	5.949679
0.4	1.032000	1.116552	1.08193	1.590140	7.878913
0.5	1.050000	1.186213	1.129726	1.339844	9.763859
0.6	1.072000	1.275504	1.189836	1.188200	11.59578
0.7	1.098000	1.387101	1.263298	1.094373	13.36728
0.8	1.128000	1.52434	1.351365	1.038230	15.07237
0.9	1.162000	1.691303	1.45551	1.008863	16.7065
1.0	1.200000	1.892929	1.577441	1.000000	18.26651
1.1	1.242000	2.135136	1.719111	1.007925	19.7505
1.2	1.288000	2.424965	1.882737	1.030440	21.15775
1.3	1.338000	2.770744	2.07081	1.066305	22.48855
1.4	1.392000	3.182272	2.286115	1.114926	23.74404
1.5	1.450000	3.671031	2.531745	1.176167	24.92605
1.6	1.512000	4.250414	2.811121	1.250235	26.03696
1.7	1.578000	4.935993	3.128005	1.337606	27.07956
1.8	1.648000	5.745796	3.486527	1.438982	28.05692
1.9	1.722000	6.700636	3.891194	1.555257	28.97231
2.0	1.800000	7.824449	4.346916	1.687500	29.82908
2.1	1.882000	9.144683	4.859024	1.836944	30.63061
2.2	1.968000	10.69271	5.433288	2.004975	31.38024
2.3	2.058000	12.50428	6.075939	2.193131	32.08125
2.4	2.152000	14.62002	6.793689	2.403100	32.7368
2.5	2.250000	17.08594	7.59375	2.636719	33.34993
2.6	2.352000	19.95403	8.483856	2.895975	33.92351
2.7	2.458000	23.28287	9.472284	3.183011	34.46029
2.8	2.568000	27.1383	10.56787	3.500123	34.96284
2.9	2.682000	31.59408	11.78004	3.849768	35.43356
3.0	2.800000	36.73272	13.11883	4.234568	35.87471
3.1	2.922000	42.64624	14.59488	4.657311	36.28839
3.2	3.048000	49.43703	16.2195	5.120958	36.67656
3.3	3.178000	57.21878	18.00465	5.628647	37.04104
3.4	3.312000	66.11746	19.96300	6.183699	37.38349
3.5	3.45000	76.27230	22.10791	6.789621	37.70549
3.6	3.592000	87.83693	24.45349	7.450111	38.00848
3.7	3.738000	100.9805	27.01458	8.169066	38.29378
3.8	3.888000	115.8889	29.80682	8.950585	38.56263
3.9	4.042000	132.7661	32.84663	9.798973	38.81616
4.0	4.200000	151.8352	36.15124	10.71875	39.05544

TABLE B.2 Isentropic Flow Table ($\gamma = 1.33$)

M	T_t/T	P_t/P	ρ_t/ρ	A/A^*	MFP
0.1	1.00165	1.006667	1.005008	5.866473	1.948722
0.2	1.00660	1.026867	1.020134	2.984731	3.887850
0.3	1.01485	1.061210	1.045682	2.047994	5.808022
0.4	1.02640	1.110733	1.082163	1.598603	7.700335
0.5	1.04125	1.176934	1.130309	1.345408	9.556535
0.6	1.05940	1.261825	1.191076	1.191701	11.369180
0.7	1.08085	1.367996	1.265667	1.096361	13.131770
0.8	1.10560	1.498695	1.355549	1.039139	14.838800
0.9	1.13365	1.657933	1.462473	1.009100	16.485840
1.0	1.16500	1.850604	1.588502	1.000000	18.069450
1.1	1.19965	2.082638	1.736038	1.008192	19.587250
1.2	1.23760	2.361167	1.907859	1.031584	21.037740
1.3	1.27885	2.694731	2.107152	1.069083	22.420300
1.4	1.32340	3.093514	2.33755	1.120283	23.735060
1.5	1.37125	3.569617	2.603185	1.185281	24.982790
1.6	1.42240	4.137367	2.908722	1.264569	26.164780
1.7	1.47685	4.813677	3.259422	1.358968	27.282780
1.8	1.53460	5.618456	3.661186	1.469590	28.338890
1.9	1.59565	6.575067	4.12062	1.597817	29.335450
2.0	1.66000	7.710851	4.645091	1.745290	30.274980
2.1	1.72765	9.057716	5.242796	1.913907	31.160140
2.2	1.79860	10.652800	5.922827	2.105829	31.993620
2.3	1.87285	12.539190	6.695248	2.323484	32.778140
2.4	1.95040	14.766800	7.571165	2.569583	33.516400
2.5	2.03125	17.393210	8.562810	2.847130	34.211040
2.6	2.11540	20.484740	9.683623	3.15944	34.864630
2.7	2.20285	24.117550	10.94834	3.510155	35.479640
2.8	2.29360	28.378900	12.37308	3.903265	36.058460
2.9	2.38765	33.368470	13.97545	4.343125	36.603330
3.0	2.48500	39.199920	15.77461	4.834484	37.116410
3.1	2.58565	46.002460	17.79145	5.382498	37.599730
3.2	2.68960	53.922700	20.04859	5.992764	38.055210
3.3	2.79685	63.126560	22.57059	6.671339	38.484630
3.4	2.90740	73.801420	25.38399	7.424767	38.889690
3.5	3.02125	86.158430	28.51748	8.260109	39.271970
3.6	3.13840	100.435000	32.00197	9.184969	39.632940
3.7	3.25885	116.897400	35.87076	10.20752	39.973990
3.8	3.38260	135.844000	40.15964	11.33655	40.296400
3.9	3.50965	157.608000	44.90705	12.58147	40.601370
4.0	3.64000	182.561300	50.15419	13.95235	40.890010

TABLE B.3 Normal Shock Properties for $\gamma = 1.4$

M_1	M_2	P_2/P_1	ρ_2/ρ_1	P_{t2}/P_{t1}	T_2/T_1	$s_2 - s_1$
1	1	1	1	1	1	0
1.1	0.91177	1.245	1.169082	0.998928	1.064938	0.147349
1.2	0.84217	1.513333	1.341615	0.992798	1.127994	0.927027
1.3	0.785957	1.805	1.515695	0.979374	1.190873	2.635738
1.4	0.739709	2.12	1.689655	0.958194	1.254694	5.372076
1.5	0.701089	2.458333	1.862069	0.929787	1.320216	9.134332
1.6	0.668437	2.82	2.031746	0.8952	1.387969	13.87009
1.7	0.640544	3.205	2.197719	0.855721	1.45833	19.50258
1.8	0.616501	3.613333	2.359223	0.812684	1.531578	25.94507
1.9	0.595616	4.045	2.515679	0.767357	1.607916	33.10892
2	0.57735	4.5	2.666667	0.720874	1.6875	40.90801
2.1	0.561277	4.978333	2.811902	0.674203	1.77045	49.26107
2.2	0.547056	5.48	2.95122	0.628136	1.85686	58.09292
2.3	0.534411	6.005	3.084548	0.583295	1.946801	67.33484
2.4	0.523118	6.553333	3.211896	0.540144	2.040332	76.92463
2.5	0.512989	7.125	3.333333	0.499015	2.1375	86.8064
2.6	0.503871	7.72	3.44898	0.460123	2.238343	96.93014
2.7	0.495634	8.338333	3.558991	0.42359	2.342892	107.2514
2.8	0.488167	8.98	3.663551	0.389464	2.451173	117.7306
2.9	0.48138	9.645	3.762864	0.357733	2.563207	128.3329
3	0.475191	10.33333	3.857143	0.328344	2.679012	139.0276
3.1	0.469534	11.045	3.946612	0.301211	2.798603	149.7874
3.2	0.464349	11.78	4.031496	0.276229	2.921992	160.5887
3.3	0.459586	12.53833	4.11202	0.253276	3.049191	171.4107
3.4	0.4552	13.32	4.188406	0.232226	3.180208	182.235
3.5	0.451154	14.125	4.26087	0.212948	3.315051	193.046
3.6	0.447413	14.95333	4.329621	0.195312	3.453728	203.8297
3.7	0.443948	15.805	4.394864	0.179194	3.596244	214.5743
3.8	0.440732	16.68	4.45679	0.16447	3.742604	225.2695
3.9	0.437742	17.57833	4.515586	0.151027	3.892813	235.9063
4	0.434959	18.5	4.571429	0.138756	4.046875	246.4771

TABLE B.4 Normal Shock Properties for $\gamma = 1.33$

M_1	M_2	P_2/P_1	ρ_2/ρ_1	P_{t2}/P_{t1}	T_2/T_1	$s_2 - s_1$
1	1	1	1	1	1	0
1.1	0.911378	1.239742	1.175051	0.998912	1.055054	0.135695
1.2	0.840904	1.502318	1.355527	0.992645	1.108291	0.920104
1.3	0.783604	1.787725	1.539547	0.97882	1.161202	2.668293
1.4	0.736191	2.095966	1.725404	0.956865	1.214768	5.495886
1.5	0.696397	2.427039	1.911577	0.927246	1.269653	9.414983
1.6	0.662601	2.780944	2.096738	0.891011	1.326319	14.38354
1.7	0.633609	3.157682	2.279751	0.849487	1.3851	20.33198
1.8	0.608523	3.557253	2.459664	0.804087	1.446236	27.17807
1.9	0.586655	3.979657	2.635697	0.756175	1.509907	34.83537
2	0.567466	4.424893	2.807229	0.706992	1.576249	43.21805
2.1	0.550528	4.892961	2.973779	0.657607	1.645368	52.2437
2.2	0.535497	5.383863	3.134994	0.608903	1.717344	61.83473
2.3	0.522097	5.897597	3.290627	0.561579	1.792241	71.91917
2.4	0.510097	6.434163	3.440525	0.51616	1.870111	82.43091
2.5	0.499309	6.993562	3.584615	0.47302	1.950994	93.30964
2.6	0.489575	7.575794	3.722889	0.4324	2.034923	104.5007
2.7	0.480762	8.180858	3.855392	0.394436	2.121927	115.9547
2.8	0.472758	8.808755	3.982211	0.359174	2.212026	127.6274
2.9	0.465467	9.459485	4.10347	0.326596	2.30524	139.4788
3	0.458807	10.13305	4.219316	0.296632	2.401585	151.4735
3.1	0.452709	10.82944	4.329917	0.269176	2.501074	163.5798
3.2	0.447111	11.54867	4.435455	0.244097	2.603717	175.7694
3.3	0.44196	12.29073	4.536121	0.221252	2.709524	188.0173
3.4	0.437211	13.05562	4.632111	0.200487	2.818504	200.3013
3.5	0.432822	13.84335	4.723624	0.181647	2.930662	212.6017
3.6	0.42876	14.65391	4.810859	0.164578	3.046006	224.9013
3.7	0.424993	15.4873	4.894012	0.149132	3.16454	237.1848
3.8	0.421492	16.34352	4.973275	0.135168	3.286269	249.4387
3.9	0.418235	17.22258	5.048837	0.122553	3.411197	261.6515
4.0	0.415198	18.12446	5.120879	0.11116	3.539327	273.8127

APPENDIX C

TABLE C.1 International Standard Atmosphere

Z (m)	P (kPa)	T (K)	ρ/ρ_o	a (m/s)
0	101.325	288.2	1.0000	340.3
1,000	89.880	281.7	0.9075	336.4
2,000	79.500	275.2	0.8217	332.5
3,000	70.120	268.7	0.7423	328.6
4,000	61.660	262.2	0.6689	324.6
5,000	54.050	255.7	0.6012	320.5
6,000	47.220	249.2	0.5389	316.5
7,000	41.110	242.7	0.4817	312.3
8,000	35.650	236.3	0.4292	308.1
9,000	30.800	229.9	0.3813	303.8
10,000	26.500	223.5	0.3376	299.5
11,000	22.700	216.7	0.2978	295.2
12,000	19.400	216.65	0.2546	295.1
13,000	16.580	216.65	0.2176	295.1
14,000	14.170	216.65	0.186	295.1
15,000	12.110	216.65	0.159	295.1
16,000	10.350	216.65	0.1359	295.1
17,000	8.850	216.65	0.1162	295.1
18,000	7.565	216.65	0.0993	295.1
19,000	6.467	216.65	0.08489	295.1
20,000	5.529	216.65	0.07258	295.1
21,000	4.727	217.6	0.06181	295.7
22,000	4.042	218.6	0.05266	296.4
23,000	3.456	219.6	0.0449	297.1
24,000	2.955	220.6	0.03832	297.7
25,000	2.608	221.6	0.03272	298.4
26,000	2.163	222.5	0.02796	299.1
27,000	1.855	223.5	0.02392	299.7
28,000	1.595	224.5	0.02047	300.4
29,000	1.374	225.5	0.01753	301.1
30,000	1.185	226.5	0.01503	301.7

Note: Density of air at sea level = 1.225 kg/m^3.

APPENDIX D: PROPELLANT TABLES

TABLE D.1 Selected Properties of Hydrocarbon Fuels

Fuel	Fuel	M_w (kg/kmol)	$\bar{h}_f^o$ (kJ/kmol)	HHV (kJ/kg)	LHV (kJ/kg)	Boiling Temp. (°C)	h_{fg} (kJ/kg)
CH_4	Methane	16.043	−74.831	55.528	50.016	−164	509
C_2H_2	Acetylene	26.038	226.748	49.923	48.225	−84	—
C_2H_4	Ethane	28.054	52.283	50.313	47.161	−103.7	—
C_2H_6	Ethane	30.069	−84.667	51.901	47.489	−88.6	488
C_3H_6	Propene	42.080	20.414	48.936	45.784	−47.4	437
C_3H_8	Propane	44.096	−103.847	50.368	46.357	−42.1	425
C_4H_8	1-Butene	56.107	1.172	48.471	45.319	−63	391
C_4H_{10}	*n*-Butane	58.123	−124.733	49.546	45.742	−0.5	386
C_5H_{10}	1-Pentene	70.134	−20.920	48.152	45.000	30	358
C_5H_{12}	*n*-Pentane	72.150	−146.440	49.032	45.355	36.1	358
C_6H_6	Benzene	78.113	82.927	42.277	40.579	80.1	393
C_6H_{12}	1-Hexene	84.161	−41.673	47.955	44.803	63.4	335
C_6H_{14}	*n*-Hexane	86.177	−167.193	48.696	45.05	69	335
C_7H_{14}	1-Heptene	98.188	−62.132	47.817	44.665	93.6	—
C_7H_{16}	*n*-Heptane	100.203	−187.820	48.456	44.926	98.4	316
C_8H_{16}	1-Octene	112.214	−82.97	47.712	44.560	121.3	—
C_8H_{18}	*n*-Octane	114.230	−208.447	48.275	44.791	125.7	300
C_9H_{18}	1-Nonene	126.241	−103.512	47.631	44.478	—	—
C_9H_{20}	*n*-Nonane	128.257	−229.032	48.134	44.686	150.8	295
$C_{10}H_{20}$	1-Decene	140.268	−124.139	47.565	44.413	170.6	—
$C_{10}H_{22}$	*n*-Decane	142.284	−249.659	48.020	44.602	174.1	277
$C_{11}H_{22}$	1-Undecene	154.295	−144.766	47.512	44.360	—	—
$C_{11}H_{24}$	*n*-Undecane	156.311	−270.286	47.926	44.532	195.9	265
$C_{12}H_{24}$	1-Dodecene	168.322	−165.352	47.468	44.316	213.4	—
$C_{12}H_{26}$	*n*-Dodecane	170.337	−292.162	47.841	44.467	216.3	256

Heat of formation, and higher and lower heating values all at 298.15 K and 1 atm; boiling points and latent heat of vaporization at 1 atm.

TABLE D.2 Physical Properties of Typical Liquid Propellants

Propellant	Chemical Formula	*MW* (g/mol)	*Cp* (J/kg·K)	ρ (kg/m³)	μ × 10⁻³ (kg/m·s)
Liquid fluorine	F_2	38.0	1541 (85 K)	1636 (69.3 K)	0.305 (77.6 K)
			1495 (69.3K)	1440 (93 K)	0.397 (70 K)
Hydrazine	N_2H_4	32.05	3081(293 K)	1005 (338 K)	0.97 (298 K)
			3174 (338 K)	952 (350 K)	0.913 (330 K)
Liquid hydrogen (LH_2)	H_2	2.016	7327[b] (20.4 K)	71 (20.4 K)	0.024 (14.3 K)
				76 (14 K)	0.013 (20.4 K)
Monomethylhydrazine	CH_3NHNH_2	46.072	2922 (293 K)	879 (293 K)	0.885 (293 K)
			3077 (393 K)	857 (311 K)	0.40 (344 K)
Nitric acid[a] (99% pure)	HNO_3	63.016	176 (311 K)	1549 (273 K)	1.45 (273 K)
			682 (373 K)	1476 (313 K)	
Nitrogen tetroxide	N_2O_4	92.016	1566 (290 K)	1447 (293 K)	0.47 (293 K)
			1871 (360 K)	1380 (322 K)	0.33 (315 K)
Liquid oxygen (LOX)	O_2	32.00	1675 (65 K)	1140 (90.4 K)	0.87 (53.7 K)
				1230 (77.6 K)	0.19 (90.4 K)
Rocket fuel (RP-1)	Hydrocarbon $CH_{1.97}$	~175–225	1884 (298 K)	580 (422 K)	0.75 (289 K)
				807 (289 K)	0.21 (366 K)
Unsymmetrical dimethyl-hydrazine (UDMH)	$(CH_3)_2NNH_2$	60.10	2814 (298 K)	856 (228 K)	4.4 (220 K)
			2973 (340 K)	784 (244 K	0.48 (300 K)

Source: From Sutton, G.P. and Biblarz, O.: *Rocket Propulsion Elements*. 7th edn. John Wiley & Sons Inc. New York. 2001. Copyright Wiley-VCH Verlag GmbH & Co. KGaA. Reproduced with permission.

[a] RFNA (Red fuming nitric acid) has 5%–20% dissolved NO_2 with an average molecular weight of about 60 g/mol, and a density slightly higher than pure nitric acid.

[b] At boiling point.

TABLE D.3 Theoretical Performance[a] of Typical Liquid Rocket Bi-Propellants

Oxidizer	Fuel	Flow Type	Mixture Ratio by Mass or Volume (v)	I_{sp} (s)	T_c (K)	ρ (g/cm³)	C^* (m/s)	M_w (kg/mol)	γ
Oxygen	Ammonia	Frozen	1.30 (0.98)	267	3177	0.88	1641	23.4	1.22
		Equilibrium	1.43 (1.08)	279	3230	0.89	1670	24.1	
	Hydrazine	Frozen	0.74 (0.66)	301	3285	1.06	1871	18.3	1.25
		Equilibrium	0.90 (0.80)	313	3404	1.07	1892	19.3	
	Hydrogen	Frozen	3.40 (0.21)	387	2959	0.26	2428	8.90	1.26
		Equilibrium	4.02 (0.25)	390	2999	0.28	2432	10.0	
	RP-1	Frozen	2.24 (1.59)	286	3571	1.01	1774	21.9	1.24
		Equilibrium	2.56 (1.82)	300	3677	1.02	1800	23.3	
	UDMH	Frozen	1.39 (0.96)	295	3542	0.69	1835	19.8	1.25
		Equilibrium	1.65 (1.14)	310	3594	0.98	1864	21.3	
Fluorine	Hydrazine	Frozen	1.83 (1.22)	334	4553	1.29	2128	18.5	1.33
		Equilibrium	2.30 (1.54)	363	4713	1.31	2208	19.4	
	Hydrogen	Frozen	4.54 (0.21)	398	3080	0.33	2534	8.9	1.33
		Equilibrium	7.60 (0.35)	410	3900	0.45	2549	11.8	
Nitrogen tetroxide	Hydrazine	Frozen	1.08 (0.75)	283	3258	1.20	1765	19.5	1.26
		Equilibrium	1.34 (0.93)	292	3152	1.22	1782	20.9	
	50% UDMH, 50% Hydrazine (A-50)	Frozen	1.62 (1.01)	278	3242	1.18	1652	21.0	1.24
		Equilibrium	2.00 (1.24)	288	3372	1.21	1711	22.6	

Source: From Sutton, G.P., and Biblarz, O.: *Rocket Propulsion Elements*. 7th edn. John Wiley & Sons Inc. New York. 2001. Copyright Wiley-VCH Verlag GmbH & Co. KGaA. Reproduced with permission.

[a] Condition: $P_c = 6.89$ MPa and $P_e = 0.1$ MPa. Optimum expansion ratio and frozen flow through the nozzle.

TABLE D.4 Theoretical Performance of Typical Solid Rocket Propellant Combinations

Oxidizer	**Fuel**	I_{sp} **(s)**[a]	T_c **(K)**	C^* **(m/s)**	M_w **(g/mol)**	γ
Ammonium nitrate (AN)	11% binder + 7% additives	192	1282	1209	20.1	1.26
Ammonium perchlorate (AP) 78%–66%	18% organic polymer + 4%–20% aluminum	262	2816	1590	25.0	1.21
Ammonium perchlorate (AP) 84%–68%	12% polymer binder and 4%–20% aluminum	266	3371	1577	29.3	1.17

Source: From Sutton, G.P. and Biblarz, O.: *Rocket Propulsion Elements*. 7th edn. John Wiley & Sons Inc. New York. 2001. Copyright Wiley-VCH Verlag GmbH & Co. KGaA. Reproduced with permission.

[a] I_{sp} and C^*: $P_c = 6.89$ MPa and $P_e = 0.1$ MPa. Optimum expansion ratio and frozen flow through the nozzle.

Index

D

E

F

G

H

J

L

M

N

O

P

T

V